Editing Technical Writing

EDITING TECHNICAL WRITING

Donald C. Samson, Jr.

New York Oxford
Oxford University Press
1993

Oxford University Press

Oxford New York Toronto
Delhi Bombay Calcutta Madras Karachi
Kuala Lumpur Singapore Hong Kong Tokyo
Nairobi Dar es Salaam Cape Town
Melbourne Auckland Madrid

and associated companies in
Berlin Ibadan

Published by Oxford University Press, Inc.,
200 Madison Avenue, New York, New York 10016

Library of Congress Cataloging-in-Publication Data
Samson, Donald C.
Editing technical writing / Donald C. Samson, Jr.
p. cm. Includes bibliographical references and index.
ISBN 0-19-506351-1
1. Technical editing. 2. Technical writing.
I. Title.
T11.4.S36 1993 808′.0666—dc20 92-13961

2 4 6 8 9 7 5 3 1

Printed in the United States of America
on acid-free paper

To my wife, Joy,
and my son, Devon—
my fishing partners

Acknowledgments

I am grateful to several people for their help with this text: my editors at Oxford University Press, Elizabeth Maguire and Susan Hannan; my copyeditor, Ginger Wineinger, who offered valuable suggestions about organization as well as sound revisions; the company representatives who granted permission to reproduce copyrighted material; the former students who allowed me to adapt their work for editing exercises; and the technical editors and teachers of technical editing who responded to my questionnaire about the text.

Contents

Introduction

Editing Technical Writing serves two audiences: students in technical communication courses and professionals in business and government who want to learn how to edit technical writing. Users of this text should have completed an upper division technical writing course or worked with technical documents in business or government; editing follows writing in the creation of technical documents and requires familiarity with types of technical documents and principles of effective technical writing. *Editing Technical Writing* describes what technical editors do and how they do it. It is not an introductory text in technical communication; however, it is appropriate for college students as early as their junior year.

Editing Technical Writing is different from such valuable editorial guides as *Words into Type, The Chicago Manual of Style,* the Council of Biology Editors' *CBE Style Manual,* the American Institute of Physics *Style Manual,* and other reference books. This text addresses many of the topics covered in them; however, it is not designed to take their place. Editors should have them readily available. This text refers readers to discussions in other guides for expanded coverage of some topics, such as preparing book manuscripts. Technical editors (and students of technical editing) should know those publication guides, which focus less than this book on helping beginners learn to edit technical material.

Editing Technical Writing also differs from other books on editing such as Karen Judd's *Copy Editing,* a useful textbook, and Judith Butcher's *Copy-Editing: The Cambridge Handbook.* They concentrate more on editing trade and scholarly books than does *Editing Technical Writing,* which focuses less on book-length manuscripts to be published as trade books, library references, or textbooks and more on documents related to specific projects, such as final project reports. This text extends Judd's and Butcher's discussions and examines the many roles technical editors have in creating technical documents. In addition, this book presents practice documents to be edited, guidelines for editing, and fuller discussions of editing grammar and punctuation.

Throughout *Editing Technical Writing* are exercises that students can use as self-tests; answer keys are provided for checking their work. In some exercises, especially editing and proofreading exercises, there is no single correct answer. A grammatical error often can be corrected in several ways, and there may be more than one way to change a sentence or a graphic that needs improvement. Good editors exercise their judgment as well as their communication skills, which is one reason editing technical material can be fascinating. So for some exercises in this text (and most editing situations in business and government), students should remember that many of the answer keys provide *suggested* responses, and other correct options may be possible.

Technical editors work with technical writers and staff in many fields to produce a variety of technical documents: project reports, journal articles, hardware and software user documentation and other instructions, proposals, manuals, video scripts, important correspondence, annual reports, and materials for briefings, among others. In this text, the terms *document* and *publication* refer to any print or on-line publication involving text and/or graphics. The text and graphics may be submitted to the editor written and drawn on paper; the editor is then working with a *hard copy* version. Or the material may be stored on floppy disks or loaded in a mainframe or minicomputer; the document is then referred to as *on disk,* whether it is on a floppy diskette or on tape. A *manuscript* is the original form of the document submitted to the editor. Technical editors rarely work from a writer's handwritten draft. The original form of the document as it was delivered to the editor—whether hard copy or the computer file on disk or on a network—is the manuscript. In this text the term *PC* stands for personal computer, whether manufactured by IBM, Apple Computers, or any other firm, as it does in several dictionaries of computer terminology listed in the bibliography.

Technology constantly changes, so much of the material that technical editors work with will also change. Technical developments make technical editing a fascinating, challenging field for people with communication skills and an interest in science and technology. Technical editors are intermediaries between technical writers and their audiences, helping communicate technical information to audiences who may need to understand it to address important issues in health, environmental studies, politics, warfare, and other areas. As technology becomes more complex, society will need more people who can explain it.

As private companies try to control their overhead expenses by reducing the number of support staff, including publication specialists, technical staff are often made responsible for producing their own documents and illustrating them. When this happens, the assistance of technical editors is all the more important to create an effective document.

There is also a growing demand for technical editors to help writers prepare documents for the expert audiences who read technical journals and attend professional and scientific meetings. Many colleges, universities, and professional schools hire editors to increase the acceptance rate of manuscripts written by faculty and staff. However, such editors should be familiar with the particular science or technology in which they edit. Without such knowledge, it is difficult to ensure technical accuracy of the documents. Writers of professional articles will be reluctant to stake their reputations and possibly their careers on the work of editors who do not understand what they are editing.

Technical editors' duties vary considerably from organization to organization. Some technical editors spend most of their time editing text and graphics for documents, rarely scheduling and coordinating production activities such as typing and printing, whereas others mostly coordinate production work and rarely edit text and graphics. Some edit the work of trained technical communicators; others work with documents written by technical staff untrained in technical writing. Consequently, no textbook can say: "This is what all technical editors do and how they do it." In this text I examine the editor's role in the different steps involved in creating a technical document, to give students and professionals an understanding of *all* possible duties, even if their future editorial work requires only some of the activities.

In large organizations (companies or government agencies), most technical documents are written by technical staff, not publications department staff. All other publication activities—from typing through printing and binding—are usually the publications department's responsibility. This is the setting examined in this text. Editors often plan, coordinate, and edit the writers' work, and supervise document production until the document is delivered to the customer (see Chapter 2, "Creating Technical Documents").

The smaller the organization, the greater the likelihood that the editor will have a broad variety of responsibilities. For example, in a small publications group, an editor might write the text of some documents (such as newsletters), design the graphics and layouts, produce the printing masters with desktop publishing software and a laser printer, and reproduce the document on a photocopier. For editors with design and layout responsibilities, the books on desktop publishing and graphics listed in the bibliography at the end of the text will be valuable. In the setting described in this text, publications department support staff rather than editors have these responsibilities, as describe in Chapter 2.

Prospective technical editors often ask how important it is to have technical training—a difficult question to answer definitively. The materials that technical editors are called upon to work with range from highly complex research reports to nontechnical materials such as press releases or consumer information. For editors with a background in science or technology in addition to good communication skills, there are substantially more opportunities in technical editing.

Experienced editors often stress the importance of a technical background. Alberta Cox has said, "the importance of technical competence in science and engineering is obvious." Lola Zook suggests that "the technical editor *must learn how to work with technical material*" (her emphasis). Also, Dr. JoAnn Hackos states that students of editing need "basic familiarity with the technical subject matter and language." In a recent survey, technical communication professionals were asked what they thought students of writing and editing should study to prepare for a career in technical communication. The respondents stressed the value of technical expertise: 70 percent recommended study of computer science, 40 percent physics, and 40 percent engineering.

Some technical communicators have suggested that technical expertise might limit editors' effectiveness, making it harder for them to remember how little a lay reader might know about a subject. This raises an important issue: Editors must take into account how much or how little their readers know about the subject under discussion in a document. (The audiences of technical documents are discussed in Chapter 4, "Editing Text.") Keeping a lay reader's limited knowledge of a subject and its terminology in mind is important for editors working with documents intended for a general audience. However, serious problems can arise when editors do not understand technical material any better than lay readers do. In such cases, no excellence in communication skills will enable them to "translate" the material to a lay audience, and any attempt to communicate it to an expert audience might introduce laughable errors in content. Revising the presentation of technical material requires fundamental understanding of what is being edited.

Also, an editor cannot check the accuracy of a document without some technical expertise. Don Bush and others have argued that a technical editor's first priority should be the technical accuracy of the document. The Board of Editors in the Life Sciences considers manuscript editors to be "concerned not only with the form but with the intellectual content of a manuscript." As computerized editing becomes more refined, editors will have more opportunity to improve a documents' content. Dr. Carolyn Rude has noted that as computer software is used more frequently to check grammar, punctuation, and spelling, editors will "spend more of their time working with the substance of documents beyond the sentence level—completeness of information, logic of argument, appropriateness for audience, organization, and format." Without some technical background, editors will be limited in their ability to address these aspects of the presentation and the technical accuracy of the document. Such editors might not even be able to proofread effectively.

Editors are also needed to work with nontechnical documents such as annual reports and some policies and procedures, and to coordinate the production of documents. However, many English and journalism students who do not have technical training can qualify for such work. In the realm of technical editing, the lack of training in science and technology will be severely limiting.

When students do develop technical expertise to add to their editing skills, they should try to acquire a sufficiently broad background in science and technology. Students who concentrate their scientific study in one field may find in their careers as editors that they cannot move into editing in other fields. Prospective editors should grasp the scientific principles that underlie the technology of many fields, such as the principles of physics.

Few technical experts begin work in their field as technical editors. However, many technical staff become interested in communication and eventually become technical writers and editors. Just as technical experts are not expected to be expert writers or editors, beginning editors are not expected to be technical experts. However, editors often are expected to have some scientific background and some interest in the technology. If they do, most technical experts will be more receptive to them and more likely to cooperate in the editorial process. Technical writers trust knowledgeable, interested editors with their writing (on which their reputations are sometimes staked), and they allow such editors to edit their writing comprehensively.

Dr. Hackos has observed that beginning editors need confidence, "an affinity for technical materials and a willingness to tackle everything." Understanding technical principles can increase editors' confidence, improve their ability to edit technical material, and help them learn more about the field. While students or working professionals study technical editing, they should also develop their understanding of science and technology through their outside reading and perhaps through formal study in college and university courses.

Editing Technical Writing does not require that readers possess a technical background. Most of the materials in this text, which are drawn from different fields, are not too technical for students still developing technical knowledge. The aim of this book is not to challenge readers to understand the science or technology in an editing exercise but to develop the skills they need to edit technical documents and to encourage them to develop technical expertise.

REFERENCES

Board of Editors in the Life Sciences. *Certification Program for Manuscript Editors. (Highlands, NC, 1991, 3.*

Bush, Don. "Content Editing: An Opportunity for Growth." Technical Communication 28, 4 (Fourth quarter, 1981): 15–18.

Cox, Alberta. "The Editor as Generalist as Well as Specialist." In *Technical Editing: Principles and Practices,* edited by Lola Zook, 8. Washington: Society for Technical Communication, 1975.

Hackos, JoAnn. "A Graduate Editing Course with a Research Component." In *Teaching Technical Editing* edited by Carolyn Rude, 41. Lubbock, Tex.: Association of Teachers of Technical Writing, 1985.

Rude, Carolyn. "The Rhetorical Basis of Substantive Editing." *Proceedings of the 37th International Technical Communication Conference,* CC 20–22. Washington: Society for Technical Communication, 1990.

Samson, Donald. "Humanities Education and Technical Communication." *Proceedings of the 35th International Technical Communication Conference,* RET 134–137. Washington: Society for Technical Communication, 1988.

Zook, Lola. "We Start with Questions: Defining the Editing Curriculum." In *Teaching Technical Editing,* edited by Carolyn Rude, 4. Lubbock, Tex.: Association of Teachers of Technical Writing), 1985.

Editing Technical Writing

CHAPTER ONE

Writing, Editing, and Proofreading

Writing, editing, and proofreading are all parts of the writing process. Writing is a series of activities that begins with thinking about a document before it is written and concludes with the document in its final form, whether printed or on line.

The writing process is recursive. That is, writing activities are not completed one after the other in a fixed sequence—first the writing, then rewriting, then editing, and then proofreading. Writers often make lists, write down ideas, and even draft complete sentences as they plan. They may revise a sentence before they go on to write the next one, and revise part of a draft before they finish it. Many teachers of writing have discussed the recursive nature of the writing process; a good introduction appears in Erika Lindemann's *A Rhetoric for Writing Teachers.*

Although dividing the writing process into discrete sequential steps misrepresents the process, identifying the different steps in creating (writing) a document can help technical editors (and writers) better understand what happens. These steps, depicted in Figure 1.1, are summarized in this chapter, then discussed more fully in Chapter 2, "Creating Technical Documents."

Technical editing students need to be familiar with writing, editing, and proofreading because technical editors are called on to perform all three. An editor's primary responsibility may be to edit technical material, but this often requires extensive writing and rewriting, and editors frequently are asked to proofread other editors' work and sometimes their own.

Technical editing differs from other kinds of editing, just as technical writing differs from other types of writing. To clarify these differences, the first part of this chapter defines technical editing and discusses its function in

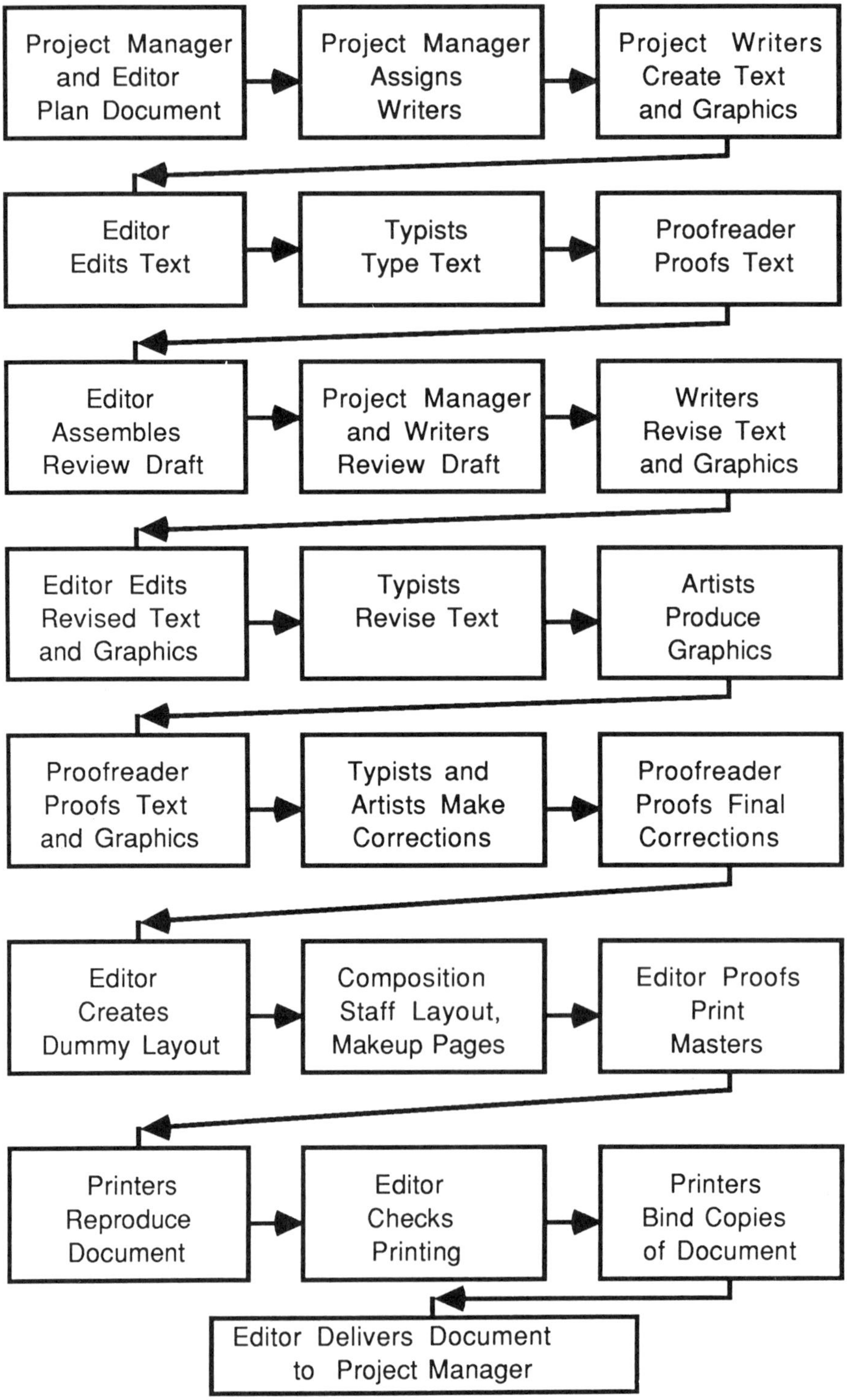

Figure 1.1. Procedure for creating a technical document.

creating technical documents. Then the discussion examines the separate activities of the writing process: writing, editing, rewriting, and proofreading.

TECHNICAL EDITING

To *edit* a document is to revise it, usually for publication in printed or online form or for oral presentation. Editing involves making changes designed to help the writer communicate effectively by making the document more appropriate for the audience, more accurate, and/or correct.

The term *technical* is more difficult to define. It is often used broadly in relation to technology, which is the application of scientific principles, usually for commercial purposes. To most people, "technical" usually means "related to science or mathematics." For our purposes, this loose definition is better than a more limited one because the term is applied to a very broad range of materials, from descriptions of experiments in physics to instructions for assembling bicycles.

Technical editing, then, is the revising of a document that presents material related to science or technology, to make it communicate more effectively. Some documents are much more technical than others, such as reports with formulas and equations or instructions to perform complex procedures. Other documents are not particularly technical, such as annual reports. Technical documents all have in common, however, an aim to communicate technical information clearly. Most such documents are designed to inform or persuade, but much technical writing is also intended to entertain, such as articles in *Natural History* and *Smithsonian.* Technical writing designed for a general audience is often called *scientific* or *science* writing. This distinction is difficult to maintain, as articles about a certain technology frequently discuss the science involved and vice-versa, and much writing about technology is for a general audience. "Technical writing" as used in this text applies to documents that convey scientific and technical information, whether or not there is an attempt to entertain the audience.

Technical documents often differ in style and format from other nonfiction prose. Technical style favors simplicity and conciseness in sentence structure, specificity in diction, streamlined and readily apparent organization of content, and heavy reliance on graphics. Also, technical documents are often prepared in specific formats. For example, one popular technical proposal format calls for text and graphics to be printed on facing pages, with no mixing of the two on the same page.

WRITING AND EDITING

To be effective, editors must understand the difference between writing and editing. On one level, the difference between writing and editing is simple: Writing produces the words in a document, and editing changes them. Many teachers and communication professionals distinguish between the writing activities that produce a first draft of a document—*invention* (coming up with what to say) and *composing* (figuring out how to say it and getting the words down)—and editing activities: *revising, rewriting,* and *proofreading.* However, these activities often overlap. For example, both the writer and the editor must consider the audience's familiarity with the material and interest in it, and as discussed before, writers often revise as they compose.

The relationship between writers and editors often links their activities. For example, editors are sometimes asked to help writers develop their abilities, especially in companies and government agencies where the writers are engineers, scientists, technicians, and other business specialists rather than communicators by profession. Editors (the communication experts in these settings) often provide other professionals with guidelines for better writing. These guidelines, many of which are summarized in the next section, can help people hone their writing skills as well as clarify some of the differences between writing and editing.

WRITING WELL

Professionals in business and government may be technical writers with training in technical communication; most often, however, they are subject-matter experts who may have had some writing courses in college but are not trained technical writers. They often (correctly) view writing as a responsibility secondary to their technical work.

Writers in business and government often do not consider "writing" to include the activities that take place before words are written down, such as defining the topic, gathering materials, and deciding what to say and when. Editors can help by emphasizing how important these preparatory activities can be to writing well. Editors should also distinguish the activities that take place after the writer has put words on paper or disk—rewriting (or revising), editing, and proofreading—from the actual writing. Editors should distinguish these actions by limiting the term *writing* to a writer's thinking, planning, and writing down his or her ideas.

Most people who think they can't write *can* write well; they just can't (or don't) edit. To them and to many English teachers, misspelled words, grammatical errors, and awkward sentences are signals of an inability to write well. That is not true: These are weaknesses in editing, not in writing.

Writing well requires a writer to think clearly, to be thorough, and to say what he or she has to say in a simple, straightforward way. Good writing is impossible without good thinking—without focusing on a topic, defining it, and developing it logically and fully. Readers expect writers to have something to tell them, something they need or want to know, and readers want to be able to get that information without undue effort.

Technical professionals write when they have information to communicate to readers. Communicating orally is sometimes impossible, and it is usually inadvisable to rely on oral communication. Most people don't listen well, and those who do can only process so much information at a time. (Consequently, people who present information orally in a briefing usually supplement their talk with printed materials.) In an academic setting, students may be forced to write whether or not they have anything to say. However, working professionals are expected to have important information to present.

Communicating well requires an ability to construct unified, coherent sentences and paragraphs. In addition, the graphics in a technical document are often more important than the text. Most readers tend to "look at the pictures" first, and some don't look at much else. A good course in technical writing (which students should complete before they undertake a study of technical editing) should provide guidelines for diction, sentence and paragraph structure, rewriting, and effective graphics for technical documents.

As a summary of principles of effective writing, consider the following eight guidelines.

1. Writers should know exactly what their subject is. If they don't define their topic carefully before they write, they may produce a draft that is rejected by their supervisor because it presents extraneous or wrong information. If writers are not sure what they are to write about, they should ask for clarification of the assignment.
2. Writers should understand how much their readers know about the topic *and* how interested they are in knowing more. If writers are unsure of their audience, they should try to learn more by consulting their supervisor or other writers and by examining other documents written for that audience.
3. Writers should understand the purpose of what they are writing. There is a great difference between a report on the results of an experiment and a proposal designed to convince a company or agency to hire a particular firm to undertake a project. To target the subject appropriately, the writer should understand fully the situation that has prompted the writing of the document.
4. Writers should recognize the important differences between technical writing in a professional setting and student writing. In business, writers are assigned projects based on their work experience. They are expected to write reports on what they know or have done in their work. Technical writers know their subject matter; consequently, invention (thinking of what to say) is not nearly the problem it is for students. Writers in business and government are no longer writing for professors who know more than they do about the topic, and they are no longer writing to show a professor how much they have learned. Rather, they are writing for readers who know far less about the topic than they do, and their aim is to inform those readers rather than to demonstrate their own expertise.
5. To plan the content of a document, rather than constructing a formal outline writers should simply list the main points or pieces of information and number the items according to the most logical order. Unless chro-

nological order is logically required (as it would be in a set of instructions) or is advantageous, writers should discuss the topics in order of importance. Writers should not use climactic order (saving the most important point for last), because most readers of technical documents want the material presented to them quickly, the most important points first. Climactic order can be used to advantage, however, if writers must develop an argument with an unpopular conclusion.

6. Writers should write about each item on their planning list in a separate paragraph, being careful to include all crucial details and appropriate explanations. "Overwriting" the first draft helps avoid the most common weakness in technical documents: failing to include needed information or explanation. Therefore, writers should try to include what might seem to be too much explanation and information in the first draft. They or the editors should decide later, in the editing stage, what should be cut out.
7. When they *write,* technical writers should pay no attention to spelling, grammar, punctuation, and sentence structure problems such as fragments. They should not concern themselves with the possible reader responses to each statement. Writers should just say what they need to say. When they have finished writing the first draft, they should set it aside for a day or two, if time permits, before they review it and revise its organization, accuracy, and content. This review is not for clarity, conciseness, and correctness at the sentence level; rather, it involves rewriting or revising to improve the document's content and organization to make it appropriate for the audience and purpose.
8. Writers should develop the habit of asking supervisors and co-workers to review each first draft, being careful to inform reviewers that it is a draft to be reviewed for content and organization. Reviewers should be asked to make any necessary additions, deletions, and changes on their copy. Many reviewers only look for typos when they review a document, even a first draft, so writers should try to cultivate reviewers who will read each draft carefully, think about its content and organization, and then make intelligent, substantive recommendations.

After writers have examined the reviewers' comments and revised the draft appropriately, they should submit the draft to an editor (or another writer) to be edited.

EDITING WELL

Editing well requires examining large aspects of a document, such as the organization and content, as well as smaller parts such as individual sentences and graphics. (See Chapter 4 for a full discussion of editing text and Chapter 5, "Editing Graphics.") Experienced editors sometimes can address both these "global" and "local" issues at the same time as they edit a document. Many editors, however, work more effectively when they first read the document only to familiarize themselves with its subject, audience, and purpose. They examine organization and content on their second reading, and address phrasing, grammar, punctuation, and so forth on a subsequent reading. Beginning editors should follow this procedure if possible when editing a short document or sections of a long one: first reading it, then addressing global matters in a *macro-edit,* and finally correcting sentence-level matters in a *micro-edit.*

The macro-edit

In a macro-edit, editors examine the content, organization, and logic of a document to see how well it addresses its subject, audience, and purpose. No sentence-level *line editing* (correcting of grammar, format, or style) takes place in the macro-edit. Necessary revisions to individual sentences or graphics are ignored for the moment in favor of larger issues such as clear thesis or purpose statements, sound and easy-to-follow organization, a thorough summary and introduction, and accurate and complete content. An effective macro-edit requires familiarity with the type of document, the subject matter, and often the company or agency producing the document. For a beginning technical editor, macro-edits are much more difficult than micro-edits.

The micro-edit

The micro-edit is what most people call *copyediting* or *line editing:* the close editing of text and graphics word by word and line by line to produce a well-written document that is rhetorically sound, grammatically correct, and consistent with company or agency standards. Principles of good style, consistency, and correctness enter the writing process here. To make sure that documents are clear, correct, and concise, many editors complete the micro-edit in two steps. In the *complexity edit,* editors make sure that the discussion in the document is presented at an appropriate level of complexity for the audience. In the *correctness edit,* editors make the document consistent with standards and correct in sentence structure, grammar, spelling, and punctuation. (These two steps are discussed at length in Chapter 4.)

The complexity edit. To communicate well, writers and editors should avoid *unnecessary* complexity in word choice, sentence length and structure, and paragraphing.

Editors should make sure the wording is appropriate for the particular reading or listening audience. For a nontechnical audience, this often requires eliminating (or limiting and defining) unfamiliar technical terms, acronyms, and abbreviations, and substituting simpler words. Wherever possible, editors should make the writing more forceful and emphatic by using verbs that state action to describe key concepts. Also, editors should ensure that important ideas are emphasized by stating them in main, not subordinate, clauses. Examining the diction of a document in a complexity edit is a good time for editors to eliminate sexist, ageist, or otherwise offensive language from the document.

In the complexity edit, editors should remember that the longer or more complex a sentence is, the more difficult it will be for readers to understand. Most sentences should conform to the customary (and simplest) pattern of subject–verb–object or subject–verb–complement. Also, editors should limit the number of sentences with long phrases before the main clause subject or between the main clause subject and verb. Unnecessary subordinate clauses should be eliminated to simplify sentence structure, and unless there is a particular reason to use passive voice, active voice should be used to make sentences direct and emphatic.

Paragraphing should also receive editors' attention in the complexity edit. Editors should control paragraph length, especially in correspondence, so that important ideas or pieces of information are not buried in long blocks of text. Editors should check the beginnings of paragraphs for clear statements of their topic. Where appropriate, editors should mark groups of information to be set off in bulleted or numbered *(displayed)* lists.

The correctness edit. Once editors have finished the complexity edit of a document, they complete the correctness edit. The correctness edit ensures that the document is correct in content as well as in presentation and compliant with company or agency standards. If the complexity edit produced substantial changes to the document, those revisions should be incorporated into the document and the text or graphic reprinted before the correctness edit takes place.

Careful correctness editing is very important. Many managers in business and government agencies are adept at spotting errors or inconsistencies in content, grammar, sentence structure, formatting, and spelling. Many consider such errors to be evidence of ignorance or sloppiness. If a document isn't exact, the writers, editors, and company or agency that produced it can look bad. Benjamin Franklin, himself a scientist and one of the great technical writers the United States has produced, said: "Trifles make perfection, but perfection is no trifle."

Effective correctness editing requires a strong background in grammar and punctuation (see Chapters 11 and 12). Editors should examine a document sentence by sentence, checking each for subject–verb agreement, pronoun case, grammatical completeness, and so forth (see Chapter 11). Only then should editors review the punctuation in the sentence, employing the guidelines in Chapter 12.

Once each sentence of a technical document has been edited for correctness, or while editors are editing for grammar and punctuation, they should check the document for consistency with the company or agency style guide. Usage, punctuation, and format should comply with the style guide or deviate from it only for a good reason. (Beginning editors who wish to deviate from the recommended style guide should do so only with their supervisor's permission.)

Macro- and micro-editing often require more than one pass through a document. To be most effective, many editors work through a document three times, once to complete a good macro-edit, once again for the complexity edit, and finally for the correctness edit. Editing has traditionally been considered a one-step process, and for long documents it might have to be. However, the process requires attention to so many different aspects of the work that most editors address them in different, well-defined steps.

Rewriting (Revising)

When writers rewrite, after they have examined the draft document and perhaps received suggestions from reviewers, they should try to "re-see" the draft from the reader's perspective. They should concentrate on changes that would make the document clearer, tighter, and generally more effective.

When rewriting, writers should address global issues first, as they are usually more crucial to the success of the document. Any problems in organization, content, or the level of the discussion must be solved before the writer attends to more local issues such as the clarity of a specific statement or the correctness of spelling and punctuation.

Rewriting is very much like editing, with one obvious difference: In rewriting, authors rework their own writing, not someone else's. Many writers (as well as editors) find it much easier to edit someone else's work than to revise their own. They tend to believe that they *have* been clear and thorough enough, and sometimes it is difficult for them to see how to solve a problem they created. However, revising the document is easier when reviewers have provided specific suggestions.

Because writers cannot always count on having a good editor edit their work, they should try to make what they have written as good as it can be. The better a draft is revised, the easier it will be to edit and the better its chances for success. Writers who received instruction in rewriting in a good writing course should be encouraged to apply what they learned. Review of the eight writing guidelines presented earlier in this chapter can serve as a good starting point.

PROOFREADING

The importance of proofreading is often underrated. Careful proofreading can ensure that a printed or online document is correct and attractive and that the document creates a good impression of the company or agency that produced it. Shoddy documents might suggest that the company or agency does shoddy work.

Proofreading is often assigned to new, inexperienced staff—a mistake, because proofreading is much more than checking for typographical errors. Proofreaders are expected to recognize and call attention to errors in spelling, grammar, punctuation, sentence structure, and format, and often in content as well. If they can't, they won't be able to spot errors that occasionally slip past technical editors, especially those who have had to rewrite substantial portions of a document. Relying on computerized spelling- and grammar-checking is also a mistake, because such programs will only detect some of the errors in a document. Correctness is often the most important criterion of excellence. A writer's attention to technical detail might be questioned when proofreaders have missed errors.

Proofreaders should understand what the writer or editor wants done and do only that. Proofreaders should not edit. They should mark as neatly as possible only what must be changed to correct an error.

Proofreaders should use the proper reference books and materials for the job—the appropriate technical dictionaries and company or agency style guide—and the proper pen or pencil. Few use intricate methods like reading backwards or dividing a page or screen into quarters. Most proofreaders simply read down the page line by line, holding a white sheet of paper under the line they are reading to help them focus on that line as they compare it one-to-one with the draft. Most readers find proofreading hard copy much easier than proofreading text or graphics on a monitor.

Misspellings are the most obvious errors to readers of published documents, so proofreaders should check carefully for accuracy of spelling. It is not wise to rely on spelling checkers in word-processing programs, which will only flag words not in their own dictionary. Most spelling checkers do not recognize correct words used improperly, like *there* and *their,* so proofreaders should check each word for context as well as spelling.

All proper nouns should be checked carefully because typists are often least familiar with these words, and misspelling the names of products or people can cause embarrassment. The accuracy of titles, headings, and illustration captions should be checked very carefully. Errors in these parts of a document are often overlooked by writers and editors, but because they stand out, such mistakes are often noticed quickly by readers.

When proofreading a document, proofreaders should verify the typed text against the original manuscript to make sure it is complete and correct. In the case of a review draft, only the writers' sketches of the graphics are available, so there is usually no proofreading of graphics at this stage. Proofreading a first-review draft involves *one-to-one* reading, comparing the typed version with the original manuscript letter by letter and number by number. One-to-one proofreading takes far more time than most people realize.

With subsequent drafts, proofreaders check the artists' graphics against the writers' originals and check text revisions that have been incorporated into the document by typists. Proofreading graphics and text corrections is not as time-consuming as one-to-one proofing, but accuracy becomes ever more important as a document nears publication. Before the final draft is prepared, proofreaders should ensure that the text and graphics conform to the style guide to be followed in the document.

Effective proofreading involves more than finding errors and making sure that typists and artists have made their changes correctly. It also involves checking the technical details in the document against a reliable reference. Proofreaders should watch for inconsistencies in data and for errors in labeling units of measure. Greek letters, formulas, and mathematical symbols should be checked very carefully, as many typists are unfamiliar with such notation and so find unfamiliar letters and symbols easy to misread.

Because proofreading is so important, and because it is difficult, demanding work, it is discussed in greater detail in Chapter 9. Experienced proofreaders should be encouraged to share successful proofreading techniques with beginners. Above all, editors should try to provide proofreaders appropriate facilities and enough time to do their painstaking work effectively.

CONCLUSION

Writing, editing, and proofreading are all parts of a technical editor's duties, and technical editors should know how to write, edit, and proofread proficiently. Also, they should be able to clarify for technical staff what is involved in the different tasks. Because editors are often in charge of a publications department's work on a document, as discussed in Chapters 2 and 10, they must know what the writing, editing, and proofreading of a technical document require so that they can plan and manage the process effectively.

QUESTIONS FOR REVIEW

1. What is technical editing?
2. Why is it important for editors to understand the differences between writing and editing and between editing and proofreading?
3. How do the macro-edit and the micro-edit differ?
4. What are the two parts of the micro-edit, and how do they differ?
5. How is rewriting similar to and different from writing and editing?

CHAPTER TWO

Creating Technical Documents

Technical editors work with three groups of people: clients, customers, and colleagues. Editors' clients are the writers whose work they edit. Their customers are technical staff in charge of projects that require documents, editors of journals and magazines, other publishers of technical material, and ultimately the readers of those documents. Their colleagues are the publications department staff who help them create technical documents.

Some technical editors work with a broad range of people to produce materials for a broad audience. Others work with only a few people for a narrowly defined audience. Most editorial work falls between these extremes. In nearly every setting, editors work with other publications staff to help writers present material to readers or viewers. Figure 2.1 identifies the standard roles of a publications department.

To generalize about editors' clients, customers, and colleagues is as difficult as it is to make generalizations about technical writers. Are technical writers communication or English specialists who have been trained to present technical information? Or are they technical staff with science or technology backgrounds who write as part of their professional duties? Clearly, the term applies to both. Just as in technical writing courses there are students majoring in the sciences and technology and others majoring in technical communication, in professional settings there are technical writers who are technical experts and technical writers who are trained in technical communication but are not technical experts.

CLIENTS AND CUSTOMERS

Technical editors' clients are writers. They may include scientists, engineers, technicians, mechanics, programmers, systems analysts, and specialists in such

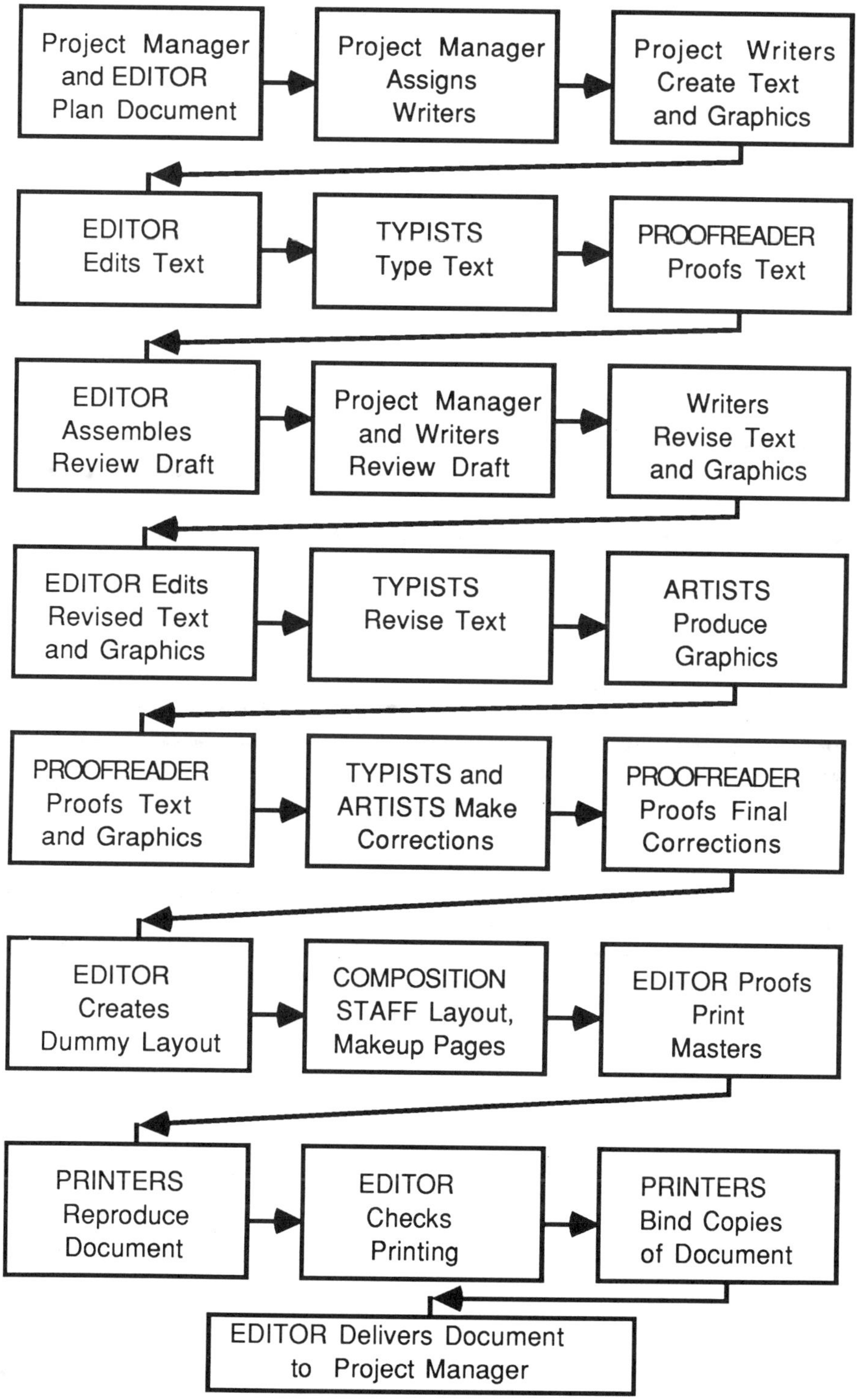

Figure 2.1. Roles of publications staff in creating a technical document.

areas of business as accounting, finance, management, information systems, planning, facilities, security, maintenance, manufacturing, contracts, and shipping.

Many of these clients write frequently as part of their regular job responsibilities; others write less often. However, they all share the need to communicate information in a written form, and depending on the situation, a technical editor might edit their writing and supervise its publication. In many research and development organizations, for example, scientists and engineers write proposals for future research, then compile progress and final reports on those projects. Often these "technical writers" are expected to write their own documents. Technical editors then make sure that their efforts' content, organization, and style are correct and appropriate for the audience.

Technical editors have two types of customers: the *primary audience* who will read the document (whether outside or inside the organization), and the *secondary audience,* the technical staff or journal editor who will review the document before it is distributed in print or on-line to the primary audience. The primary audience might be broad, such as that of a daily newspaper or a weekly news magazine, or narrow, such as the readership of a pathology journal. (Primary audiences are discussed in Chapter 4, "Editing Text.") Editors must keep the primary *and* the secondary audience in mind as they plan, edit, and produce the document. The document must satisfy both audiences by being appropriate for the primary audience in content, organization, and style.

Technical editors in trade publishing (general-interest books, newspapers, and magazines) and textbook publishing have a different relationship with clients and customers than do technical editors on the staff of companies or government agencies. Trade and textbook editors must please their customers (the magazine's or book's readers) over their clients (the writers, whom they may never meet), if a choice must be made. Technical editors in organizations outside publishing, however, need to satisfy their clients as well as their audiences. Often technical editors must cooperate fully with their clients even if this reduces the time available for editing the document for the primary audience. These editors usually work face to face with clients, a situation that requires greater interpersonal and negotiation skills than trade editing but allows for a closer collaboration.

COLLEAGUES

Editors need to understand the range of publication activities because they work closely with other publications department staff or handle all stages of document production themselves. A typical publications department organization is shown in Figure 2.2. Also, they need to be able to plan work on a document and explain that process to technical staff. Editors should know what happens in typing or typesetting text, creating graphics, laying out text

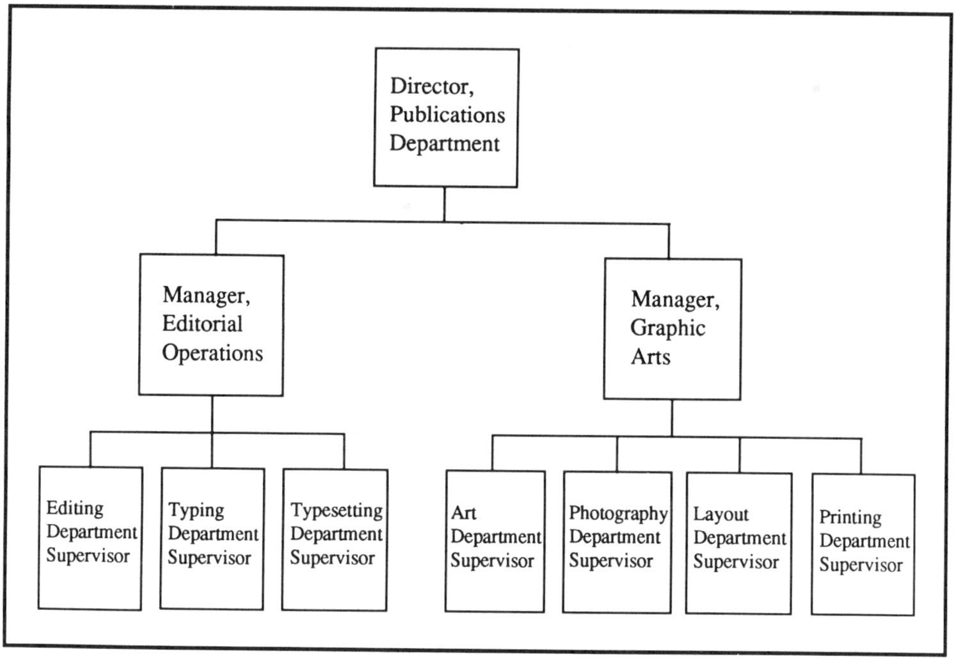

Figure 2.2. A typical publications department's organization.

and graphics, making up pages, printing and binding documents, and creating on-line documents. The most educated editors will be better able to estimate and negotiate for the time and money needed to create attractive and technically sound documents.

After the writers have produced the words and rough graphics for a technical document, editors edit the text and graphics as described in Chapters 4 and 5. Then the text goes to typists and perhaps then to typesetters, and graphics to be prepared by graphic artists are routed to them. The typed or typeset text and the writers' sketches of graphics are usually reviewed by project staff. Changes to the text are processed, and the graphics are created by hand ("on the boards") or with computerized drawing equipment. Text and graphics are proofread and corrected by typists and artists. As the corrections to the text and graphics are proofread and photographers provide any prints needed for the document, the editor creates a "dummy layout" of the document, sketching what goes where on each page or screen according to principles of aesthetics and good communication. Sometimes composition specialists create a more polished layout from the editor's dummy layout. The pages are then made up. For a printed document, makeup is done by hand or with computers. In the traditional hand method, graphics and sections of text are trimmed and then taped or pasted down on stiff paper, following the arrangement in the layout. These pages, called "mats" or "mechanicals" or "print masters," are reviewed by the editor, and any errors are corrected. The mats are then reproduced on a photocopier or used to make plates for a printing press, and the printed copies of the document are bound.

In the electronic method of makeup, the editor or a layout specialist uses desktop or electronic publishing software to merge text and graphics files to create the pages or screens of the document. A laser-printer printout of each page is used as the printing master. For an online document, the editor creates the screens by merging the files. Usually, a printout is used for proofreading before the document is reproduced electronically.

Scheduling production activities is perhaps the most crucial task in a publication project, especially for a beginning editor who has not yet learned how long staff need to perform their duties to the level of quality desired. As discussed in Chapter 10, editors must schedule the activities backwards from the completion date of the document (when the copies of the document are turned over in finished form to the client or customer). Editors should keep colleagues fully and honestly informed about the project at all stages.

PRODUCTION

Processing Text

The increased capabilities of word processing systems have been both a blessing and a curse for typists. Text revisions are far more easily made, because entire pages need not be retyped. However, because of this convenience, typists are frequently asked to make changes too often, before the document has been thoroughly edited. Also, because of its complicated equipment, typesetting can take far longer than many project managers anticipate.

To expedite these processes, the editor and the project manager should estimate the document's length and define its format carefully, specifying line length, column widths, and table style. Also, editors should submit sections of the document for typing as they become available (called *waterfalling* the text) rather than waiting until the entire text is ready. The editor and the project manager should also establish ahead of time how many drafts will be produced and how *clean* (free of handwritten corrections) the text should be for review of each draft. Above all, writers and editors should provide legible materials for the typists.

In most organizations, typed text is stored on disks in a central word processing group. If the text is to be typeset, the revised disks are loaded into typesetting equipment, the files copied, and typesetting codes added. Because typeset text is more complicated to revise than word processed text, editors should try to wait until the changes from project reviews have been made to the text before the files are typeset.

Significant advances in typesetting equipment are being made each year. Editors who work closely with typesetters and graphic designers should be familiar with the books on typesetting and design listed in the bibliography and should keep abreast of recent developments by reading computer journal articles on typesetting and desktop publishing in, for example, *Publish!*.

Processing Graphics

Graphic artists draw by hand or with computerized equipment rough or finished graphics for a document. Having well-edited submissions from the editor increases greatly the accuracy and speed of the artists' work. However, finished graphics can take significant production time, especially for complicated art (e.g., scenarios, complex flow or organization charts, or isometrics).

Before the artist begins, the art supervisor or coordinator should be informed of the standards for the document (page or screen size, line thickness, type sizes and faces, etc.) and of the approximate number of easy, medium, and hard pieces of artwork. Also, the number and difficulty of any classified or proprietary graphics should be estimated, as they often take longer to prepare because of security measures. Computerized equipment allows new originals to be generated easily for printing, but minor changes to a graphic are often far slower done on automated equipment than manually.

For a technical document editor, the main pitfall with graphics is that writers try to convey more information than the graphics' design will bear, making them weak and difficult for readers to interpret (see examples, Chapter 5). Each graphic, even a large table of data, should be designed and controlled so that its message is clear. Also, each graphic should have a caption that identifies clearly its subject or message. This caption can help the artist understand what to emphasize in the graphic, thus saving revisions and money.

The cost of producing graphics for a technical document is high, whether computer-assisted equipment is used or the graphics are drawn on the boards. Consequently, editors should not have graphics prepared by artists before reviewers of the first draft and the project manager are sure that the graphic will be used unchanged in the final document. Editors are often responsible for adhering to a publications department budget for creating a document, and they can conserve money, time, and the energies of artists and proofreaders by having graphics created only if they will be used in the final document.

Photography

Large organizations often maintain a photography laboratory and staff photographers as part of the publications department. The photo lab is usually responsible for retrieving and printing existing negatives, as well as photographing staff, equipment, facilities, and processes for the organization's documents. In smaller companies or agencies, technical editors may be responsible for photography themselves, or they may contract with an outside photographer (which may be very costly).

The growing emphasis on visual communication, the refinement of printing techniques, the development of scanning electron microscopes to nearly one-angstrom resolution, and the growing ability to create special effects with photographic images have led to increased use of photographs in technical documents and presentations. Editors should schedule work by this publications support group early in the work on a document and estimate carefully the time and cost of photo sessions and lab work.

Laying Out a Document

A layout is the arrangement of text and graphics on a page of a document or a screen of an on-line document. Layout must be considered very early in a publication's design to ensure that the final document fits the chosen design and length. When the writers prepare for the planning conference and write their first drafts, they often work within space budgets established by the project manager. The editor is often consulted to help estimate the page length of each section. The editor should *size* (estimate the size of) the text and graphics for each section. Layout should be structured in advance by planning and controlling the content of the document—the amount of discussion (see Chapter 4 for further discussion).

Laying out a page is similar to putting together a puzzle, but this puzzle allows more than one solution. Usually, however, one page arrangement is preferable to the other possibilities, especially when compared to the rest of the document. Principles of effective page design include placing graphics above text, balancing text and graphics on each page and on facing pages, and using *white space* (nonprinted areas) to avoid *gray pages* (pages with an unbroken block of text).

Making Up Pages

Page makeup is the pasting or taping of finished graphics and sections of text on *page frames* (stiff sheets of paper used for platemaking or direct printing). Page makeup can be done by hand or electronically.

The traditional hand method is still widely used and will continue to be for small documents and in organizations with limited computer equipment. In hand layout, composition staff use art knives or razor blades to cut out and trim graphics and blocks of text. They then follow the layout to arrange the text and graphics on each page frame and use rubber cement, wax, or tape to attach them. Any needed security or disclosure markings are affixed to each page *(mat)*. After the editor has reviewed the made-up pages, they are numbered.

In the electronic (computer) method, graphics and blocks of text are placed with desktop or electronic publishing software, and each page is printed on a laser printer. The master can be duplicated directly on a photocopier, or it can be used to make plates for offset printing. New originals for pages made up electronically can be generated easily, but minor changes can take time, and they can be made only by a trained operator. Last-minute corrections cannot be made as easily with automated equipment.

Hand page makeup has disadvantages and advantages. Because the elements of each page are glued, waxed, or taped in place on the mats, they can come loose, and their edges can create shadow lines, called *cut lines*, upon printing. However, hand makeup has an important advantage: Editors can make minor changes to a mat quickly and easily by typing the change on adhesive paper and *cutting it in*—attaching the trimmed correction over what needs to be changed.

Editors should take great care with the mats or *mechanicals* (printing masters), especially those made up by hand. It may be very difficult to generate a replacement for a lost or damaged mat, especially one with graphics, and having to replace mats might jeopardize the printing schedule for the document.

Printing

An organization's printing resources may range from a small photocopier to a print shop capable of full-color work, from a secretary who makes photocopies to professional printers, separators (who create the negatives required for color printing), strippers (who mount the negatives used in printing on special sheets of paper), binders, checkers, and other staff.

A document can be printed in black ink on white paper by various methods. Conventional office copiers can provide sufficiently good black-and-white (and even color) copies of drafts for review. For the final document, however, offset printing should be used for pages with photographs that require *screens* (plastic overlays that will produce shaded areas on the printed page; see Chapter 5).

In offset printing, a mat or a laser printout is used to produce a printing plate, which is attached to one roller of the offset press. As the press operates, the image from the printing plate is transferred to a rubber sheet on another press roller and then to the paper. Offset printing reproduces a sharper printed image of photographs than photocopying or stencils. Also, offset printing is less expensive than photocopying when many copies are being made, and a variety of paper stocks can be used. Offset printing is a longer process than photocopying: printing plates must be made, and *make-ready* (preparing the press for printing) is more complicated than turning on a photocopier. However, there is greater control of print quality.

Printing black ink on paper is considered *one-color* printing. Many small print shops can print two- and three-color work by making a different plate for the material to be printed in each color and running the pages through presses once for each ink color. However, it is sometimes difficult to get accurate *registration* (correct alignment of press runs so that the colors don't overlap or seem crowded). Fast-drying inks must be used to prevent smearing on subsequent press runs.

Full-color *(four-color)* printing is the most complex, time-consuming, and costly method. It is generally used only for marketing materials, document covers, and other high-visibility documents. Four-color printing requires more skill because color separations must be made and aligned on press, and most print shops have far more one- or two-color presses and operators. It can be difficult to estimate how long four-color work will take and how much it will cost (see Chapter 10).

Binding

Once printed, the copies of a technical document must be collated (put in the proper order). This can be done by machine if the pages are all one standard size. Over- or undersized pages often must be interleaved by hand.

Library binding

In library binding, commonly used for hardbound books, groups of pages called *gatherings* are stitched together with thread, then stacked in order and glued down the left side. A hard cover is glued on afterward.

Perfect binding

Perfect binding, used for most paperback books and for a growing number of journals such as *The New England Journal of Medicine* and *Audubon,* is similar to library binding, but the gatherings are held together by the glue along the spine, not stitching. The cover is glued on, as with library binding. Both methods hold the pages securely, but with both the document may not lie open flat. Pages to be inserted in a library- or perfect-bound book must be glued *(tipped)* in. Pages to be removed must be cut out, although they can sometimes be pulled out of a perfect-bound book.

Most technical documents are bound by other methods, commonly saddle-stitching, stapling, looseleaf binding, and mechanical binding.

Saddle-stitching

In this process, the cover and pages are printed on sheets twice as wide as the finished document. For example, the pages for a document 8.5 inches wide and 11 inches tall are printed at 17 inches wide by 11 inches tall. After the pages are gathered, they are folded once and stapled (not stitched) at the fold. *Science, Scientific American,* and *Technology Review* are bound this way. With saddle-stitching, the pages are held securely, and the document can be folded open flat. However, the thickness of the volume is limited, and it is very difficult to insert or remove sheets or sections.

Stapling

The document pages (including the covers) are stapled once in the upper left corner or several times down the left side. The pages are held securely with this method, but the thickness of the document is limited and it is very difficult to insert sheets or sections or remove them without tearing them out. A stapled document will not lie open flat.

Looseleaf binding

Pages are drilled with holes to fit a two-, three-, or five-ring binder. Binding is quick, the document will lie open flat, and sheets or sections can be inserted or removed easily. This method is effective for manuals that are updated with new pages or are produced piecemeal. However, pages punched for looseleaf binding can tear at the holes and fall out. Also, the added expense of the binders must be considered, including a cover or label.

Mechanical (plastic or spiral) binding

Mechanical binding is done with plastic or wire. In plastic *(comb)* binding, a row of slots is punched along the left side of the cover and text pages, and a plastic spine is attached by passing each *finger* through the appropriate slot. The document will lie open flat if a large enough plastic binder has been used (better too large than too small). Sheets or sections can be inserted or removed relatively simply with a hand-operated machine. Pages do not tear as easily at the slots as those punched for looseleaf binding.

Similarly, in spiral binding, a row of holes is drilled along the left side of the covers and pages, and a coiled wire is fed through the holes, creating a bound document similar to a student's spiral notebook. Spiral binding is trickier than plastic binding. Inserting pages is more difficult, but they are held more securely than with plastic binding, making the method appropriate for documents that will not be revised or will be handled roughly. The term "spiral binding" is sometimes used for plastic binding, so check which method is intended.

Checking a Document

Once a sample of a bound document has been page-checked to verify the collating, the remaining copies can be bound. For classified or proprietary *(company-sensitive)* documents, each page should be checked to make certain none are missing, duplicated, marked incorrectly for classification, or unintentionally blank. For an on-line document, random disks should be checked to see that the document copied correctly.

In production work on a document, there is no substitute for exactness. If readers examine a faulty document or view a screen that has errors, they might question the quality of work done by the organization that produced it. Doing production work perfectly takes time, and making sure that the production work has been done perfectly takes time. Adequate time for production work by an editor's colleagues in the publications department must be built into the schedule for creating a document, because the quality of the document will suggest to its readers the quality of the work the organization does.

THE EDITOR'S ROLE

A technical editor's work on a document often begins long before the writing. In some companies, government agencies, and academic settings, editors do not become involved until the writers have finished the first draft of the document (and sometimes the final draft). Often, however, editors are part of the initial planning for a project and the documents related to it.

Companies and agencies that want successful documents usually have teams at work long before the writing begins, especially for proposals and software manuals. Technical, financial, and managerial staff plan the work, often with an editor representing the publication aspects.

A technical editor is often the publications department's *lead* (manager) on a project, estimating the staffing, scheduling, and costs and supervising production work. When the publications department provides writers for a project, the editor often supervises their work as well.

Some experts estimate that many technical editors spend no more than 20 percent of their time actually editing text and graphics. The rest of the time, they design and plan documents; budget, schedule, and supervise editing and production activities; and interact with technical staff, especially project managers. Editors' roles in business and government agencies vary widely from company to company and agency to agency, but technical editing work is similar across companies and agencies, so it is valuable to examine editors' different responsibilities, many of which are represented in the list in Figure 2.3.

This section of Chapter 2 examines the many different activities that might make up an editor's role in creating a technical document.

Editors' work can be broken down into three main activities, which coincide with the stages of work on a technical document: planning publications department work; editing and proofreading; and supervising and coordinating production work.

The Planning Stage

As editors begin work on a project, they should learn about the technology involved by studying pertinent reports. Also, in many technical documents, especially proposals, there is a fair amount of *boilerplate,* discussions of experience or capability that are repeated in many types of company documents. Editors can work with company information specialists to gather boilerplate and useful ideas and materials such as related photographs.

In most large companies and agencies, technical communicators move from project to project, occasionally finding themselves working with unfamiliar technology. However, examining recent similar documents can provide background and discussion that would be appropriate to include, such as descriptions of previous experience. Editors should study such reports before discussing the technology with technical staff, so they can establish themselves

EDITORIAL EVALUATION AREAS

Area I: Degree of Supervision Required

BEHAVIOR - Attitude toward one's self and one's responsibilities. Perception of editorial role. Cost consciousness. Commitment to better work habits.

WORK HABITS - Ability to be self-starter and use time effectively. Tendency to be thorough, orderly, consistent, dependable, inquisitive, and flexible. Ability to work well with others.

Area II: Quality of Editing Work

BASIC SKILLS - Ability to research, plan, organize, and execute.

COMMUNICATION SKILLS - Creativity, clarity, brevity, simplicity, completeness, continuity, honesty. Outlining and summarizing skills. Writing style. Skill in grammar and spelling.

KNOWLEDGE - Familiarity with technical content, graphics, media, company (goals, history, and markets), product, customer. Basic intelligence and comprehension.

IMPACT ON CONTENT - Control of strategy, organization, points to be stressed, emergence of sales message, responsiveness, completeness of technical discussions.

EXCELLENCE - Conformance to requirements.

Area III: Management Traits

ASSERTIVENESS - Ability to get things done despite all obstacles.

LEADERSHIP - Ability to evaluate, select, organize, motivate, direct, and train others, and to successfully delegate.

MANAGEMENT SKILLS - Perspective, objectivity, judgment, self-confidence; ability to perform successfully under pressure, make decisions quickly and decisively, and deal successfully with people at all levels.

ACCEPTABILITY - Respect of others. Absence of adverse personality traits or other weaknesses that offset professional and management skills.

Figure 2.3. Editorial evaluation areas. (Source: Martin Marietta Corporation, © 1987, used with permission.)

as technically competent. Editors should also find out the document's purpose and the nature of its primary and secondary audiences.

Technical editors are in an excellent position to catch errors and inconsistencies—they may be the only people who read the *entire* document with an eye toward the accuracy of all details. Individual reviewers often concentrate only on parts of a long document.

Also, editors must understand the technology to catch problems in content and to make the document accessible to a lay audience. Technical writers' reputations rest in part on their accuracy, and they don't want to risk having a well-intentioned but ignorant editor damage their work. An aerospace editor who changes "pitot valve" to "pilot valve" or "pivot valve" would be laughed at and perhaps never trusted again. Editors should use the planning stage to get educated, by reading and by talking with technical staff.

In the planning stage, editors should also learn about the people involved in the project, especially those who will be writing sections of the document. Other publications staff can often offer advice, such as which writers will take a second draft to produce a good discussion, and what particular reviewers will insist on. Editors can then plan assignments for assistant editors, typists, and proofreaders and decide which sections should not be edited heavily for draft review because of expected rewrites.

Editors should meet with the project manager to review publications standards, decide how the publications department will produce drafts, and establish the printing requirements for the document. Printing requests can include much more than drafts: Outlines, marketing statements, reviewers' comments, and other materials are frequently reproduced for the writers.

Editors should offer to distribute document format and style guidelines to the writers. Critical sections, writers, and technical sources should be identified with the help of the project manager, and the ornateness of the document and its production schedule should be established. Editors should determine what format (size, shape, and layout of material on the pages) is to be followed or adapted: Will fold-outs be used? How large? Will turn pages be acceptable? How fancy is the publication to be? Many business managers believe that expensive, flashy documents are more likely to impress and be successful, so they want glossy or laid paper and color printing. Often they do not know that color printing will take much longer and cost much more.

The best way editors can keep a document from being unnecessarily elaborate is to provide an estimate of the cost for an appropriately attractive document. Such estimates are usually high enough that managers make the right choice.

If necessary, editors should arrange facilities and equipment for publications production, especially for documents containing proprietary or classified materials. Safes and door locks might have to be installed before work can begin, and computers and other equipment might have to be approved by security staff.

In the planning stage, editors should also consult the heads of publications department support groups to arrange staffing, establish standards and schedules, and determine costs. These groups must be informed of the document's estimated size, schedule, and complexity, and of any special features, such as company-sensitive or classified materials or deviations from departmental standards. Leaders of the support groups can determine how the document production schedule fits with other jobs scheduled in their shops at the same time, and plans can be made to handle the workload.

Editors in charge of technical documents must consider *shoploading* to make certain that the artists, typists or typesetters, composition specialists, and printers will not be too busy to support the effort properly. Determining the priority of this job within the support groups is important as well. Editors should diplomatically inform the project manager if the document does not have top priority.

During planning, editors should estimate their need for publications staff, negotiate for those staff, and assign them. Editors should also estimate publications department costs for the document and inform the project manager. Unfortunately, there is no easy formula for such estimating. A publications department should have its own cost standards for editing (and for writing, if publications staff write documents), typing or typesetting, art work, page makeup, printing, and binding. Editors should not believe others' opinions of how easy any document will be to create. Some technical staff base such estimates on work on previous documents and not on the particular problems represented by the technology or staff involved in the present project. Some technical staff assume that graphics and even sections of text can be lifted *(picked up)* from related documents, simplifying the publication effort. Editors should suspect facile estimates and plan document production schedules carefully.

In meetings with the heads of support groups, editors should make certain that all the text and graphics standards for the document are understood. After each meeting, editors might meet with support group staff to discuss the project with them, especially if the document they will be working on is a major publication. If editors do this, support group staff will know more about the project and be more enthusiastic about working on it.

For major documents requiring a number of writers, a document planning conference often takes place at the end of the planning stage. The editor may write or edit an outline that clearly and logically divides the content of the document, given its audience and purpose. The project manager may distribute marketing information, copies of earlier documents, or other materials that the writers and the editor can use as references. This conference can help editors identify better the purpose and audience of the document.

If a document involves many writers, the editor may distribute document guidelines to the writers and support group leaders, as mentioned above. In them, acronyms should be defined and preferred abbreviations listed. These guidelines should include the agreed-upon text and graphics standards, along with any exceptions to the company or agency's style guide for art sizing, line length, typefaces and sizes, spacing, boxing and ruling of tables, use of callouts or legends, and so forth.

The final deadline—delivery of the document to the project manager or other client—is the most important for editors. Working back from that due date, there are many other milestones: binding; printing; sign-off; layout and page makeup; art and text corrections; review of the final draft; preparation of final text and art; and completion of editing. Some documents may require other work before they are delivered, such as packing, labeling, and page-checking. Given the tight schedules of most technical document efforts, each milestone must be met or the final submission can be jeopardized. No matter how well editors have planned, coordinated activities, and negotiated, however, last-minute changes to a document often produce overtime work for many publications staff.

Editors should be honest with support group staff, who in all fairness should know in advance what is required of them and must be told promptly of any changes to that work. Just as it is unwise for a teacher to change a major assignment after students have begun work on it or for a contract officer to change the scope of work on a project while bidders are preparing proposals to undertake it, an editor should avoid suddenly altering the work to be done by the support groups. Support groups usually work under the pressure of deadlines, and the most willing staff have been fully and honestly informed about the project in the planning stage.

The Writing and Editing Stage

When the writing of a technical document begins, editors' duties change. Once sections of the document have been written and submitted, editors determine how appropriate the content and organization are for the purpose and audience; they examine the graphics for appropriateness and balance; they check each section for logic and completeness of evidence to support claims; and they identify sections that need fuller development (see Chapters 4 and 5).

To hold down costs, editing for a review draft often differs from editing for the final draft. On preliminary drafts, editors usually perform no more than the macro-edit (see Chapter 4). Despite the fact that some reviewers might be distracted by typos, misspellings, and other sentence-level problems in review drafts, editors should not pay much attention to such problems. Rather, they should concentrate on the organization and content of the document.

Preliminary drafts of a technical document are for review of content and organization, and they need not be highly polished, as sections are often revised radically. At this point, editors should:

1. Make sure the numbers of sections match the numbering in the document outline, and make sure the graphics are keyed to the proper sections

2. Read each section for content
3. Add a sentence to clarify the organization of each section, if necessary
4. Examine the content for appropriate development of main points for the particular primary audience
5. Examine the graphics for variety, clarity, and support of the main points of each section, and clean up graphics quickly for printing of the draft

The project manager, technical staff, and the editor should check the draft for technical accuracy and completeness. Throughout the creation of the document, the project manager, editor, and writers need specific guidance and information from reviewers. As one author once wrote in a note on a first draft: "Comments are appreciated but are not too helpful. If something is wrong, fix it. If something needs to be added, add it."

If the main idea of a section of a document is not clear or the main points do not develop directly the main idea, the editor should discuss the section with the writer and clarify the purpose and main points. Similarly, if a graphic is not clear or is unnecessarily complex, the writer should be consulted on how it might be revised. Fine points should not be debated or explored for a review draft. Fine-tuning should wait until the draft of the document has been reviewed and revised.

A technical document that requires more than two preliminary drafts is in trouble. Excessive drafts waste time and energy in production and technical review. They can be avoided by careful staffing and a good planning conference.

For the final draft, editors should edit the document thoroughly in the macro- and micro-edits described in Chapter 4, focusing especially on the following:

- Achieving the document's purpose
- Developing ideas clearly and thoroughly
- Supporting points adequately
- Making the discussion appropriate for the particular audience
- Ensuring accuracy in content and correctness in presentation

The graphics in each section must be checked to make sure that they develop clearly and forcefully the thesis of the section. Many technical documents, especially marketing materials and computer documentation, have been criticized for containing graphics that serve no purpose.

If the document includes borrowed materials (usually illustrations) that are copyrighted, as journal articles and especially books often do, the editor may be responsible for arranging reprint permission. If the materials are from United States government publications, permission to reprint is not required, but the source of the material should be acknowledged. Beginning editors should be aware that arranging permission to reprint often takes months and costs hundreds of dollars—and sometimes permission is not granted. Correspondence with the original publishers should begin early.

While editors complete their activities in the writing and editing stage, keeping support groups informed about the document, production activities begin.

The Production Stage

Editors facilitate production work by providing a dummy, a tentative layout of the document for the composition specialists, by coordinating work among the groups, and by being available to answer questions. During production, editors may also become the primary proofreader, checking revisions to text and graphics and examining the printing masters created in page makeup.

If editors are themselves responsible for creating the document on-line or with a PC, they may type all text and even create the graphics with desktop publishing software. The page can then be printed out on a laser printer, or the disk or tape can be loaded into typesetting equipment for production of a high-quality printing master. When editors do their own production work, they should avoid proofreading their own work. It is difficult for editors who have worked very closely with a document to see problems in it. Editors who do their own design and production work on documents should have other editors or proofreaders check their work carefully to ensure clarity and correctness.

Before a document is printed, editors should check the printing masters carefully for errors in the final corrections, to ensure that captions match graphics properly, that the masters are of good quality and free of smudges or loose pieces of text or graphics, if made up by hand, and that each page has been numbered properly and labeled with proper classification or disclosure labels. This review is the last step before the printing masters go to press and often involves the project manager and perhaps the print shop supervisor as well as the editor. Writers and technical staff should not take part, because they often cannot suppress the desire to make last-minute changes that might delay printing. Changes at this point often threaten the completion deadline, and only very important concerns raised by high-level managers should be allowed to jeopardize a final deadline.

Even after a document is printed and the copies are bound, or copied electronically onto disk or tape, editors may have added responsibilities. If the copies of the document are to be shipped, the editor might be responsible for packing. Packing a document is not usually an editorial task in a large organization but might be in a small firm or agency. Editors should not underestimate the importance of careful packing and shipping, which can ensure that the document arrives on time and in good condition. There is no sense in investing time and money to produce an attractive document only to have it folded, scuffed, torn, or stained in shipping, or to create a bad impression by having it arrive late.

As early as possible editors should establish a project's packing and shipping requirements, which usually include:

- Packing date and time
- Classification or sensitivity information
- Number of copies
- Bound size of copies
- Delivery date, place, and time

Packing and shipping requirements should be discussed with the print shop, mail room, or other appropriate group. Odd-sized materials may require special packaging ordered in advance, such as engineering drawings sent in reinforced cardboard shipping tubes and computer disks sent in plastic mailers.

If the mailing will be only a few copies of a small document, stiff cardboard sheets should be used between copies, inside a mailing envelope slightly larger than the contents. The envelope should be securely taped and properly labeled with addresses inside and out. If the mailing will involve larger documents or many copies of small documents, the copies should be packed in reinforced cartons with dividers to separate the sets. Stiff cardboard sheets should be used to protect the covers of documents facing the walls of the carton. Shrink-wrapping each copy can protect its appearance, but it can produce an ugly package that creates a bad first impression. Shipping cartons should be just large enough to hold all the documents; any extra space should be stuffed with packing filler so that the contents will not shift in handling.

If the document contains classified or sensitive materials, additional labeling and wrapping may be required, as dictated by the company or agency's contracts or security staff. Packing and marking standards should be followed carefully. Editors may have to fill out document control or transfer forms, using the copy numbers stamped on the covers of the documents, reference numbers such as the project contract number, and other identifying information. Such paperwork should be filled out painstakingly, even though it may have to be completed quickly to meet the schedule. Errors could result in security violations.

Editors often deliver technical documents in-house. If documents are intended for an external audience, editors may still be in charge of delivery. Especially for time-sensitive mailings such as contract reports, proposals, and other marketing materials, editors must be careful to arrange on-time delivery. How and when documents reach a customer suggest the agency's way of doing business, just as the documents themselves do.

When a document is delivered, editors are often quickly assigned to another project. However, there is still one essential piece of work to do: *cleaning up.* Editors prepare materials for archives, return or destroy drafts, return borrowed materials, and catalog originals of graphics. Cleaning up a publication project well can simplify work on future publication efforts.

Archiving—the cataloguing and storing of materials used to prepare the document—is important. In cooperation with information specialists, editors often write abstracts of the document for the organization's library or information center; index and catalog it; and label originals of graphics with the name, date, and reference number of the document, and prepare them for storage.

Not all materials used to create a document are saved. Editors should return writers' manuscripts or dispose of them properly, especially classified materials. Borrowed materials such as specifications, standards, and other documents should be returned promptly.

Cleaning up a document effort often involves meeting separately with the supervisors of support groups. When support work did not go as planned, editors should discuss the problem with the appropriate supervisor in a *debriefing*. The best approach editors can take is: "What can I do to make things go more smoothly next time?" Sometimes the problem is something the editor did or did not do. Consequently, editors need to be careful to explore the problem and learn from it, accepting suggestions from support group supervisors.

When the document has been completed, editors should try to make certain that publications department staff are recognized for their contributions. In large organizations, many employees feel that their work isn't recognized or rewarded sufficiently, or that the credit for a successful document is given only to supervisors and managers. Editors should thank support group staff individually for their efforts, perhaps singling out those whose work was exceptional in a memo or report that becomes part of the employee's personnel file.

CONCLUSION

Technical editing students should be able to distinguish between clients, customers, and colleagues, and they should understand an editor's roles in creating technical documents. Editing technical documents involves much more than line-editing text and graphics, because most technical editors have planning and supervisory responsibilities as well.

QUESTIONS FOR REVIEW

1. Who are editors' clients and customers? How do the two groups differ?
2. Who are editors' colleagues? What are their main duties in preparing a technical document?
3. How do editors' responsibilities change when they use desktop or electronic publishing software to produce a document?
4. What are editors' primary responsibilities in the planning stage of a document?
5. What are editors' primary responsibilities in the writing and editing stage of a document?

6. What are editors' primary responsibilities in the production stage of a document?
7. How does editing a document for review of the first draft differ from editing the final document?
8. What five main goals should editors keep in mind as they edit the final draft of a document?
9. Who should be excluded from the final pre-printing review of a document, and why?
10. What might cleaning up a publication effort involve?

CHAPTER THREE

Using Editing and Proofreading Symbols

Editing and proofreading symbols, shown in Figure 3.1, are nearly universal, regardless of subject matter. These symbols are so widely used that charts of them occur in dictionaries intended for the general public. Most of the symbols are used for both editing and proofreading, but some proofreading marks are not used in editing.

Editing symbols are used on paper *(hard copy)* to indicate to a typist, typesetter, or graphic artist the changes the editor wants made when the text of a document is typed on disk for the first time, when the disk version of text is revised, and when graphics are created or revised.

Proofreading symbols are used at different points in creating a document. As discussed in Chapter 2, it is important to remember that "proofreading" is done at many points in creating a technical document, not only at the galley or page proof stages. When the text has been put on disk, the disk version is proofed against the original manuscript, and when graphics are created they are proofed against the original sketches. When the text and graphics have been corrected following final review, and the document is about to *go to press,* the printing masters are proofread.

Editing symbols are used in the text and in sketches for graphics, but usually not in the margins of the document. Proofreading marks are made in the text (or on a copy of a graphic) *and* in the left and right margins of computer processed text and graphics or a set of proofs. If a change is called for in a line, the proofreader indicates what is to be done in the left or right margin, whichever is closer to the point of the change. If a number of changes are to be made in the line, the marks in the margin are separated by slashes.

Having proofreading marks in the margin is very important. When proofs are corrected, typists or typesetters do not read every line of the document.

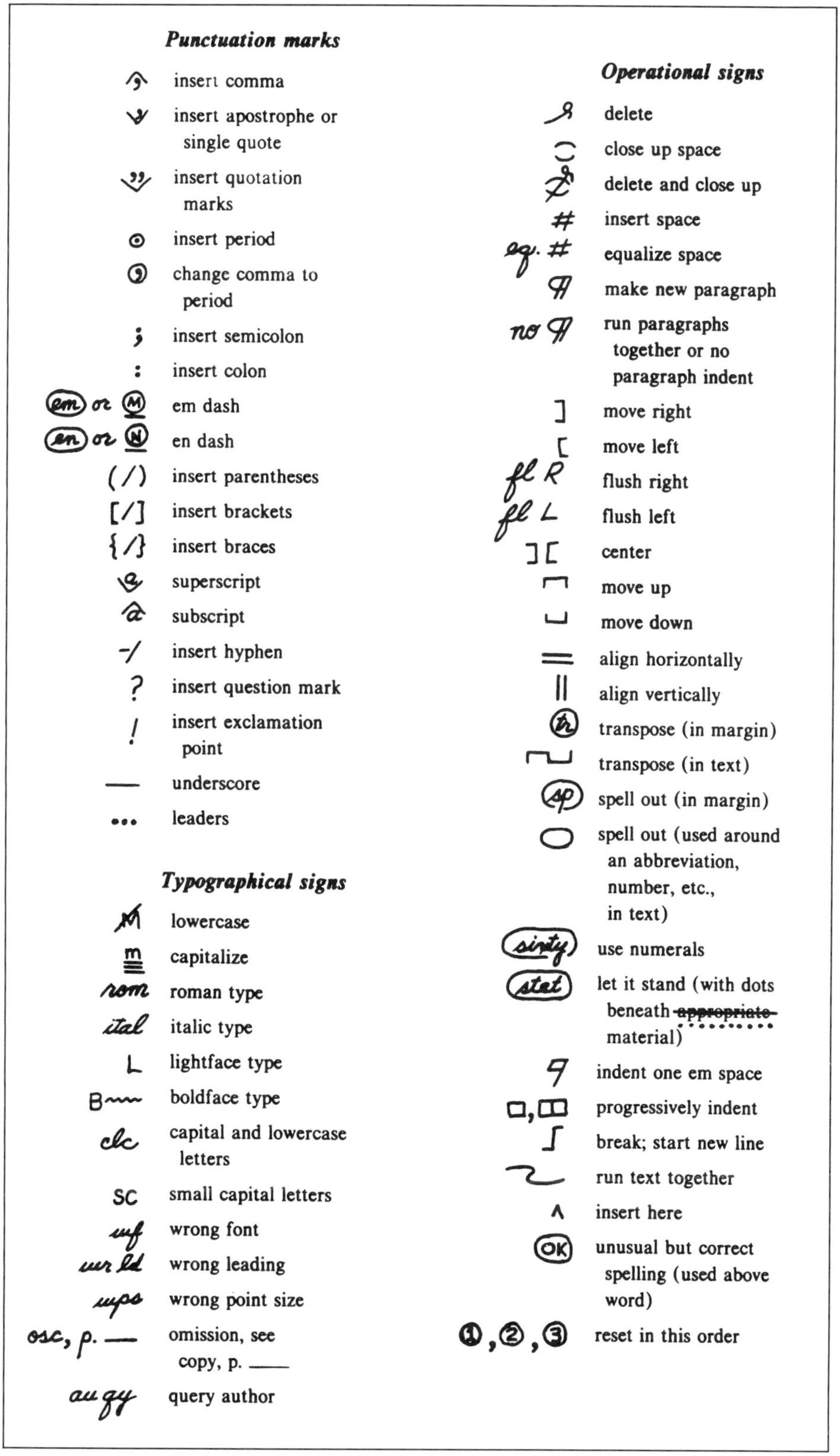

Figure 3.1. Editing and proofreading symbols. (Source: Martin Marietta Energy Systems, *Document Preparation Guide,* p. 6–1. Martin Marietta Energy Systems, © 1983, used with permission.)

Rather, they look in the margins for proof marks that indicate changes are needed nearby. Marking changes in the nearer margin reduces the time it takes them to find and make the changes.

Editors should know what stage the document is in, because editing marks are not always used on new text and graphics. If the text is new (not on disk) and the graphics have not been prepared, editors do not need to mark additions or deletions with formal symbols; they can merely add what is needed or cross out what should not be typed or drawn. When typists or artists work with new material, they type every letter or draw every line as it appears in the manuscript. Consequently, editors should simplify their marking of the material and avoid using marks that aren't necessary.

If the text of a document has been put on disk and the graphics prepared, some editing may still be required, especially when revisions are generated by reviews. At this point, editors must indicate additions and deletions with the proper marks. If a comma is added without using a caret (^), the typist might think the comma was already there and not add it to the disk version. Similarly, material to be deleted from the disk version of the text or graphic should be marked with a delete sign (ꝯ) to signal it clearly.

Proofreaders should always mark each change to text with the appropriate symbol in the margin. If there is room, proofreaders should use the symbol in the line as well, but if the text is tight, a caret or a slash should be used in the text to indicate where to make the change indicated in the margin.

Proofreaders should *never* mark on original graphics. Changes to graphics should be marked only on photocopies or on vellum (tissue) overlays.

The following discussion of the editing and proofreading symbols is task-oriented, focusing on what the editor or proofreader wants to do (for example, add text or change punctuation). For each task, editorial marking is discussed first, then using proofreading symbols.

TO ADD TEXT OR PUNCTUATION: ^

Editing New Text or Graphics

To add words or punctuation to new text or unprocessed graphics, simply add the material. Only if there might be a question where the addition belongs should a caret be used to indicate the place.

Editing Processed Text or Graphics

To add text or punctuation to material that has been typed or drawn, use a caret (^). The tip of the caret should just touch the bottom of the line at the point where the addition should be inserted. Write the letter, word, or words above the caret if there is room. If not, write the addition in the closest margin, circle it, and draw a line from the circle to the top of the caret.

Proofreading

To add letters, words, or punctuation to proofs, place a caret in the text where you want the text or punctuation added. Make sure there is a ∧ in the margin to signal the addition. Print the punctuation mark in the margin. If it is a period, circle it. If it is a single or double quotation mark, invert the caret underneath it to prevent confusion with adding a comma. If it is some other mark of punctuation, place the caret over it. Parentheses ( ), brackets [ ], and braces {} to be added should be drawn in the margin with a slash separating them. Remember, a slash between symbols, letters, or words in the margin indicates that two or more separate changes are to be made in the line.

Examples:

Editing new text:

Recognition often comes late to scientists , as it did to Gregor Mendel and Barbara McClintock.

Editing text on disk:

Recognition often comes late to scientists∧ as it did to Gregor Mendel and Barbara McClintock.

Proofreading text:

Recognition often comes late to scientists ∧ as it did to Gregor Mendel and Barbara McClintock.

TO REMOVE TEXT OR PUNCTUATION:

Editing New Text or Graphics

To delete text or punctuation from new material, simply cross it out or use correcting fluid ("white-out").

Editing Processed Text or Graphics

To delete material from text or copies of graphics that have been processed, use a dele (or), which rhymes with "really." The second symbol above, slightly more complex, is often used, but the simpler first symbol is sufficient. The dele should be made neatly through the letter(s) or lines to be deleted. The dele can also be used to delete a block of material by circling it and drawing a dele through the upper right of the circle. The dele is often used in conjunction with the *close-up* described below.

Proofreading

Draw a dele through the material to be removed from the text or the graphic, and draw a dele in the margin to signal the deletion. If the type is too tightly spaced for clear marks within the text body, use a circled note in the margin and mark the material to be deleted with a dele.

Examples:

Editing new text:

Recognition often comes late to scientists, as it did to Gregor Mendel and Barbara McClintock.

Editing text on disk:

Recognition often comes late to scientists, as it did to Gregor Mendel and Barbara McClintock.

Proofreading:

Recognition often comes late to scientists, as it did to Gregor Mendel and Barbara McClintock.

TO CHANGE TEXT OR PUNCTUATION

Editing New Text or Graphics

To change letters, words, or punctuation in new material, simply cross out neatly what you don't want and write in the correction.

Editing Processed Text or Graphics

To make changes in text or copies of graphics that have been processed, draw a dele through the letters, words, or punctuation to be deleted and write above the material or in a circled note in the margin what is to be added in its place.

Proofreading

Draw a dele through the material to be removed and draw a dele in the margin to signal it. Place a caret in the text at the point of insertion and write the change over a caret in the nearer margin. Again, if the text is tightly spaced, the change can be clarified by a circled note in the margin.

Examples:

Editing new text: Gregor Mendel

Recognition often comes late to scientists, as it did to ~~Isaac Newton~~ and Barbara McClintock.

Editing text on disk: Gregor Mendel

Recognition often comes late to scientists, as it did to ~~Isaac Newton~~ and Barbara McClintock.

Proofreading: Gregor Mendel

Recognition often comes late to scientists, as it did to ~~Isaac Newton~~ and Barbara McClintock.

TO CLOSE UP SPACE: ⁀

Editing

To indicate that space should be deleted between letters, words, or lines of text in new or processed text or graphics, use the close-up or *ligature* (⁀).

When a dele has been used to remove a word or a line, or when typed text has uneven word spacing because right justification was used, editors do not have to indicate a close-up. The typist or typesetter will assume there is to be no blank space.

If a deletion forms two words that could be separate or joined as one, editors should indicate whether or not the space is to be closed up. For example, marking "wavellength" by calling for the deletion of one "l" could result in "wave length" or "wavelength." If the editor intends "wavelength," the original should be marked "wavellength." However, marking "wavelength" would call for two words. If there might be confusion, the editor or proofreader can write (one word) or (two words) in the margin.

Proofreading

Proofreaders should err in the direction of overindicating rather than underindicating what is to be done about letter spacing and word spacing, to reduce time wasted when typesetters or artists have to ask about the change or correct their misinterpretation.

If typists or typesetters are in doubt about whether terms should be joined as one, hyphenated, or typed as two words, they usually consult a reference work such as Silverthorn's *Word Division Manual*, the U.S. Government Printing Office (GPO) *Style Manual* or its *Word Division Supplement*, or the company or agency's style guide. The guide should list the preferred treatment of troublesome compounds that are commonly used in the organization's documents—for example, "wire wrap" versus "wirewrap" in an electronics firm's style guide. A sample of instructions from the GPO *Style Manual* section on handling compounds appears in Figure 3.2.

TO ADD SPACE IN TEXT: | and

Two symbols are used to indicate that space should be added in processed text and graphics: | and #. The vertical line indicates that one space should be added between the letters or symbols divided by the mark. The pound sign is used to indicate that a space should be added between the two lines separated by it.

Editing New Text or Graphics

Editors should count on typists and artists to add space as necessary, so symbols for adding space are usually not needed in new material. However, in rare situations where confusion might exist, the editor should mark for space as done for processed material. For example, if the typists would probably not know that "wire wraps" is to be two words, the editor can save them time by marking "wire|wraps." If the text sent to be typed is single-spaced, editors

Solid compounds

6.8. Print solid two nouns that form a third when the compound has only one primary accent, especially when the prefixed noun consists of only one syllable or when one of the elements loses its original accent.

airship	cupboard	footnote
bathroom	dressmaker	locksmith
bookseller	fishmonger	workman

6.9 Print solid a noun consisting of a short verb and an adverb as its second element, except when the use of the solid form would interfere with comprehension.

blowout	holdup	setup	*but* cut-in
breakdown	makeready	showdown	run-in
flareback	markoff	throwaway	tie-in
giveaway	pickup	tradeoff	
hangover	runoff		

6.10. Compounds beginning with the following nouns are usually printed solid.

book	house	school	way
eye	mill	shop	wood
horse	play	snow	work

Figure 3.2. Sample guidelines for handling compound terms. (Source: Government Printing Office *Style Manual,* p. 74.)

should write "Please double space" at the top of the first page and circle the request if double-spaced text is desired.

Editing Processed Text or Graphics

Use the space symbols | and # as needed to indicate space to be inserted in the text or callouts (wording) on a graphic. For graphics, a circled instruction for the artist is preferable to the symbol alone.

Proofreading

Proofreaders should use the symbols as described above, marking in the text (or on a copy of the graphic) and in the margins.

TO CAPITALIZE A LETTER OR WORD: ≡ and =

Editing and Proofreading

To capitalize a letter or word, underline it three times for full-sized capitals or twice for *small capitals,* which are used in abbreviations such as BC and PM.

Examples:

The motor runs on dc power provided by batteries.
Archimedes developed the principle in the third century bc.

Editors and proofreaders use the underlining for capitals similarly in text and copies of graphics, but proofreaders should also write (caps) in the margin.

TO LOWERCASE A LETTER OR WORD: /
Editing and Proofreading

Indicate that a capital letter should be lowercase by drawing a slash through it. To lowercase a word, draw a slash through the first letter and extend the top of the slash horizontally over the word.

Examples:

> Power is measured in Watts.
> Maximum speeds for aircraft are often indicated in MACH numbers.

TO BEGIN A NEW LINE:
Editing and Proofreading

To indicate that one line should end at a particular point and the text continue on the next line, insert the line break mark after the last word you want on the line.

Example:

> Recognition often comes late to scientists, as it did to Gregor Mendel and Barbara McClintock.

In proofs, write (tr) in the margin.

TO CONTINUE A LINE:
Editing and Proofreading

To indicate that text should continue on the same line, draw an arrow to connect the following text.

Example:

> Recognition often comes late to scientists, as it did to Gregor
> Mendel and Barbara McClintock.

TO BEGIN A NEW PARAGRAPH: (¶)
Editing and Proofreading

To indicate the beginning of a new paragraph in new or processed text, insert (¶) (the paragraph sign inside a circle).

Example:

> Recognition often comes late to scientists, as it did to Gregor Mendel and Barbara McClintock. (¶) Some scientists, however, are recognized and rewarded earlier in their careers, as Stephen Hawking has been.

TO CONTINUE A PARAGRAPH: no ¶
Editing and Proofreading

To indicate that new or processed text should be added to the end of the previous paragraph, insert no ¶ at the beginning of the paragraph to be run in.

Example:

> Recognition often comes late to scientists, as it did to Gregor Mendel and Barbara McClintock.
>
> no ¶ Some scientists, however, are recognized and rewarded earlier in their careers, as Stephen Hawking has been.

Sometimes typists prefer a *run-in* line in place of the no ¶ symbol.

Example:

> Recognition often comes late to scientists, as it did to Gregor Mendel and Barbara McClintock.
>
> Some scientists, however, are recognized and rewarded earlier in their careers, as Stephen Hawking has been.

TO WRITE OUT A NUMBER OR ABBREVIATION:
Editing and Proofreading

To indicate that a number or an abbreviation should be written out as a complete word, circle it. Most style guides recommend writing the numbers ten and below as words and 11 and above as numerals. Also, as most style guides recommend writing out numbers that begin a sentence, large numbers should be repositioned. Conversely, circle a written-out number if it should be a numeral.

Examples:

> 2 motors will be shipped for testing; the other 23 will be stored until after testing.
>
> Twenty-three motors will be stored while the two other motors are tested.

The second sentence will have an awkward combination of written-out and numerical forms, so it might be better to rephrase the sentence: "While we test 2 of the motors, we will store the other 23."

TO TRANSPOSE TEXT:
Editing New Text or Graphics

To reorder symbols, letters, or words in new material, it is often best to cross out the unwanted version and write the correction above it.

Example:

> Recognition often comes [late] to scientists, ~~late,~~ as it did to Gregor Mendel and Barbara McClintock.

Editing Processed Text or Graphics

To transpose symbols, letters in a word, or words in a line of text, it is sometimes proper to strike the material and rewrite it, as above. Often, however, rewriting is unnecessary. Editors should mark only the symbols, letters, or words to be moved with a transposition symbol (∿). To move text from one line to another, it is usually easier for typists if editors circle what is to be moved and draw an arrow to where it should be inserted.

Example:

> Recognition comes often to scientists late, as it did to Gregor Mendel and Barbara McClintock. Some scientists are recognized and rewarded early in their careers, as Stephen Hawking has been.

The transposition symbol must be drawn carefully, with the ends hooking into the line, as in the example. If it is drawn sloppily, the typist might revise the sentence incorrectly—here, to read, "Often recognition comes. . . ."

Proofreading

Proofreaders should use the transposition symbol within a line and in the margin, or use the circle and arrow. They should not rewrite text, as typesetting the rewrites may result in new errors. In tight text, proofreaders can draw a line through what is to be changed and write the correct reading in a circle in the margin.

TO CENTER ON THE PAGE:] [
Editing and Proofreading

Material to be centered on a page, such as first-order heading or a caption, is marked with reversed brackets:] [.

Examples:

>]CHAPTER 5[
>]Figure 5.1. Glen Canyon Dam[

Proofreader should also make these marks in the margin.

TO MOVE MATERIAL: [,] , ⊔, ⊓, and (ALIGN)
Editing New or Processed Text or Graphics

To align text, use a bracket to indicate the direction of movement.

Example:

> Editors can reduce the time and cost of having text typed and graphics drawn for a technical document by making sure that their copymarking follows standard symbols used in nearly all organizations.
> [Typists and artists are familiar with these conventional symbols and can interpret them effortlessly if they are drawn neatly.

In this example, the typist would align the second paragraph with the first, retaining the indentation. Brackets are used in the same way to indicate movement in other directions.

Parallel lines can be drawn alongside a body of text and the text it is to be aligned with. However, this may confuse the typist about what should be aligned with what.

Example:

> Typists and artists are familiar with these conventional symbols and can interpret them effortlessly if they are drawn neatly. If they are not drawn neatly, however, time will be wasted when the typist has to ask the editor what he or she wanted or when the editor has to correct the typist's interpretation.

Proofreading

Proofreaders should use parallel lines as in the example above, with (align) in the margin. If there is any doubt about what is to be aligned, the proofreader should explain in a circled note in the margin.

Example:

> Typists and artists are familiar with these conventional symbols and can interpret them effortlessly if they are drawn neatly. If they are not drawn neatly, however, time will be wasted when
>
> the typist has to ask the editor what change was wanted or when the editor has to correct the typist's interpretation.
>
> (align right column with left)

TO INDICATE ITALICS: ——
Editing and Proofreading

Editors and proofreaders should underline material to be italicized with one straight line. Proofreaders should also write (ital) in the margin.

Example:

> Crotalus abyssus abyssus is native to the Grand Canyon.

TO INDICATE BOLDFACE TYPE: ~~~~
Editing and Proofreading

To indicate boldface type, editors and proofreaders should draw one wavy line underneath. Proofreaders should also write (bold) or (bf) in the margin.

Example:

> NEVER operate this equipment with the shield off.

Boldface type is used in mathematical material to indicate vectors (line segments with direction). To indicate that x is a vector to be boldfaced, it should be marked x.

TO INSERT A DASH: — or –

Editing and Proofreading

On typewriter and word processor keyboards, two hyphens with no space before or after them indicate a dash. As discussed in Chapter 12, editors and proofreaders working with typeset material should distinguish between em and en dashes, which have different uses. To indicate the type of dash desired, print a line or double hyphen with an M or N below it. Editors as well as proofreaders should write — or – in the margin alongside the line in the text containing the dash to avoid confusing it with a hyphen.

Example:

> Mr. Johnson wants these changes made to the proposal--if it's not too late to make them.

TO INSERT A HYPHEN: =

Editing New Text or Graphics

In new text, editors should add only the hyphens needed in compounds (sometimes called *hard* hyphens) according to the style guide in use. In some situations, especially those involving chemical nomenclature, many style guides call for the copymarking of all hard hyphens with an equals sign: = . Most typists and artists can be counted on to hyphenate words correctly, so editors should mark hyphenation in new text or graphics only when there might be confusion.

Example:

> The I-beam testing will be completed next week.

Editing Processed Text or Graphics and Proofreading

Editors and proofreaders call for the insertion of a hyphen with a caret at the point of insertion and an equals sign above it. (Proofreaders should also write (hyph) in the margin.)

Example:

> The I=beam testing will be completed next week.

TO INSERT SUBSCRIPTS AND SUPERSCRIPTS: ∧ and ∨

Editing New Text or Graphics

Editors inserting subscripts and superscripts in new text or graphics need only print them where needed. If the typist or artist might be confused, however, the editor should use carets as described below.

Editing Processed Text or Graphics

In processed text or graphics, editors should use carets to mark subscripts and superscripts.

Example:

$$abc = h_z y + \{j^2 [p_z - k_z (h + y) \times q]\}^4$$

Proofreading

In tight text, proofreaders should correct with a dele within the line and a caret and subscript or superscript in the margin. If a number or letter in the text needs to be sub- or superscripted, the proofreader should mark it with a caret and print the correction in the margin.

TO OVERRIDE A CHANGE: . . . (stet)
Editing and Proofreading

Editors and proofreaders occasionally make changes they later decide against. In new text, white out the change; in processed text or graphics, neatly cross out the change, then put a row of dots under the original and print (stet) in the nearer left or right margin.

Example:

(stet) Recognition ~~often comes~~ late to scientists, as it did to Gregor Mendel and Barbara McClintock.

TO SIGNAL A QUESTION FOR THE AUTHOR: (QA)
Editing and Proofreading

When editors or proofreaders have a question for the author, they should write QA and the question (concisely stated, circled) in the margin.

Example:

Edison never had a rival in the development of the use of electricity (QA: what about Tesla?)

COPYMARKING EXERCISE

Copymark the following discussion of galley and page proofs, which is new text not yet on disk. Edit lightly—the paragraphing needs some attention, and there are many small errors, but only one sentence needs major revision.

Galley proofs are different from page proofs, and
the proof-reader or editor who checks these proofs must
understand the differences between them.
Galley Proofs are pulled (printed) after the text of
the the manuscript has been tpyeset. Often the galleys
are on long stripes of paper that holds far more than
the text of one page in the final print version. The
galleys do not have graphics- figures or tables, in
them.

They may be pulled on cheap paper, and be faint in
places or poorly alined, but there is no need to mark
this. Galley proofs should be checked character by
character against the copy the typesetter used.
Changes should be made in the text (if possible) and
marked in the margin, or marked in the text and
written in the margin if there is not room in the
text.

Page proofs are pulled after the correction to the
galleys have been made by the typesetter, and each page
of the job has been made up. Each page of the proofs
correspomds to one page of the finished job. Page
proofs contain graphics that must be proofread. The
quality of photographs may be poor in page proofs, but
the appearances of all other graphics and the text
should be good, or marked for fixing.

Page-proofs should be checked against the changes
in the galleys, to make sure that all errors were
corrected. Any line in which a correction was called
for should be checked agianst the original manuscript
(ms) to ensure that no new errprs were introduced in
the correction of the errors in the galleys.
[There is no need to check the rest of the page proofs
against the ms, but the ms is usually returned with
them, for reference.] Graphics should be checked to
insure that each is labelled, placed and titled
properly. Running heads and page numbers should be
checked. Long or short pages should be marked with the

simplest possible fixes. As with the galleyproofs, changes should be marked in the text and written in the margin. Or made in the text (if possible) and marked in the margin.

When proofreading, marks should be made only on the proofs, not on the original manuscript. Also, no copy editing should be done, because changes at this point in the process publication are to costly. In reading galleys or pageproofs, the editors aim is to speed up, not hinder the process.

Answer Key

Galley proofs are different from page proofs, and the proof reader or editor who checks these proofs must understand the differences between them.
Galley Proofs are "pulled" (printed) after the text of the the manuscript has been tpyeset. Often the galleys are on long stripes of paper that holds far more than the text of one page in the final print version. The galleys do not have graphics figures or tables in them.
They may be pulled on cheap paper and be faint in places or poorly alined, but there is no need to mark this. Galley proofs should be checked character by character against the copy the typesetter used. Changes should be made in the text (if possible) and marked in the margin, or marked in the text and written in the margin if there is not room in the text.

Page proofs are pulled after the correction to the galleys have been made by the typesetter and each page of the job has been made up. Each page of the proofs correspomds to one page of the finished job. Page proofs contain graphics that must be proofread. The

quality of photographs may be poor in page proofs, but
the appearances of all other graphics and the text
should be good, or marked for fixing.
Page proofs should be checked against the changes
in the galleys, to make sure that all errors were
corrected. Any line in which a correction was called
for should be checked aganst the original manuscript
(ms) to ensure that no new errors were introduced in
the correction of the errors in the galleys.
There is no need to check the rest of the page proofs
against the ms, but the ms is usually returned with
them, for reference. Graphics should be checked to
insure that each is labelled, placed and titled
properly. Running heads and page numbers should be
checked. Long or short pages should be marked with the
simplest possible fixes. As with the galley proofs,
changes should be marked in the text and written in the
margin. Or made in the text (if possible) and marked
in the margin, or

When proofreading, marks should be made only on
the proofs, not on the original manuscript. Also, no
copy editing should be done, because changes at this
point in the process publication are to costly. In
reading galleys or page proofs, the editors aim is to
speed up, not hinder the process.

Discussion of Possible Copymarking

The copymarking of this passage is based on the guidelines for editing text in Chapter 4, for grammar in Chapter 11, and for punctuation in Chapter 12.

Paragraph 1

The hyphen in "proof-reader" should be removed, and a ligature used to indicate that "proofreader" is one word. Other than that, the first paragraph is correct as it stands. The statements could be presented in two sentences. However, the writer has chosen to combine the related ideas in one sentence that is not overly long or complex, so the editor should let the sentence stand.

Paragraph 2

The beginning of this paragraph should be marked for indenting. The "P" in "Proofs" should be lowercase, as in paragraph 1, and "pulled" should be put in quotation marks to indicate that the word is being focused on as a term. The second "the" after "of" should be deleted, and the spellings of "typeset" and "strips" should be corrected. The "s" in "holds" must be deleted so the verb agrees with its subject, "strips." The hyphen after "graphics" and the comma after "tables" should be marked as one-em dashes, to set off the phrase "figures or tables." (Parentheses could be used in place of the dashes.) Commas should not be used in place of the dashes after "graphics" or "tables" because a comma after "graphics" would seem to set up a series of three components.

Paragraph 3

The first sentence should be run into paragraph 2 because it continues the focus on defining galley proofs. A new paragraph 3 should begin with the second sentence, as the writer's focus shifts to what proofreaders do with galley proofs. In this new third paragraph, the parentheses around "if possible" could be dropped, but the editor should assume the writer has a reason for deemphasizing the phrase by putting it in parentheses. There is no need to break the last sentence of the paragraph into two, although it is somewhat complex. The editor should not mark the alignment problem; as this is new text, the typist will automatically fix the problem.

Paragraph 4

"Correction" should be changed to "corrections" to agree with the plural verb "have been made." The comma after "typesetter" should be deleted because the following "and" does not join two main clauses—they are subordinate clauses introduced by "after." The spelling of "corresponds" should be corrected, and the "s" dropped from "appearances." The comma after "good" should be removed because "or" joins two adjectives completing the statement about "appearance," not two main clauses.

Paragraph 5

The hyphen in "Page-proofs" should be deleted for consistency with the earlier usage. The comma after "galleys" is removed to emphasize why the page proofs should be checked against the corrections to the galleys. The spellings of "against" and "errors" should be corrected, and "ensure," meaning "to make sure," should be substituted for "insure." The abbreviation "ms" should be capitalized. Parentheses should be used in place of brackets, and the comma after "them" is not needed. The American spelling is "labeled" (British is "labelled"), and most style guides would add a comma after "placed" to emphasize the three separate modifiers in that series. "Galleyproofs" should be marked as two words. The last sentence of the paragraph should be revised to parallel the final sentence of paragraph 3. As making changes in the proof and marking them in the margins is preferable to writing changes in the margin, that alternative should be presented first. A comma should be used after "margin" and before "or" to help the readers separate the two alternatives; without the comma, readers might read the sentence as "made in the text and marked in the margin or marked in the text."

Paragraph 6

There is no need to mark the close-up of the space between paragraphs 5 and 6. This is new text, so the typist will close such a space unless instructed not to. The first sentence has a dangling participle, "proofreading"; who is doing the proofreading is not made clear in the sentence. The error is best corrected by interpreting "proofreading" as a modifier of "marks" and deleting "When" and the comma. The error could be corrected by inserting "proofreaders" in front of "marks" and changing "marks" to "mark," and deleting "should be made," but that would create awkward repetition and reduce the emphasis on the marks. "Copyediting" is properly one word, but it could be interpreted as two, so a ligature should be used. The words "process" and "publication" should be transposed. The correct word "too" should be substituted for "to," and space should be inserted between "pageproofs." An apostrophe must be added to "editors," and a comma should be added after "hinder" to set off the nonrestrictive phrase "not hinder."

Minimizing Copymarks

Editors can simplify typists' work by minimizing the number of editing marks on a manuscript. One way to accomplish this is to use "white-out" for covering errors. Instead of copymarking the manuscript as above, the editor could use white-out for many of the deletions and spelling corrections, providing the typist with a cleaner copy.

Using white-out does take longer than using standard editing marks. For a small document with a tight typesetting deadline or budget, however, the extra editing time spent using white-out might save valuable time later, if fewer corrections are needed. Editors should decide on a case-by-case basis which method is better.

PROOFMARKING EXERCISE

The processed text of the discussion of galley and page proofs appears below, in a galley proof. Compare it to the edited version of the discussion and mark the proof as needed.

Galley proofs are different from page proofs, and the proof reader or editor who checks these proofs must understand the differences between them.

Galley Proofs are pulled (printed) after the text of the the manuscript has been typeset. Often the galleys are on long strips of paper that hold far more than the text of one page in the final print version. The galleys do not have illustrations-figures or tables-in them. They may be pulled on cheap paper and be faint in places or poorly aligned, but there is no need to mark this. Galley proofs should be checked character by character against the copy the typesetter used. Changes should be made in the text (if possible) and marked in the margin, or marked in the text and written in the margin if there is not room in the text.

Page proofs are pulled after the correction to the galleys have been made by the typesetter, and each page of the job has been made up. Each page of the proofs corresponds to one page of the finished job. Page proofs contain illustrations that must be proofread. The quality of photographs may be poor in page proofs, but the appearance of all other illustrations and the text should be good or marked for fixing.

Page proofs should be checked against the changes in the galleys to make sure that all errors were corrected. Any line in which a correction was called for should be checked against the original manuscript (MS) to ensure that no new errors were introduced in the correction of the errors in the galleys. There is no need to check the rest of the page proofs against the MS, but the ms is usually returned with them for reference. Illustrations should be checked to insure that each is labelled, placed, and titled properly. Running heads and page numbers should be checked. Long or short pages should be marked with the simplest possible fixes. As with the galley proofs, changes should be made in the text (if possible) and marked in the margin, or marked in the text and written in the margin.

Proofreading marks should be made only on the proofs, not on the original manuscript. Also, no copy editing should be done, because changes at this point in the publication process are to costly. In reading galleys or page proofs, the editors' aim is to speed up, not hinder the process.

The proofmarked version appears below.

Galley proofs are different from page proofs, and the proof reader or editor who checks these proofs must understand the differences between them.

Galley Proofs are pulled (printed) after the text of the the manuscript has been typeset. Often the galleys are on long strips of paper that hold far more than the text of one page in the final print version. The galleys do not have illustrations figures or tables in them. They may be pulled on cheap paper and ~~be~~ faint in places or poorly aligned, but there is no need to mark this. Galley proofs should be checked character by character against the copy the typesetter used. Changes should be made in the text (if possible) and marked in the margin, or marked in the text and written in the margin if there is not room in the text.

Page proofs are pulled after the correction to the galleys have been made by the typesetter and each page of the job has been made up. Each page of the proofs corresponds to one page of the finished job. Page proofs contain ~~illustrations~~ that must be proofread. The quality of photographs may be poor in page proofs, but the appearance of all other ~~illustrations~~ and the text should be good or marked for fixing.

Page proofs should be checked against the changes in the galleys to make sure that all errors were corrected. Any line in which a correction was called for should be checked against the original manuscript (MS) to ensure that no new errors were introduced in the correction of the errors in the galleys. There is no need to check the rest of the page proofs against the MS, but the ms is usually returned with them for reference. ~~Illustrations~~ should be checked to insure that each is labelled, placed, and titled properly. Running heads and page numbers should be checked. Long or short pages should be marked with the simplest possible fixes. As with the galley proofs, changes should be made in the text (if possible) and marked in the margin, or marked in the text and written in the margin.

Proofreading marks should be made only on the proofs, not on the original manuscript. Also, no copy editing should be done, because changes at this point in the process publication are to costly. In reading galleys or page proofs, the editor's aim is to speed up, not hinder the process.

CHAPTER FOUR

Editing Text

Editing text well first requires attention to underlying theoretical issues of communication: an awareness of the audiences of technical documents, and familiarity with theories of readability, legibility, and usability.

THE AUDIENCES OF TECHNICAL INFORMATION

The world is full of people who read about technical subjects. They read Steven Hawking's *A Brief History of Time* (which has sold more than 1 million copies), *National Geographic* articles on gorillas, magazine articles and U.S. government publications on health issues, installation instructions for VCRs and other electronics, advertisements for equipment and machinery, and so on. Some of these "technical readers" are well educated in science; some are not. Some are interested in science or in technology (the application of the principles of science) for its own sake. Others read technical material only to find out how to hook up a garage-door opener or to decide which outboard motor to buy. Some readers digest information quickly, whereas others take longer to grasp concepts.

The more homogeneous a technical document's primary audience, the easier it is for writers and editors to make sure the material will be interesting and presented at the proper pace, with the appropriate language and degree of explanation. For example, if marketing staff estimate that most users of a set of instructions for installing a garage-door opener have at least a high school education, middle incomes, and limited technical experience, but they are motivated to install the opener, the instructions can be written in matter-of-fact statements with nontechnical terms. Simple drawings of parts and descriptions of tools would be appropriate. If the instructions are written for professional carpenters, they can be shorter, and writers can assume familiarity with the equipment and procedures involved.

When the audience is broader (such as readers of a word processing program manual), the more difficult it becomes for writers and editors to present material for a range of readers with different computer experience and reading comprehension skills.

Some technical writing textbooks address the problem of audience analysis by positing specific types of audiences, usually some variation on the following: lay readers, technicians, executives, experts, and a combined audience of some or all of the four. This classification rightly recognizes that readers have different degrees of familiarity with a particular technical subject, from lay readers with minimal knowledge to experts who know all about it. Also, this system recognizes that readers will have different uses for the information in any given technical document, from technicians who manufacture parts to executives who make decisions about plant expansion.

However, such a classification system is often based on a questionable premise: that level of education differentiates the audience. Lay readers supposedly have high school degrees; technicians have associate's degrees or perhaps two years of BS study in science or technology; executives have MBAs; and technical experts have PhDs (or at least an MS). Such a division is simplistic, because there are too many technical experts with only BS's (and perhaps with just high school degrees), too many lay readers with PhD's, and too many successful executives without MBA's to distinguish audiences on the basis of education.

Rather than dividing readers by education, editors should ask two questions: "How much do the readers know about the topic?" and "How interested are they in the topic?" A philosophy professor reading about the Hubble telescope may know very little about it but be very interested. An optics technician may know a great deal about the Hubble but not be interested if he or she has read about it many times.

The philosophy professor uneducated in astronomy but interested in the Hubble telescope is an example of a *lay* audience, as is a high school student uneducated in astronomy and not interested in the Hubble. As lay readers, they have a limited knowledge of the subject, whatever their degree of interest. That degree of interest is very important for the writer, however, who can vary his or her techniques, as discussed below.

The optics technician reading about the Hubble is an example of an *expert* audience, as a physicist or an amateur astronomer might be. These expert readers are probably familiar with the subject, whether it is central to their profession or just a hobby. However, that does not mean that these readers would necessarily be interested in a discussion of the Hubble.

Assuming that an advanced degree in science or technology characterizes the expert reader of technical documents is specious. Many experts in a scientific field are more widely known for their work in another field. For example, Vladimir Nabokov, the famous novelist, was an accomplished lepidopterologist who held a research fellowship in entomology at Harvard University while he was publishing poetry, essays, stories, and novels, before he went on to teach Russian literature at Cornell University. Also, some experts in science do not have advanced degrees in the field. An assistant curator of the Museum of the Rockies at Montana State University, who did not have a college degree, was awarded National Science Foundation and MacArthur Foundation grants for study of dinosaur nesting sites. He discovered 400 dinosaur eggs, including one with the embryo intact, and has been recognized as an authority on dinosaur nesting.

The *lay* audience and the *expert* audience, then, represent the extremes of a continuum; there is a broad group between them that makes up the *middle* audience.

The Lay Audience

Lay audiences don't know a great deal about the subject, so writers and editors should provide background information, avoid unfamiliar terms (or define them if they are essential), and avoid references to unfamiliar processes and equipment. They should write in a simple style, use easily understood comparisons, and avoid equations and mathematical formulas. A slow pace, developing each point before moving on to the next, will give the audience time to absorb each point. Explanations should be overdone rather than underdone. The lay audience might not be motivated to read the article or document, so writers and editors should consider ways to pique their curiosity—such as how the information could save them money, warn them of health hazards, or reveal something interesting about their friends or family. For a lay audience, writers should use simple graphics: informal tables, photographs, drawings, maps, cartoons, flow charts, and line, bar, and circle graphs.

The Middle Audience

Middle audiences are the hardest to reach because of the variation in how much they know about the subject. Writers should again provide background information, although less than with the lay audience, use simple comparisons, and avoid unfamiliar terms or lengthy discussions of unfamiliar processes and equipment. Keep the prose simple and the pace moderate, avoid complex equations and formulas, and explain terms and processes. Graphics for a middle audience might include informal and simple formal tables, line and bar graphs, photographs, perspective drawings, organization charts, schedules, simple logic diagrams, and more complex circle graphs, maps, and flow charts than for the lay audience.

The Expert Audience

Expert audiences do not need background information unless a new idea or unfamiliar process is being discussed. Even then, little background information will be required. For the expert audience, writers and editors should keep the prose simple (as with the other audiences), but equations and formulas can be presented as necessary. Expert readers will expect to see how a writer's results were arrived at, so data and its manipulation should be shown in full. The pace should be rapid, and explanations given only for obscure equipment or processes. Graphics for experts can include formal tables, photographs, perspective and isometric drawings, and more complex line graphs, flow charts, maps, and schedules than for a middle audience. Exploded view, cutaway, and cross-section drawings; schematics; wiring and block diagrams; scatter charts; and graphs with semilogarithmic and logarithmic scales are also appropriate.

The Combined Audience

In addition to the lay, middle, and expert audiences, the *combined* audience is composed of two or all three of the other audiences. Many technical documents are read by combined audiences; however, it is rare that all of the audiences read the same sections. For example, a long article on the Stealth bomber might have lay readers (such as a senator unfamiliar with avionics and aerospace engineering), middle readers (such as a pilot, or a reader of *Aviation Week and Space Technology*), and expert readers (such as a designer of jet aircraft engines or radar-absorbing structures). The senator might not attempt to read the technical sections, while the designer probably would not read introductory discussions of the technology or the purpose of the aircraft. For such a *mixed* or *combined* group, writers and editors should identify clearly the subject matter of document sections in the table of contents so readers can identify the sections of interest. Within each section, the discussion should be appropriate for its audience. The summary or introduction should provide a clear, specific overview of the entire document because most readers of a document read the summary or introduction.

Determining Audience

Writers need to know who their readers are before they can write appropriately on a subject, and they often come to editors for that information. Earlier documents prepared for the audience (or a similar one) can be examined to get a sense of how to approach the task, but only if those documents were successful. Editors need to know whether readers will use the document for entertainment, for helping them do their job better, for learning how to do something new, or for other purposes. Also, editors should read outside publications targeted at their readers to see how diction, quotation, sentence length and complexity, graphics, and patterns of organization have been used. Try to involve future readers in reviews of document drafts, as is done in usability testing (described below). Most of all, editors must keep the audience's knowledge and interest (motivation) in mind. Editors should never assume readers know enough to understand—or are interested enough to struggle through—a discussion that does not present interesting, accessible material. Editors should help writers ask questions about their readers, not make assumptions about them.

Finally, editors must distinguish between primary and secondary audiences (see Chapter 2). In most situations, documents for external audiences will be reviewed in-house before they are printed and distributed to the outside audience. The in-house reviewers are the secondary audience; the outside readers (such as the firm's customers or the buyers of a magazine) are the primary audience. Editors must satisfy their secondary audience of the quality of a document as they try to make it communicate effectively with the primary audience. Usually this is not a problem when the secondary audience is familiar with the primary audience. Sometimes, however, editors find that their secondary audience (in-house reviewers) want changes in the document that might make it less effective for the primary audience. In this situation, editors should present their case (not argue it) and follow their supervisor's instructions. They should always remember that editors exist to help writers—but the boss is always the boss.

READABILITY, LEGIBILITY, AND USABILITY

To be effective, technical documents should be readable, legible, and usable. More and more software packages contain formulas designed to measure readability; computer equipment manufacturers are increasingly concerned with legibility; and technical communicators are becoming more concerned with how easy their documents are to use.

Readability

"Readability" means "understandability." The more readable a document is, the more easily it can be understood by a broader audience, ranging from those who read well to those who read poorly. Readability testing of materials for adult readers originated in the 1930s and was popularized in the 1940s by Rudolph Flesch. He developed a formula that could measure how easily readers could understand a document. Flesch's formula and other readability indexes use representative samples from a document, passages of at least 100 words chosen from various places. Sentence length and the proportion of polysyllabic words are measured, and a formula is used to calculate how easily the document can be understood. The basic assumption is that longer sentences and words make writing harder to understand.

Many readability formulas exist today, and editing students should be familiar with some popular ones, such as Flesch's Reading Ease Formula, Robert Gunning's Fog Index, and J. Peter Kincaid's adaptation of Flesch's formula, the Flesch-Kincaid Index. They are applied here to the passage below (part of which appears in its final format in Figure 4.1, in the discussion of legibility).

> IMPORTANT NOTICE TO ALL POLICYHOLDERS
>
> When reviewing your enclosed auto policy renewal, you will notice your policy term is now six months rather than twelve months. The premiums shown on your declarations page are one-half the amounts that would have been charged for a twelve-month term.
>
> At ———— we are continually looking for ways to provide our policyholders quality coverage and service at a competitive price. Controlling our operating costs is one way we can continue to offer you the best rates possible. In making this change, we will be better able to control our costs while improving our ability to respond to the needs of our policyholders.
>
> We believe this change will benefit you, and we look forward to continuing to serve you in the years ahead.
>
> P.S.: The payment plan you have selected will have no significant changes. You will continue to have the same number of payments due each year. We have added a nominal service charge to our direct bill installment plan. If a service charge has been applied, it will be noted separately on your bill.

Flesch Reading Ease Formula calculation

The Flesch formula calculates a score of 0–100 for a passage. The higher the score, the greater "reading ease" of the passage. A passage that scores 90–100 is rated as "very easy," requiring only a fourth grade education; a passage that scores 0–30 is rated "very difficult," requiring a college education. The calculation is based on average sentence and word length:

$$\text{Reading ease} = 206.835 - (1.015\text{ASL} + 0.846\text{ASW})$$

where ASL = average sentence length and ASW = average number of syllables per 100 words.

The sample has 178 words in the text (not counting "P.S.") in ten sentences. The average sentence length is 17.8 words, and the average number of syllables per 100 words is 155. Therefore,

$$\text{RE} = 206.835 - (1.015 \times 17.8 + 0.846 \times 155) = 57.6.$$

On the Flesch scale, a passage that measures 50–60 is rated "fairly difficult" and requires "some high school" education.

Gunning Fog Index calculation.

The Gunning Fog Index is also based on average sentence length, but Gunning attempted to refine measuring the difficulty of the diction by focusing on "hard" words of three syllables or more, not counting proper nouns or those created by combining words, forming plurals, or conjugating verbs.

$$\text{Fog index} = 0.4\ (\text{ASL} + \text{TSW})$$

where ASL = average sentence length and TSW = number of trisyllabic words per 100.

In the passage, "reviewing," the name of the company (two words, three syllables each), "controlling," "operating," "improving," "continuing," and "selected" would not be counted as hard words, so there are 20 words of three or more syllables among the 178 words in the text, or 11.2 per 100. Therefore,

$$\text{Fog index} = 0.4\ (17.8 + 11.2) = 11.6.$$

According to the Fog index, 11.6 years of education (in essence, a high school education) would be required to understand the passage.

Flesch-Kincaid Index calculation

The Flesch-Kincaid Index measures the grade level of education required to understand the document by counting words and syllables per sentence:

$$\text{GL} = 0.39\text{NWS} + 11.80\text{NSW} - 15.59$$

where NWS = average number of words per sentence and NSW = average number of syllables per word.

The average sentence length is 17.8 words, and there is an average of 1.55 syllables per word. Therefore,

$$\text{GL} = 0.39 \times 17.8 + 11.80 \times 1.55 - 15.59 = 9.64.$$

According to the Flesch-Kincaid index, a tenth grade education would be required for understanding the passage.

Interpreting readability scores

These three readability formulas and others are based on the assumption that writing is more easily understood when short words are used in short sentences. However, that is not necessarily true. For example, the sentence "We measure work in joules and force in newtons" won't mean much to most readers. It isn't very readable even if it gets a good readability score. Therefore, readability theorists recommend that passages of at least 100 words be chosen from different sections of a document to get a representative estimate.

Another problem with the use of such formulas is determining what the mathematical result means. Keyed to educational levels, the results become more questionable as students' reading skills decline and as schools produce disparate reading abilities among students in the same grades, sometimes by as much as a year and a half. Readability formulas do not take into account the readers' familiarity with the *meaning* of the long and short words in the text. Finally, the formulas do not take into account graphics or design formats, which can significantly affect a document's difficulty.

Readability testing can have real value for editors, however. Many word processing programs can point out lengthy sentences and identify unfamiliar words. Editors can use such programs to identify overly long sentences and jargon quickly. The readability scores can encourage technical writers to control sentence length and diction. Quantification appeals to many technical writers, who may become more interested in the theory of readability when readability scores are used to show the difference that revising writing can make.

Legibility

Legibility is the measure of how easily the letters and graphical elements of a document can be identified visually by a reader. Consider text printed on dark green paper, or very tiny labels on a graphic, or indecipherable handwriting. The words might be understandable if they could be made out at all by the reader. A similar problem in legibility occurs with typefaces that are hard to read, such as script and decorative faces such as Bobo Bold and Rope.

Legibility can be a problem in a number of ways. Handwritten text and graphics may be illegible, complicating typing and production of graphics. Tight leading (pronounced *ledding*)—the space between lines—and letter or word spacing may create legibility problems. Also, some typefaces are less legible than others, such as sans-serif typefaces. (A "serif" is the small horizontal line at the bottom or top of vertical parts of letters.) Text in all caps is harder to read than text in initial caps or lowercase. The greatest legibility problem for technical editors, however, is controlling type *point size.* In many kinds of technical documents, especially in page-limited proposals and other tight layouts, technical staff often want to reduce graphics or use a smaller point size to squeeze more material into the space available. To ensure legibility, editors should try to hold firm to at least 10-point text and 8-point callouts in graphics.

As an example of how legibility can affect readability, the text in Figure 4.1 is sufficiently readable: It requires some high school education. The typeface itself is legible; each letter can be made out clearly. The somewhat long lines (50 to 55 characters per line is best) are spaced sufficiently far apart that the reader's eyes can track properly from the end of one line to the beginning of the next. However, the text is difficult to read because of its

> WHEN REVIEWING YOUR ENCLOSED AUTO POLICY RENEWAL, YOU WILL NOTICE
> YOUR POLICY TERM IS NOW SIX MONTHS RATHER THAN TWELVE MONTHS. THE
> PREMIUMS SHOWN ON YOUR DECLARATIONS PAGE ARE ONE-HALF THE AMOUNTS
> THAT WOULD HAVE BEEN CHARGED UNDER THE TWELVE MONTH TERM.

Figure 4.1. The effect of tight word spacing and all capitals on legibility. (Source: Merastar Insurance Company, Chattanooga, Tennessee, © 1987, used with permission.)

capitalization and tight word spacing. Text in all caps is less legible than text in initial caps or lowercase because the differences in the characters don't stand out as much. Also, the word spacing is too tight. There is not enough space between words to make them stand out clearly as separate words.

To improve the legibility of a document, editors should keep in mind the following guidelines for type styles and faces:

- Initial cap and lowercase type are easier to read than all caps.
- Standard typefaces are easier to read than italic, script, outline, condensed, expanded, or other unusual typefaces.
- Serif typefaces are easier to read than sans-serif faces in text, but sans-serif faces work well for headings and callouts in graphics and presentation materials.
- In most printed documents, text should be in 10- to 12-point type, headings in 14- to 18-point type, graphics callouts in 8- or 10-point type, and transparencies in 24- to 36-point type. (Point sizes are illustrated in Figure 4.4 below.)
- Text kerned properly, with enough spacing between letters in a word so that they don't touch each other, is easier to read than type that is too tight or too loose.

Editors should remember the following guidelines for line and word spacing:

- Type set with regular (equivalent) word spacing is easier to read than unevenly spaced words.
- Type set with a small amount of extra leading is easier to read than type set solid (no extra leading).
- Short lines are easier to read than long lines.
- Ragged right text is easier to read than right-justified text, unless the typesetter can provide nearly equivalent word spacing.

The following guidelines for paper and ink are also helpful in increasing legibility.

- Type on smooth-finish paper is easier to read than type on glossy (coated) or textured paper.
- Text on off-white paper is easier to read than that on bright white or colored paper.
- Black ink on near-white paper is easier to read than white or colored ink on colored paper.
- Fast-drying ink should be used on coated (glossy) paper to prevent smearing.
- Bright inks (black or colored) are effective in short documents such as brochures or advertising materials but are harder to read than medium black ink in longer documents.

Generally, legibility is less a measure of reading ability and familiarity with the material and more a matter of the physiology of the eye, which can identify characters more easily when they are big enough and in sufficient contrast to their background. Editors can anticipate problems in legibility (as well as readability) if they know their audience. For example, if the document is intended for elderly people, larger type and line spacing and shorter lines could be used.

Usability

A document's usability is the ease with which readers can use it, especially instructions. For example, a software manual will be very usable if readers can easily find what they are looking for, understand the discussion, learn quickly how to operate the software, and encounter no gaps in the information or vague or confusing steps.

In a sense, usability subsumes readability and legibility. To be usable, a document must be readable (so it can be understood) and legible (so the characters that make up words and numbers can be identified). However, usability is a matter of the document's content and organization as well. To help readers use the document, editors consider the format and the layout of the pages or screens. They examine the indexing, graphics, and other devices that might facilitate the document, such as tabs and color-coding.

Usability is influenced by more than design considerations, however. A document's usability can be increased by making its organization tight and apparent. The content must also be controlled so readers aren't misled by unclear discussion or by incorrect inferences. Only the information needed to perform the activity under discussion should be presented, clearly and sufficiently developed.

Usability is not measured with numbers or levels. As with legibility, the aim is to achieve the greatest usability for the particular audience. Maximum usability has been achieved when readers have no trouble using the document efficiently. Usability testing (which is also called *document validation, document verification,* or *protocol analysis*) is becoming common in the computer industry, and manufacturers in other industries will follow, in part to avoid lawsuits. In usability testing, readers who resemble the primary audience of the document try to use the document. They report on any problems they encounter so that the problems can be resolved before the document is released to the primary audience.

Usability design incorporates the results of usability testing, but it should also implement comments from users of similar documents produced by the company or agency. Usability testing helps writers and editors uncover problems they may not see because they are so close to the document and thoroughly understand its subject. Often the writers and editors are an expert audience, and if the document is intended for a lay or middle audience, the editors should try to arrange for document validation by a lay or middle audience. To keep up to date on usability testing, editors should review issues of *Technical Communication* and *IEEE Transactions on Professional Communication,* and proceedings of conferences such as the Society for Technical Communication annual conference.

Editors must ensure that they edit text properly to make documents readable, and design documents properly so they are legible and usable, to communicate effectively with their audience. Editing text requires concern for readability and legibility, and designing many documents and laying out their text and graphics requires concern for usability.

EDITING TEXT

Editing technical text involves several of the steps in document production discussed in Chapter 2 and presented in Figure 4.2. Editing and copymarking text can be approached in different ways. Some editors mark mechanical errors and typos as they read through a document for the first time; others do no sentence-level editing until the second reading. Some read the beginnings of sections first, to get an overview of the document. Some editors must work section by section because they never have all of the document in their hands at one time until it is printed and bound.

In many companies and agencies, and in the setting described in this text, documents are submitted to editors for macro-editing before draft review, as described in Chapter 2. After the review and subsequent revisions, editors do more macro-editing and complete micro-editing to prepare the final version.

For a draft review, editors should complete the following tasks:

- Review the purpose and audience of the document
- Review the standard for the document
- Read the section or complete document carefully
- Macro-edit the text
- Confer with the writer(s) if necessary
- Have the text typed, proofread against the original manuscript, and merged with rough graphics for draft printing

After the document has been reviewed by the writers, editors should:

- Complete the macro-edit
- Micro-edit for complexity
- Micro-edit for correctness
- Edit the graphics and submit them to the artists
- Submit text revisions to the typists or typesetters
- Copyfit the text for layout
- Have text and graphics proofread
- Prepare a dummy layout of text and graphics

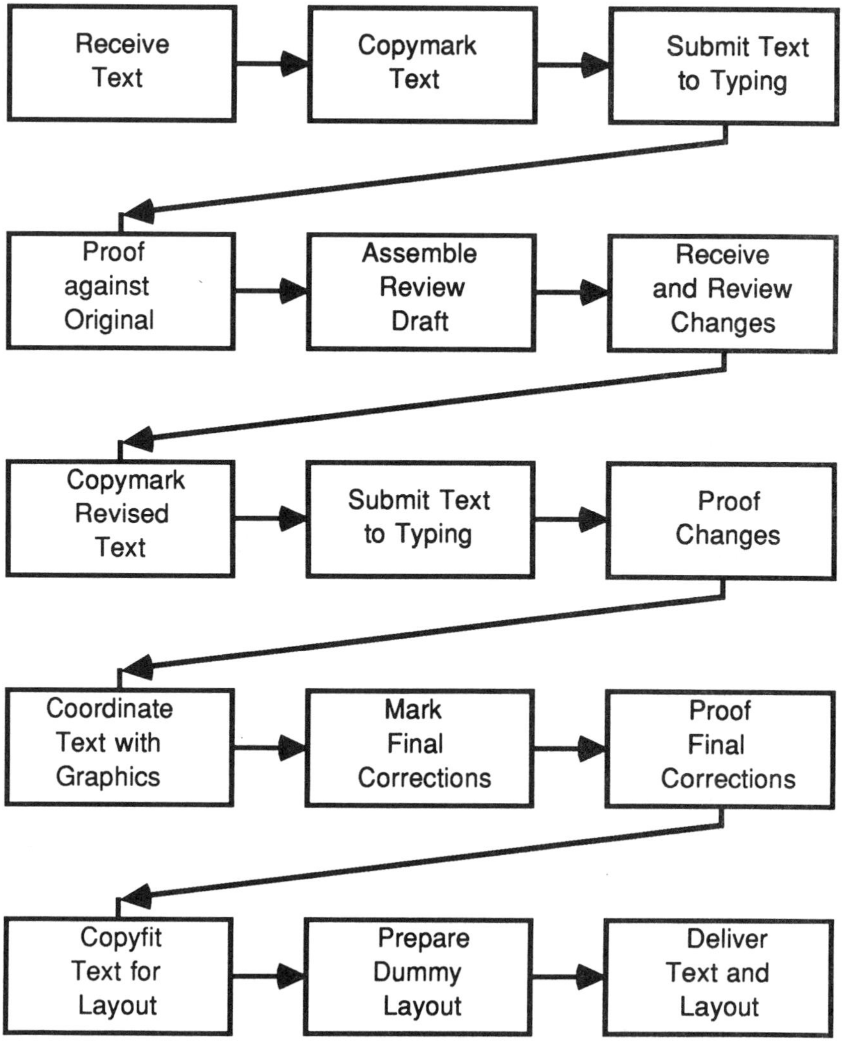

Figure 4.2. Steps in editing and producing text.

The Macro-edit

Before editing begins, editors must schedule the submission of *write-ups* (text and graphics) with the project manager and encourage him or her to enforce the schedule. When writers' submissions are late, editing time shrinks because subsequent deadlines must be met. The worst mistake an editor can make is miss a deadline; consequently, scheduling writers' submissions and editing of them is very important. Once write-ups are received, the macro-edit can begin.

In a macro-edit, editors examine the document's overall content, organization, and logic. They consider whether the pace of the discussion is appropriate for the audience, and they examine how well the document accomplishes its purpose.

In a macro-edit, editors rarely address grammar or style. Sentence-level revisions are usually ignored because subsequent review of the draft might eliminate whole sections. Instead, editors focus on larger issues such as clear thesis or purpose statements, sound and apparent organization, thoroughness of summary and introduction, and appropriateness and accuracy of content. An effective macro-edit requires familiarity with the type of document, the subject matter, and sometimes the company or agency producing it. For a project that will not be reviewed in draft, many editors complete the macro-edit before they begin the micro-edit.

For a beginning technical editor, macro-edits can be difficult. Instead of facing sentence-level questions such as how to punctuate a particular sentence, the editor must address larger issues about organization and content. Try to keep the following questions in mind:

- Will the purpose of the document be clear to the audience?
- Is the background information provided in the introduction sufficient to orient the readers?
- Is the subject covered adequately?
- Is the organization sound, and will it be apparent to the audience?
- Are headings used where needed to clarify the document's organization?
- If the document is not argumentative, do the most important points and most significant details come first?
- Are important points clear? Are facts and details used as supporting evidence?
- Does the summary include all of the most important points?
- Does the content fulfill the purpose of the document?

Before marking the manuscript, find out whether the text is on disk—this determines how the manuscript is marked, as discussed in Chapter 3. If the text is not on disk, any reorganizing can be done with scissors and tape, and many changes can be made without editing marks. If the text is on disk, all changes must be indicated with proper symbols on the manuscript.

Review the standard carefully before your pen touches paper, whether it is a company style manual, a field-specific guide like the Council of Biology Editors' *Style Manual,* or some other guide. In some situations, such as proposals to the Department of Defense, a document might be rejected if it fails to meet requirements stated in the standard or other instructions. In less extreme cases, deviating from the standard might inconvenience reviewers of a journal article or users of a manual accustomed to seeing certain types of documents arranged and developed in certain ways.

Editors should consider the audience and purpose of the document (see above). Are they a lay, middle, or expert audience, or a mix? Is the document designed to inform or persuade? How motivated will the audience be to read it? If editors are unsure how to answer any of these questions, they should ask their supervisor, the project manager, other editors, or anyone else who might be able to provide the information. Someone will know—or the document needs to be rethought. Editors should not edit until they have these questions answered.

If the document is relatively short and is unclassified, many editors photocopy the writers' manuscripts, keeping them unmarked (but page numbered) and editing on a copy. Filing the original write-up can be useful for later reference, if text or graphics are lost or if a writer wants to review the manuscript to answer a technical question.

Be careful not to mark on the originals of any graphics submitted, because they may be *repro* (originals that will be printed from later). Instead, graphics should be saved for the micro-edit, when photocopies or transparent overlays attached to them will be copymarked. In the macro-edit, however, editors should make sure all graphics are numbered and labeled properly with captions.

Next, many editors read through the entire document (or section) to get a sense of its content and organization, then use a fine-tipped red or green pen to mark the text. (Red and green stand out better for typists than black or blue.) They might read the write-up first for its organization and edit appropriately, for example, adding headings. Many read for content next, adding a sentence of explanation or defining terms. Other editors address organization and content at the same time.

As editors macro-edit the text, they should be sensitive to possible changes in meaning, keeping in mind that changing a writer's meaning without consulting the writer is the second-worst error an editor can commit. Examine the graphics provided by the writer to make sure they are suited to the material and the audience. Any comments or questions for the author should be written in the margin and circled.

If there is time before the text must be submitted to the typist as new text or a revision, many editors try to meet with the writer to solve any problems in organization or content. Often the writer can approve quickly a proposed change or provide necessary explanations. Editors should remember that many writers are sensitive about their work, especially if they are not confident. If part of the discussion is not clear, good editors take the "I'm not sure I understand you here—could you help me?" approach, which works much better than saying, "This isn't clear, and I want you to fix it."

If meeting with the writer isn't possible, editors should change the organization and content as they think necessary. As hard as it might be, try not to edit at the sentence level. You might, however, take a little time to fix outrageous typos and grammatical errors. If time permits and the draft seems to be strong enough to survive draft review without rejection, you might edit the summary or introduction and section openings for a clear focus on the subject.

Finally, editors should use the macro-edit to confirm decisions about document design and formatting made in the planning stage. For example, if a write-up has been submitted double-spaced, in 12-pitch type, with 1.5-inch margins, and the standard for the document calls for single-spaced, 10-pitch text with 1-inch margins, the editor should instruct the typists in a circled note at the top of the first page as follows: Single space, 10 pitch, 1 inch margins. Initial planning decisions regarding typeface, size, line spacing, and other design criteria can be revised as needed as write-ups are submitted. However, fixing the *specs* (specifications) before write-ups are prepared can save valuable time in editing and typing or typesetting. If there is time before printing the review draft to format the write-ups consistently, the draft will make a much better impression on reviewers (the secondary audience that approves the document).

The macro-edit can familiarize editors with the entire document. Also, it can shape the document globally, clarifying and strengthening organization and content. It is a very important step in editing a document whether or not the document will undergo draft review.

Sample of a Macro-edit

The document below is a revision of an article that appeared in an in-house monthly publication of a large government agency. Although the agency does some very technical work, many of its employees are not technical specialists. The newsletter helps keep them informed on technical subjects. This article, then, will be read predominantly by a lay audience. The editor will be able to confer with the writer before the article is submitted for publication, as it is not yet on disk.

Astronomy and Technical Development

The stars have occupied human thought for millennia. Over the generations, we have succeeded in gaining ever greater insight into the underlying forces at work in the cosmos. In the Space Station era, the family of permanent observatories in space will open the way to new, comprehensive studies of key remaining problems in astrophysics. These studies can help us understand: (1) the birth of the universe, its large-scale structure, and the formation of galaxies and clusters of galaxies; (2) the origin and evolution of stars, planetary systems, life and intelligence; and (3) the fundamental laws of physics governing cosmic processes and events. If we succeed, we will leave a legacy to rank us with the great civilizations of the past. No one knows, however, if we will be able to solve the technical problems inherent in such an effort.

For centuries, astronomy and technology have progressed hand in hand. Our study of the universe has benefitted from improvements in observational equipment. And, developments in astronomy have led to practical applications in other disciplines.

From 1500 to 1600, increasingly accurate maps of the sky for navigation were produced. Early star maps were crude renderings, but developments in telescopes by Galileo and others improved star mapping considerably.

In the next century, one of great scientific advancements in Europe including Harvey's demonstration of the circulation of blood and the establishment of the British Royal Society, Christien Huygen's developed his principle for wave motion: "Each point on a progressive wavefront can be considered as a source of secondary spherical waves; the amplitude of any future point on the wavefront can be obtained from a superposition of these secondary waves, and their envelope gives successive wavefronts" [1]. Such early studies of wave motion influenced the work of nineteenth-century scientists.

Christiaan Huygens also invented a pendulum clock that could be used for navigational time-keeping. Sir Isaac Newton developed the calculus (Leibniz developed it soon after, independently, but Newton usually gets all the credit), the three laws of motion and the law of universal gravitation to explain the motions of planets and comets.

From 1800–1900, there were increasingly sophisticated optical innovations by astronomers, including William Herschel, Fraunhofer, Lord Ross, Alvin Clark, etc. Also, increasingly sensitive photographic techniques were developed.

In this century, Lockyer discovered a new chemical element, Helium, on the sun before it was known on the Earth. Hans Bethe predicted hydrogen fusion at the center of the sun. Lyman Spitzer developed the astrophysical plasma theory, the basis of present devices for releasing energy from controlled fusion. Very long baseline radio astronomy techniques were developed for high-precision geodesy to survey the structure of the Earth. Also, techniques of celestial mechanics (precursors to the development of accurate spacecraft navigation) were developed.

The mutually beneficial interaction between astrophysics and technology continues today.

[1]Brackenridge, J. Bruce, and Robert M. Rosenberg, <u>The Principles of Physics and Chemistry</u> (New York: McGraw-Hill), p. 280.

—30—*

To prepare for macro-editing, the editor should review the goals and questions listed above. Then read the manuscript to examine its purpose, organization, content, and appropriateness for its audience, and note any major questions regarding content or organization for a conference with the writer.

The editor should notice some of the following problems in this piece: The first part of the paragraph that begins "In the next century" contains irrelevant information, and the definition of Huygens's principle is inappropriate for the lay audience. The editor should ask if there was supposed to be a section on the years 1700–1800. Also, ideas for graphics should be discussed. The editor might suggest an illustration similar to Figure 4.3. The editor might suggest moving the first paragraph to the end of the piece, as it emphasizes future developments that logically would follow those past accomplishments noted in the article. If the writer agrees, the text could be copy-marked as below before submitting it to the typist. The editor can also submit the graphic concept to the art department.

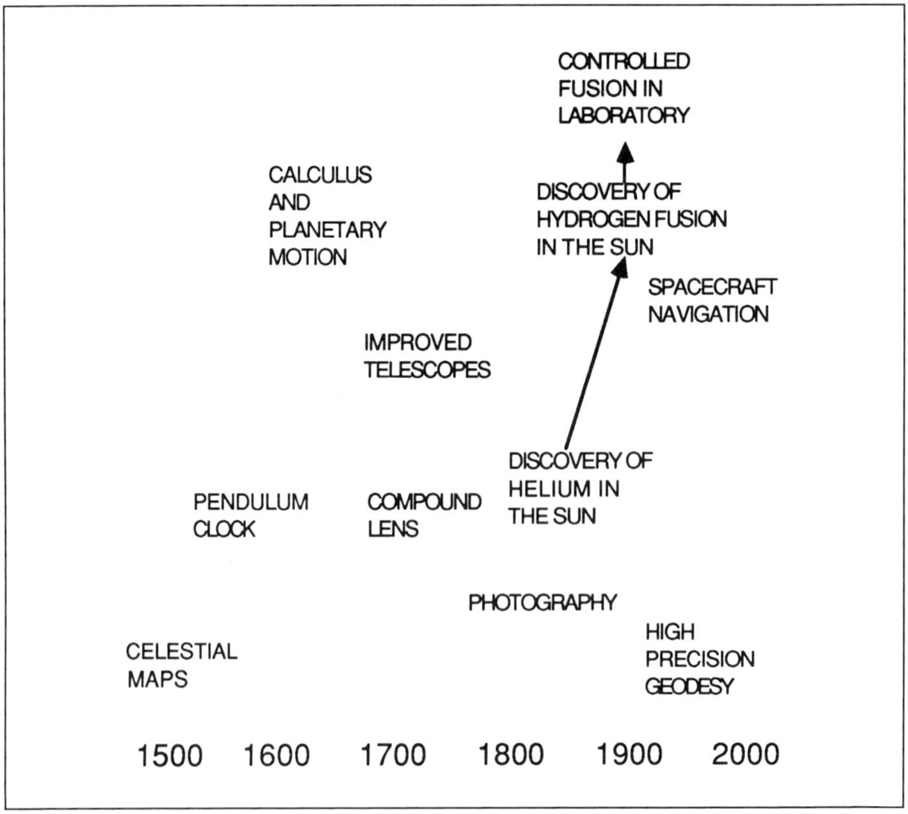

Figure 4.3. Possible illustration concept for astronomy article.

Astronomy and Technical Development

The stars have occupied human thought for millennia. Over the generations, we have succeeded in gaining ever greater insight into the underlying forces at work in the cosmos. In the Space Station era, the family of permanent observatories in space will open the way to new, comprehensive studies of key remaining problems in astrophysics. These studies can help us understand: (1) the birth of the universe, its large-scale structure, and the formation of galaxies and clusters of galaxies; (2) the origin and evolution of stars, planetary systems, life and intelligence; and (3) the fundamental laws of physics governing cosmic processes and events. If we succeed, we will leave a legacy to rank us with the great civilizations of the past. No one knows, however, if we will be able to solve the technical problems inherent in such an effort.

For centuries, astronomy and technology have progressed hand in hand. Our study of the universe has benefitted from improvements in observational equipment. And, developments in astronomy have led to practical applications in other disciplines.

Increasingly accurate maps of the sky for navigation were produced. Early star maps were crude renderings, but developments in telescopes by Galileo and others improved star mapping considerably.

In the next century, one of great scientific advancements in Europe including Harvey's demonstration of the circulation of blood and the establishment of the British Royal Society, Christien Huygen's developed his principle for wave motion: "Each point on a progressive wavefront can be considered as a source of secondary spherical waves; the amplitude of any future point on the wavefront can be obtained from a superposition of these secondary waves, and their envelope gives successive wavefronts" [1]. Such early studies of wave motion influenced the work of nineteenth-century scientists.

Christiaan Huygens also invented a pendulum clock that could be used for navigational time-keeping. Sir Isaac Newton developed the calculus (Leibniz developed it soon after, independently, but Newton usually gets all the credit), the three laws of motion and the law of universal gravitation to explain the motions of planets and comets.

From 1800–1900, there were increasingly sophisticated optical innovations by astronomers, including William Herschel, Fraunhofer, Lord Ross, Alvin Clark, etc. Also, increasingly sensitive photographic techniques were developed.

1900–N NOW

~~In this century,~~ Lockyer discovered a new chemical element, Helium, on the sun before it was known on the Earth. Hans Bette predicted hydrogen fusion at the center of the sun. Lyman Spitzer developed the astrophysical plasma theory, the basis of present devices for releasing energy from controlled fusion. Very long baseline radio astronomy techniques were developed for high-precision geodesy to survey the structure of the Earth. Also, techniques of celestial mechanics (precursors to the development of accurate spacecraft navigation) were developed.

The mutually beneficial interaction between astrophysics and technology continues today.

*

~~[1] Brackenridge, J. Bruce, and Robert M. Rosenberg, The Principles of Physics and Chemistry (New York: McGraw-Hill, 1970, p. 280.~~

—30—*

Art Dept.: I need a drawing about 4″ tall by 5.5″ wide to illustrate developments in astronomy and technology. The labels below (in Fig. 4.3) should appear in the drawing, maybe with a simple drawing to represent each development. Emphasize the sketch of the development; label each in 8 point all caps or smaller. Also, indicate relations as shown with the arrows? Dates are to be approximate; no need for exact placement. I think something sketchy rather than final would be fine—maybe hand lettering? I'll need a sketch by the end of next week (check it Wednesday?) and the finished piece by the 23rd. Call me if that's a problem.

Discussion of the Macro-editing

As the text for the article is not yet on disk, the editor might choose to physically cut out the fourth paragraph and the reference to be deleted. Marking those discussions, however, does not create a messy text. The passages should not be marked with deles because this is new text and there is no material on disk to be deleted.

Adding the headings is easy at this point; they clarify the organization of the article for review. Few other changes are made in the macro-edit. The entire article might be rejected, so there is no reason to spend time editing at the sentence level. The editor should ask the typist to print the text double spaced to facilitate micro-editing. When the typed text has been proofread against the original, it is ready for the micro-edit.

*The —30— indicates the end of the document.

The instructions to the art department could include sketches for guidance, but the directions here will be sufficient for artists in this organization.

The Micro-edit

The micro-edit is often called *copyediting* or *line editing.* In a micro-edit, editors work through text and graphics line by line and letter by letter, concentrating on making them rhetorically sound, grammatically correct, consistent with standards, and effective. Principles of good style, consistency, and correctness enter the editing process here.

Some experienced editors complete the micro-edit in one step, addressing paragraphing, sentence structure, diction, grammar, punctuation, and spelling in one pass. Beginning editors can usually accomplish the micro-edit more effectively in two steps. In the first, the *complexity edit,* the editors make sure that the content, pace, and diction are appropriate for the audience. They make sure that the document is readable—that the audience can handle the complexity of the discussion. In the second step, the *correctness edit,* the document is made consistent with standards and correct in sentence structure, grammar, spelling, and punctuation.

The complexity edit

The complexity edit should ensure that the complexity of the discussion is appropriate for the audience. There is nothing wrong with making a technical document complex—if it is to be read by an expert audience. Experts expect complexity. Writing teachers who tell students to always simplify their writing are ignoring this important fact. Technical documents have varying levels of appropriate complexity, from the instructions for lay audiences that use familiar terms, lots of explanation, and simple graphics to simplify the subject matter, to the report on AIDS research for an expert audience which uses scientific terms, avoids explanation of basic concepts or procedures, and uses complicated graphics to chart data.

The level of complexity of a document is important. If a document is too complex for a lay audience, it will not be readable. If a document aimed at technical experts is oversimplified, it might seem simpleminded or condescending. Editors' changes should make any document easier for the intended audience to understand, more emphatic, and more focused. For a lay audience, this often requires eliminating or defining unfamiliar technical terms, acronyms, and abbreviations; substituting simpler words for jargon and technical terms; simplifying sentence structure; and streamlining graphics.

The aim of the complexity edit is to make sure the text communicates its information well by achieving the right level of complexity in word choice, sentence length and structure, and paragraphing. Complexity editing usually takes place after a draft of the document has been reviewed and writers have revised their sections as necessary.

As text is submitted by writers, editors should examine the schedule for submission of the text for typing and check the revisions as they are submitted to see the extent of changes. If sections have been rewritten, they are often macro-edited and submitted to typing first, so they can be returned to the editor in time for a subsequent micro-edit. Other write-ups are often grouped by their degree of revision. Depending on the typing schedule, the cleanest write-ups might be micro-edited first, to get revisions done as soon as possible if the typists are not busy, or the tough sections might be micro-edited first, taking longer to get revisions to typing if typists are busy. Either way, editors should estimate for the typing supervisor the extent of revisions and confirm the schedule for typing work.

Once the order of sections for micro-editing has been established, editors begin by rereading them, paying particular attention to reviewers' comments and whether the writer's changes seem to satisfy them. Changes in organization and content are noted to confirm that the text satisfies the nine questions of the macro-edit (above). Check the section number for the write-up against the outline of the document, and the reference numbers for graphics against the callouts in the text. Also, check the page format to make sure it meets the standard.

After making any revisions called for in the previous steps, complexity edit copymarking of the revised text can begin.

Paragraphs. Editors should strive to control paragraph length in technical documents, especially in instructional manuals and in correspondence. The focus of each paragraph should be clear, as should transitions between paragraphs.

Shorter paragraphs, which are easier to digest, can control the pace of a discussion by breaking it into more readable portions. Short paragraphs are especially useful with lay audiences. The more complicated the information being presented and the less knowledgeable the audience, the shorter the average paragraph length should be. (However, the paragraphs should be of varying lengths to prevent monotony.) The more educated the audience, the more familiar they are with the subject, and the more interested they are, the longer the paragraphs they can handle. Even for an expert audience, it is unwise to use many paragraphs more than 20 lines long in a report and 10 lines long in a letter. Long paragraphs intimidate most readers, and sometimes individual points get lost. There is nothing wrong with a one-sentence paragraph, but a series of one-sentence or very short paragraphs can make a document sound like an elementary textbook. Short paragraphs may be appropriate for a set of instructions. Generally, variation in paragraph length is the goal.

In the complexity edit, editors should make sure that only one main idea or body of information is presented and developed in each paragraph. If a paragraph drifts into a new topic, begin a new paragraph. The main idea or topic of the paragraph should be stated early in the paragraph, unless it is something the writer needs to convince the reader of before it can be stated.

The main idea of an argumentative paragraph might be delayed until supporting detail has been presented. Most readers expect an indication of paragraph topic near its beginning, and providing that sort of introduction can help readers focus on what is important. Also, using a transition at the beginning of a paragraph to link it to the previous one can unify the discussion in the document.

Editors also check that points and details are presented in the most effective order, usually the most important first, then the next most important. Climactic order—saving the most important point for last—works well in some arguments, but generally the most important information should be presented first, when a reader's attention and comprehension are greatest. In long documents, editors should consider using the first paragraph in a main section to tell what will be covered in that section, and the last paragraph to summarize the main points of the section.

In the complexity edit, editors often add transitional words, so that the reader can see easily how the paragraph develops, as in the following example:

> Safety *officials* have identified three *reasons* for the *crash. First,* ice on the wings interfered with the operation of the ailerons. *Also,* a malfunction in the windshield defroster obscured the pilot's vision. But the most important *reason,* according to the *officials,* was pilot error: The flight recorder indicated they were unaware of their altitude until just before the *crash.*

Note how the repetition of key words and the transitions link the sentences in a tight paragraph that stays focused on the topic clearly stated in the first sentence.

Editors often mark a series of points or pieces of information to be *displayed* in a bulleted or numbered list set off from the body of the paragraph. Displayed lists (also called *stacked* or *bulleted* lists) are usually indented twice as much as a normal paragraph on both the left and right, with an extra space above and below the list. Stacked lists are especially useful for emphasizing individual points, as in the guidelines for legibility presented earlier. If the items are steps that should be followed in the order given, editors should use numbers (with or without underlining, depending on the style guide) instead of bullets.

Sentences. The complexity edit requires attention to sentence structure and length. The longer a sentence or the more complex its structure, the more difficult it may be for an audience to understand, especially a lay audience. Expert audiences are more accustomed to long, complex sentences. An interested, educated audience familiar with the subject can tolerate longer sentences than a less educated audience, an educated audience unfamiliar with the subject, or an uninterested audience. Many word processing programs can highlight long sentences so editors and writers can check them. How can editors tell what length is appropriate? The most effective method is to read what the audience reads and to seek the advice of experienced editors.

Simple sentence structure increases readability. To accomplish this, rely on the common sentence structures we use when we speak: subject–verb, subject–verb–object, and subject–verb–complement. Whenever writers use a different structure, they make a sentence more difficult. For example, "The engineer knew that the test results were faulty" is much simpler than "That the test results were faulty was known by the engineer." Both sentences say the same thing, but the first version has a more familiar structure. Some writers try to impress readers with long or complicated sentences designed to make them appear educated or intelligent. Editors should encourage writers to impress readers instead with how clearly and simply they can present technical material.

In the complexity edit, editors often eliminate unnecessary subordinate clauses (discussed in Chapter 11). Subordinate clauses can make a sentence longer as well as more complex, so they should be used sparingly. For example, "Send the report that was written by Smith to Jessica Locke, who is the supervisor of training" contains two unnecessary subordinate clauses. The sentence could be reduced to "Send Smith's report to Jessica Locke, the supervisor of training." The second version is shorter and simpler (with only one subject and verb). Important ideas should be emphasized by stating them in main clauses, not subordinate clauses.

The subject and verb of a clause should not be split unnecessarily. Sentences are easier for readers to process when they can see quickly and easily what is being said about the subject. For example, "Michelle Smith, having graduated from Radford University, went to work for a software firm" could be restated as "Michelle Smith went to work for a software firm after graduating from Radford University." This revision clarifies the sentence by putting the subject and verb together.

When sentences begin with the subject followed soon after by the verb, without too many intervening words, they are more readable. For example, "Having graduated from Radford University with a major in computer science and a minor in mathematics, and having received academic honors as a student, Michelle Smith went to work for a software firm" is much more difficult to comprehend easily than "Michelle Smith went to work for a software firm after graduating with honors from Radford University with a major in computer science and a minor in mathematics."

Editors should rely on the active voice in sentences, using passive voice when the doer of an action is deemphasized (as discussed in Chapter 11). Active voice is more forceful and direct than passive voice. For example, "Robots can perform manufacturing operations that are dangerous to humans" is more direct than "Manufacturing operations that are dangerous to humans can be performed by robots." The second sentence, with the passive verb "can be performed," is correct and might be preferred if the writer wants to emphasize the dangerous manufacturing operations. The active voice sentence has more focus on robots (the subject) than on the kind of operations they can perform. The passive voice sentence points more to "manufacturing operations" than "robots," because "manufacturing operations" is the grammatical subject of the sentence, and the real doer of the action, "robots," is tucked away in a prepositional phrase. A greater proportion of sentences with active verbs will make a discussion more direct and economical, but a variety of sentence structures and lengths can increase readers' interest as well as readability.

Editors should aim for variety in sentence structure and length, always trying to keep the sentences straightforward.

Diction. Diction (word choice) can be more important than sentence or paragraph structure in communicating effectively with an audience. Many technical documents for lay audiences fail because readers are faced with technical terms and explanations that are appropriate for middle or expert audiences but not for lay readers. However, some technical documents fail to communicate well because they patronize middle and expert audiences, avoiding technical terms that should be used for those audiences and providing explanations that aren't necessary. Making documents communicate well is in part a matter of content (addressed in the macro-edit), but it is also a matter of diction, so diction should be addressed in the micro complexity edit.

Many experienced editors can address diction as they examine the paragraphing and sentence structure in the text of a document. Beginning editors, however, should scan the text in a separate step for words that might be inappropriate for the audience. In documents for a lay audience, they should also look for terms for which clearer, simpler synonyms might be used. For lay audiences, they substitute simple, familiar words for possibly unfamiliar technical terms. For middle audiences, they use necessary technical terms but provide needed definitions. For expert audiences, they define only neologisms or familiar terms used with a new meaning. Sometimes technical jargon is appropriate, especially in documents designed to be read by technical experts. For a general audience or an audience unfamiliar with the subject matter, however, editors use more familiar, simpler terms. For any audience, editors should choose the more straightforward expression, such as "use" instead of "utilize," "improve" rather than "ameliorate," "begin" rather than "initiate," "person" rather than "individual."

To make writing more forceful and emphatic, place key ideas in verbs that express action, not in nouns. Verbs are the strongest parts of English speech. A sentence like "Implementation of the plan will begin in June" should be stated "The plan will be implemented in June" so the key idea "implement" is placed in the verb. Rather than fail to emphasize the sentence's true action by saying, "Analysis of the samples was undertaken by Smith and Company," put the action in the verb: "Smith and Company analyzed the samples." Action verbs are stronger than being or linking verbs. "The tests were a failure" is weaker than "The tests failed." Editors should change an inflated, wordy expression such as "We have made visual contact with you" to "We see you." Similarly, rephrase "The engineers were able to initiate work on a new design" as "The engineers began work on a new design."

When addressing diction in the complexity edit, eliminate any phrasing that might be construed as sexist or otherwise offensive. Editors must be sensitive to sexist language, but they should resist constructing awkward or illogical statements in trying to avoid it. For example, many people suggest replacing gender-specific pronouns such as "he" or "she" with a construct like "s/he." Such a form would be inappropriate in most technical documents. Replacing "he" or "she" with the neutral plural "they" is often recommended, but converting the focus from singular to plural can be a problem. For example, if the discussion focuses on the "typical" technical writer, using "they" to refer to that writer will confuse many readers, and changing the discussion to the "typical" technical writer*s* may subvert the author's focus.

The proper response to a passage that slights any person on the basis of sex, age, race, or religion sometimes is not to delete or revise it but to point out the problem to the writer. Sometimes a writer fails to see that a statement might be offensive, and once it's pointed out, the writer is anxious to correct it. If there is no time for a conference, the editor should edit the passage lightly to remove the potentially offensive statement(s). If the writer balks at a change, the editor might seek the counsel of a supervisor in handling the problem. Beginning editors especially should approach such situations cautiously, without crusading.

Sample complexity edit

The text below is the macro-edited version of the article discussed above. The editor has met with the writer, who agreed to the suggested cuts and indicated that there was no section on 1700–1800 planned. The macro-edited text has been typed on disk and is now ready for complexity editing. Remember, the readers are predominantly a lay audience, but some will recognize many of the scientists' names.

The editing symbols discussed in Chapter 3 have been used to copyedit the manuscript. The style guide specifies that titles and first-order headings (set flush left) are to be in bold type, so the manuscript has been marked accordingly.

Cross-fertilization
Astronomy and Technical ~~Development~~

Over the ~~For~~ centuries, astronomy and technology have progressed hand in hand. The ~~Our~~ study of the universe has benefitted from improvements in observational equipment. By the same token, ~~And~~ developments in astronomy have led to practical applications in other disciplines.

1500-1600

• Increasingly accurate maps of the sky for navigation ~~were produced. Early star maps were crude renderings, but developments in telescopes by Galileo and others improved star mapping considerably.~~

1600-1700

Christiaan Huygens invented a pendulum clock ~~that could be used~~ for navigational time-keeping. ~~Sir Isaac~~ Newton developed the calculus ~~(Leibniz developed it soon after, independently, but Newton usually gets all the credit)~~, the three laws of motion and the law of universal gravitation to explain the motions of planets and comets.

1800-1900

Increasingly sophisticated optical innovations by astronomers, including William Herschel, Fraunhofer, Lord Ross, Alvin Clark, ~~etc. Also~~ increasingly sensitive photographic techniques ~~were developed.~~

1900-NOW

~~In this century~~ Lockyer discovered a new chemical element, Helium, on the sun before it was known on the Earth. ~~Hans~~ Bette predicted hydrogen fusion at the center of the sun. Lyman Spitzer developed the astrophysical plasma theory, the basis of present devices for releasing energy from controlled fusion. Very long baseline radio astronomy techniques ~~were developed for~~ high-precision geodesy to survey the structure of the Earth. ~~Also~~ techniques of celestial mechanics (precursors to the development of accurate spacecraft navigation) ~~were developed.~~

The mutually beneficial interaction between astrophysics and technology continues today.

~~The stars~~ have occupied human thought for millennia. Over the generations, we have succeeded in gaining ever greater insight into the underlying forces at work in the cosmos. In the Space Station era, the family of permanent observatories in space will open the way to new, comprehensive studies of key remaining problems in astrophysics. These studies can help us understand:

- the birth of the universe, its large-scale structure, and the formation of galaxies and clusters of galaxies;
- the origin and evolution of stars, planetary systems, life and intelligence, ~~and~~
- the fundamental laws of physics governing cosmic processes and events.

If we succeed we will leave a legacy to rank us with the great civilizations of the past. ~~No one knows, however, if we will be able to solve the technical problems inherent in such an effort.~~

Discussion. The change in the title emphasizes the mutually supportive relationship of astronomy and technology suggested in the first paragraph. The first paragraph sets up the body of the article as an identification of developments, so the bulleted items in the chronological sections are stated in noun phrases, not in complete sentences. Paragraphs with more than one development are broken into separate bulleted items for emphasis. The headings are marked for boldface, as specified by the standard.

In the first paragraph, "For" has been changed to "Over the" to support the idea of progress from one century to the next. "The" has been substituted for "Our" because *we* have not been studying the universe (at least not in the scope of this article). "By the same token" emphasizes the simultaneity of the cross-fertilization.

In the section 1500–1600, the sentence about early star maps has been dropped as it does little more than restate the point about maps' increasing accuracy.

In the section 1600–1700, the unnecessary subordinate clause "that should be used" has been cut, as has the irrelevant detail about Leibniz.

In the 1800–1900 section, the informal "etc." has been replaced by "and many others," to stress that there were many others. The second item has been converted to a noun phrase for parallelism.

In the section 1900–NOW, the redundant "In this century" has been cut, and the separate developments have been set as bulleted items and converted to noun phrases. The addition of "a precursor for all modern fusion efforts" establishes the consequence of Bethe's prediction. The appositive "precursors to the development of accurate spacecraft navigation" has been elevated from within parentheses to parallel the statement about Bethe's prediction.

The heading "A Perspective of the Search" has been added to divide what follows from the discussion of historical developments. In this section, "Astronomical searches" has been substituted for "The stars" to clarify the focus. The second and third items in the bulleted list have been transposed on the grounds that an understanding of the origin and evolution of the stars will follow an understanding of the fundamental laws of physics governing cosmic processes and events. The final sentence has been deleted because it undercuts the positive emphasis of the article's ending.

The last two items in the "Perspective" section are transposed to help keep readers from thinking that "the origin and evolution of stars" is a restatement of "the birth of the universe," the first item in that list.

If fewer changes were called for in the complexity edit, the editor might choose to place correctness edit markings on the same pages. However, because the text here is already heavily marked, it should be submitted to typing for revisions and proofread before the correctness edit. The typing of complexity edit revisions might introduce errors that will need to be fixed in the correctness edit.

The correctness edit

Effective correctness editing requires close attention to content and presentation. Details must be accurate, and the standard for the document must be followed carefully. Correctness editing requires a strong background in grammar and punctuation (see Chapters 11 and 12).

Correctness editing is important because mistakes can be costly. A few years ago, a firm in the Pacific Northwest offered to undertake some work for another firm for $1,000. The second firm readily agreed. Then someone at the first firm realized that $1,000 should have been typed as $100,000. Only by going to court could they break the contract to do the work for $1,000.

Mistakes can also make writers and editors look foolish. Some years ago at a large company, an employee who was very sensitive about his height wrote a letter to a government official. In the last sentence, he wrote: "If you have any questions about this matter, please contact the undersized." Only after he had reviewed the letter, signed it, and given it to his secretary to mail was the error caught by his boss's secretary.

Such mistakes don't happen often, but they can be costly or even dangerous. Incorrect instructions for milling a part to a particular tolerance can result in an entire order of parts being ruined; unclear or insufficient instructions for the disposal of medical waste could lead to medical staff contracting AIDS.

In the correctness edit, editors should change only what is essential to correct errors, to make the write-up conform to the standard, or to improve the presentation of the material significantly. At this point, macro- and complexity editing should not be necessary and should not take place.

In the correctness edit, editors should read for technical accuracy. In most organizations, final responsibility for the accuracy of a technical document rests with the writer, not the editor, but good editors should catch errors. Recently, editors of several technical journals have called for greater accuracy in articles submitted for publication. More and more research organizations are having grant proposals and article manuscripts reviewed by professional editors before they are submitted, to ensure accuracy and clarity and to increase their acceptance rate.

The correctness edit is also the final check to ensure that the standard for the document has been followed explicitly for text, graphics, formats, references, and so on. Whether the standard is a company style guide, a government manual, a general guide like *The Chicago Manual of Style,* or a field-specific manual like the *ACS Style Guide,* the document should conform to it in every way.

Sentences are checked for spelling, grammar, and sentence structure. Main clauses are checked for a complete subject and verb and grammatical completeness (discussed in Chapter 11). Definite pronouns are checked for clear antecedents, agreement, and case. The pronouns "this" or "that" can be troublesome, especially when referring to an idea—they should only be used to take the place of a noun previously stated.

Examine the punctuation in each sentence, inserting, deleting, or changing it as needed. Editors must be able not only to punctuate correctly but to explain why certain punctuation is correct or preferable.

Correctness editing differs from complexity editing by focussing less on making the document appropriately readable for the audience and more on making the document technically accurate and consistent with the standard for the document.

Correctness editing is similar to proofreading in that both are designed to eliminate errors. When there are few correctness changes on a manuscript, editors sometimes mark the margin with an X to draw attention to them, similar to using proofreading marks. However, correctness editing differs by taking place before final corrections to text and graphics and before the document is laid out. Also, it is an editorial rather than a proofreading function, to be undertaken by editors more familiar with the content of the document than proofreaders usually are.

Sample correctness editing

The text below is the complexity-edited version of the article on astronomy and technology. The disk version of the complexity-edited text has been revised, and revisions have been proofread. The manuscript is now ready for correctness editing.

The style guide dictates that hyphenated terms not be capitalized, and that bulleted items can either be indented or set flush left but should have the first letter capitalized. Also, the spelling of foreign names is to be based on the *Dictionary of Scientific Biography* but with English equivalents, and titles and headings are boldface.

Because few changes are called for here, an X has been used in the margin so the typist or typesetter will not overlook a correction.

X **Astronomy and Technical Cross-Fertilization** (lc)

Over the centuries, astronomy and technology have progressed hand in hand. The study of the
X universe has benefitted from improved observational devices and techniques. By the same token, developments in astronomy have led to practical applications in other disciplines.

1500-1600

- Increasingly accurate maps of the sky for navigation

1600-1700

X X • Christiaan Huygens invention of the pendulum clock for navigational time-keeping
- Newton's development of the calculus, the laws of motion and the law of universal gravitation as a means to explain the motions of planets and comets

1800-1900

- Increasingly sophisticated optical innovations by astronomers (William Herschel, Fraunhofer, Lord Ross, Alvin Clark, and many others)
- Development of increasingly sensitive photographic techniques

1900-NOW

- Lockyer's discovery of a new chemical element, Helium, on the sun before it was known on Earth
- Hans Bethe's theoretical prediction of hydrogen fusion at the center of the sun, a precursor for all modern fusion efforts
- Lyman Spitzer's development of astrophysical plasma theory, the basis of present devices for releasing energy from controlled fusion
- Very long baseline radio astronomy techniques used in high-precision geodesy to survey the structure of the Earth
- Techniques of celestial mechanics, precursors to the development of accurate spacecraft navigation.

The mutually beneficial interaction between astrophysics and technology continues today.

A Perspective of the Search

Astronomical searches have occupied human thought for millennia. Over the generations, we have succeeded in gaining ever greater insight into the underlying forces at work in the cosmos. In the Space Station era, the family of permanent observatories in space will open the way to new, comprehensive studies of key remaining problems in astrophysics, helping us understand:

- the birth of the universe, its large-scale structure, and the formation of galaxies and clusters of galaxies;
- the fundamental laws of physics governing cosmic processes and events;
- the origin and evolution of stars, planetary systems, life and intelligence.

If we succeed, we will leave a legacy to rank us with the great civilizations of the past.

Discussion. The correctness-edited version of this text in some ways looks like the work of a proofreader. Many of the changes are ones a proofreader following the standard would be expected to catch. In correctness editing, however, there is greater leeway to make changes, as the document is not yet in the proof stage. The type of change represented here by moving the sentence about Lockyer to its proper place is much easier to fix in the editing stage. A proofreader would not be expected to know that Lockyer's discovery predated 1900, and often proofreaders are not expected to check the accuracy of content. However, editors who have worked with the material in a document should have resources at hand enabling them to check such information if they do not know it already.

In this correctness edit, several changes have made the piece conform to the style guide. In the title, the "F" in Fertilization has been lowercased in the compound, as has the "H" in Helium in the 1900–NOW section. The "W" in "NOW" and the heading "A Perspective of the Search" have been marked for boldface, and the items in the bulleted list in the final section have been marked to begin with a capital letter, all per the style guide. The period after the last item in 1900–NOW has been dropped. However, the semicolons and the period in the bulleted list in the "Perspective" section have been retained, to suggest that they form parts of one sentence. The list has been marked for flush left, to match the other and to conform to the style guide. Also, the possessive form for Huygens has been corrected. Finally, the spelling of names has been edited in conformance with the *Dictionary of Scientific Biography.*

Errors in the content have been corrected as well. Lockyer's discovery was in 1871, so the item has been moved to the preceding section.

In this example, correctness editing has concentrated more on making the format and content of the piece conform to the appropriate standards. Outside of the proper names, few errors needed to be corrected. However, a proofreader looking only for typos or errors in grammar or punctuation might have missed those errors, especially the date of Lockyer's discovery. Editors should recognize the difference between correctness editing and proofreading so that all necessary changes are made.

Following correctness editing, the text is again processed and the changes proofread. The text is copyfitted, and a dummy layout is prepared to ready the document for layout, page makeup, and printing. Before moving to discussion of those steps, however, we should consider how macro- and microediting done electronically (on-line or on a PC) might differ from the hard-copy marking process just described.

Editing and Proofreading On-line

Editing a document on a monitor rather than on paper *(hard copy)* presents special challenges for technical editors. However, on-line editing also allows editors more control over some steps.

In hard-copy editing, the disk or tape version of a file is printed out and the editor copy- or proofmarks the text and graphics. The text is returned to typists or typesetters and the graphics to artists for correction, after which a new version is printed out, returned to the editor, and proofed.

The hard-copy system has strengths and weaknesses. An important strength is that editors, usually not accomplished typists, are not responsible for the typing generated by editing or proofreading. Especially when a great deal of material must be edited, it is more cost-effective to have a trained typist do the work in less time, with fewer errors, and for less pay.

Another advantage to hard-copy editing is that editors can see a print document in the same form as its readers will, as black ink on white paper rather than less than half a page of white or colored text on a black or colored screen. (With the development of full-page monitors and more capable desktop publishing software, this is a less significant factor than it once was.) With the hard-copy system, editors mark changes for illustrations on copies or overlays, and trained graphic artists carry them out. This is the best way to produce complicated technical graphics.

Weaknesses in the hard-copy system have led to its replacement in many organizations, as more and more technical editors have PCs or terminals at their desks. First, having someone other than the editor enter corrections sometimes leads to confusion. Resolving the confusion can waste time. Also, it is easy to transfer files electronically, and the amount of paper used can be reduced with on-line editing. This reduction of paper waste is especially valuable with classified or sensitive documents, whose hard copy must be disposed of by special procedures. If editors with security clearance edit classified documents on-line, there is no reason to arrange for typists with the proper clearance. On-line editing is especially useful for text, but it should be used sparingly for revising graphics. Simple graphics can be revised by editors trained in the software, but complex ones should be altered only by professional artists.

Editing students skilled in WordPerfect, Microsoft Word, or other leading word processing programs have an obvious advantage over unskilled editorial candidates, especially in organizations that rely on on-line editing. Such organizations will use a mainframe system, stand-alone or networked PCs, or a combination of the two approaches. All three choices have advantages and disadvantages determined by the hardware and the information systems specialists who configure it. Many publications departments have found that on-line editing is feasible only when the organization makes an adequate commitment of PC equipment or mainframe time, systems analysis support, and most importantly, data recovery capability.

Copyfitting

Measuring the text of a document (or of a section) to find out how much space it will take when printed is called *copyfitting*. Once the text has been edited, typed on disk, and revised, and while it is being corrected and proofread, editors copyfit the text as one of the first steps in creating a dummy layout.

In the planning stage, tentative decisions about the final form of the document are made, including typefaces and type sizes for text, headings, and graphics that follow the instructions in the standard. Two common faces are Times Roman (for text) and Helvetica (for headings and graphics). The differences between them and the different point sizes (one point equals approximately $\frac{1}{72}$ inch) can be seen in Figure 4.4. Other characteristics of the final format are planned, such as whether the text will be justified with an even left and right margin or flush left (an even left margin and *ragged* right margin). A tentative layout might be prepared, especially for a document with length restrictions.

To prepare the dummy layout, editors must know the finished sizes of the graphics and the amount of space the text will take. To copyfit the text, determine the number of characters in the text (letters, numbers, spaces, punctuation marks, and symbols) by counting a number of random lines and multiplying by the number of lines per page, then by the total pages. The line length is set in picas (about six to the inch), and the number of characters per pica per typeface is determined by consulting a specimen book. Editors then multiply the line length in picas by the number of characters per pica to get the total typeset characters per line. This figure is divided into the estimated characters in the text to determine the approximate number of lines of typeset text.

The amount of space the text will occupy is also influenced by the *leading* (the spacing between the lines). With longer line lengths and smaller type sizes, leading should be increased for easier reading. A passage set to different leadings is shown in Figure 4.4. For example, consider a document in which the average number of characters per line is 80. With 500 lines of text, the estimated total is 40,000 characters. A typeface and size that provide 3 characters per pica set in a 33-pica line, for 99 (round up to 100) characters per line of typeset text, would require 400 lines. Consult the type specimen chart for samples of different leadings to find the total space occupied by those 400 lines of text. At the top of the first page of the text, the editor should *spec* (state the specifications for) the text, indicating the typeface, type size in points, leading in points, and line length in picas. If the text were to be set flush left in 10-point Times Roman on 12-point leading in a 33-pica line, the editor would write: $\frac{10}{12} \times 33$ Times Roman FL. Individual headings should be specked in the margin next to them. Other instructions such as indentation, boldface, italics, and flush left or centered for headings should also be indicated in a circled note at the beginning of the text.

When copyfitting is complete, editors have a good idea of how much space the text will require. They can then create a dummy layout of each page of the document.

Layout

Before the text and graphics for a document are laid out and made up into printing masters, layout staff review the editor's dummy layout. Often they

TIMES ROMAN (For Text)

Common Sizes:

6 The laws of physics must be of such a nature that they apply to systems of reference in any kind of motion. The laws of physics

8 The laws of physics must be of such a nature that they apply to systems of reference in any kind

10 The laws of physics must be of such a nature that they apply to systems of

12 The laws of physics must be of such a nature that they apply to

14 The laws of physics must be of such a nature that they

Line Spacings:

12/12

The laws of physics must be of such a nature that they apply to systems of reference in any kind of motion.

12/14

The laws of physics must be of such a nature that they apply to systems of reference in any kind of motion.

HELVETICA BOLD, ALL CAPS (For Graphics and Headings)

6 **THE LAWS OF PHYSICS MUST BE OF SUCH A NATURE THAT THEY APPLY TO SYSTEMS OF REFERENCE IN**

8 **THE LAWS OF PHYSICS MUST BE OF SUCH A NATURE THAT THEY APPLY TO**

10 **THE LAWS OF PHYSICS MUST BE OF SUCH A NATURE THAT**

12 **THE LAWS OF PHYSICS MUST BE OF SUCH A NA**

14 **THE LAWS OF PHYSICS MUST BE OF SUCH A**

20 **THE LAWS OF PHYSICS MUST**

24 **THE LAWS OF PHYSICS**

30 **THE LAWS OF PHYS**

36 **THE LAWS OF PH**

Figure 4.4. Type design.

revise the dummy to create the final layout, but it is very valuable because of the editor's familiarity with each section of the document.

Each document has a final format, such as an 8.5 by 11-inch printed page or a computer monitor screen. (In discussions of usability, "format" has been extended to include aspects such as typeface and line spacing, but it also refers to size and shape.) Within a format, different arrangements of text and graphics on the printed page or computer screen are possible. An editor creating a dummy layout for a printed or on-line document must try to accomplish the following:

- Balance text and graphics on each page (and each set of facing pages)
- Focus the reader or viewer's attention
- Fit the text and graphics comfortably within the limitations of the page or screen
- Place each graphic near what it illustrates

In addition, the editor follows these specific guidelines, whether hand or computer page makeup will be used:

- Place most graphics *above* text
- Place graphics toward the outside of the page
- Balance (mix equally) text and graphics on each page (or screen) and on facing pages
- Place any graphic that will not fit on the page with its text callout at the top of the next page
- Position captions consistently, centered or flush left (below a figure and above a table)
- Make sure photographs are cropped and sized properly so that the focal point occupies the greater part of the photograph
- Remember that readers will shift their focus from the most eye-catching graphic on a page to the next most eye-catching graphic as they read down a page
- Use margins, empty *(white)* space, headings, lists, and so forth to avoid large blocks of text—*gray pages*
- Have a consistent design throughout the document for columns, headings, etc.
- Use turn pages only when necessary
- Print the document on the front side only, or on the front and back of all pages

These principles are often adapted for different types of documents. Whatever the adaptations, editors should pursue three main aims in page layout: *consistency* in presentation, *clarity* of the organization, and *emphasis* on the most important information.

Some documents tend to rely on particular layouts. Instructions, for example, often use numbered lists of steps, different heading sizes (and perhaps fonts and colors), and lots of white space in which users can make notes. Comic strips usually print whatever the characters say or think above their heads and whatever the cartoonist says below or between the panels of the strip.

Figure 4.5 presents a sample dummy layout for the article on astronomy (the final layout is shown in Chapter 5). In the dummy, the three-column format is indicated by the broken lines, and headings are provided for reference. The note "(from bottom of Col 1)" emphasizes that the text does not flow across the top of the page but down the first column. Boxing is indicated for the graphic and the sidebar. This dummy layout may be revised by the layout staff to better match the facing page, but the dummy should be submitted as a suggestion from the editor.

CONCLUSION

Macro- and micro-editing often require more than one "pass" through a document. To edit effectively, many editors work through a document once to complete a good macro-edit, once again for a good complexity edit, and a third time for a good correctness edit. If the length of the document prohibits three passes through it, the complexity and correctness edits are often done simultaneously.

Traditionally, editing has been considered by many to be a one-step process. However, editing in well-defined steps, with clear focus on the tasks involved, can result in much better editing, especially for beginning editors.

SAMPLES TO EDIT

The three documents below, "COD Test Procedure," "Cleaning and Maintaining the Mouse," and "Planet Earth," are scrambled versions of edited texts. "COD Test Procedure" is designed for a middle audience familiar with the equipment, materials, and methods described in the instructions; the manuscript needs macro- and micro-editing. "Cleaning and Maintaining the Mouse" is designed for a lay audience; it also needs macro- and micro-editing. "Planet Earth" is designed for a lay or middle audience; it has been macro-edited but needs micro-editing. An answer key appears after the exercises.

COD Test Procedure

Macro- and micro-edit these instructions according to the guidelines in this chapter. The instructions are already on disk.

Three edits of this manuscript appear in the answer key: the macro-edited version, with the questions for the author; the complexity-edited version; and the correctness-edited version.

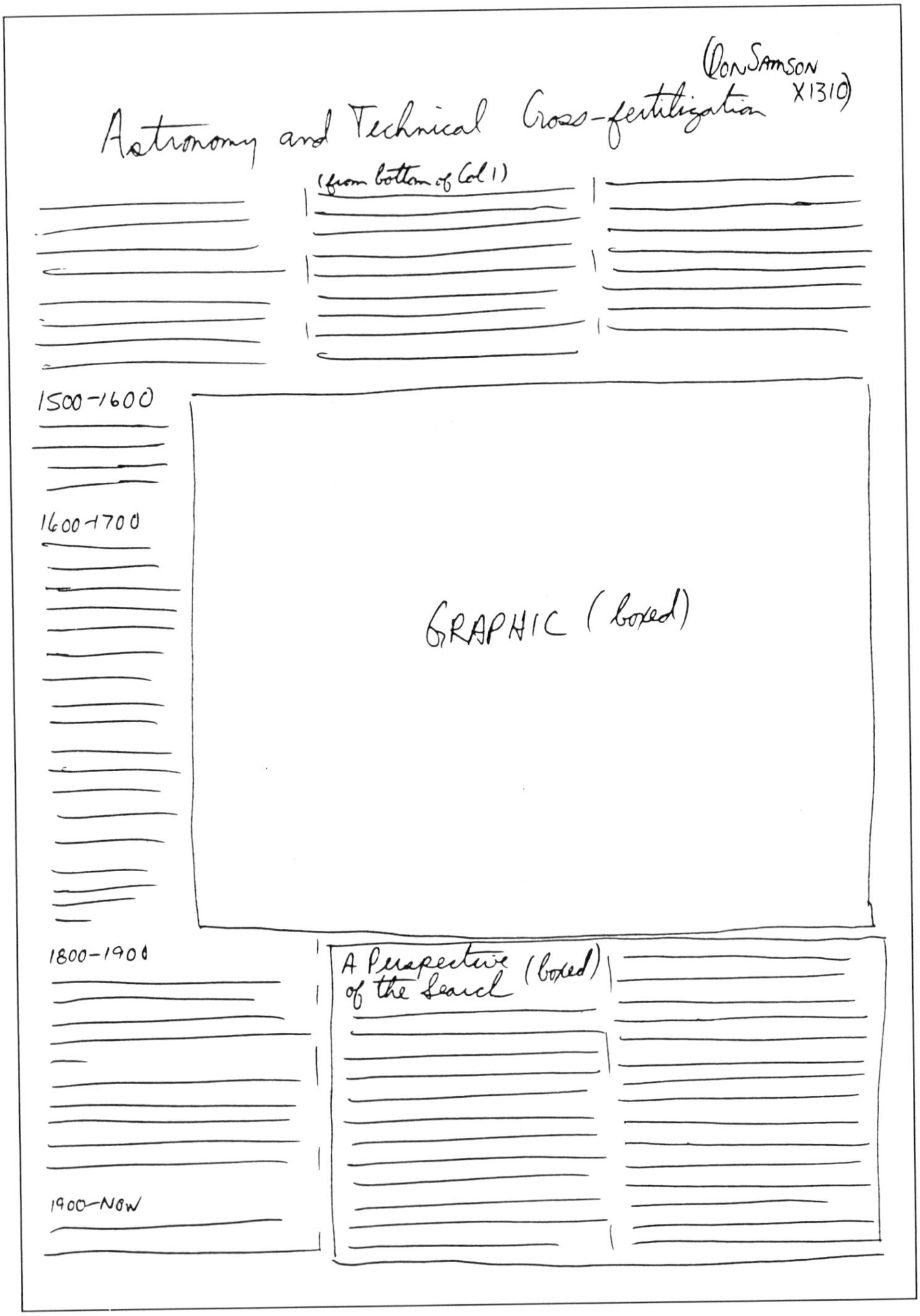

Figure 4.5. Dummy layout, astronomy and technology article.

COD Test Procedure

The chemical oxygen demand (COD) test procedure is used to measure polutional strength of industrial and domestic wastewaters. This allows measurement of a waste in terms of the total quantity of oxygen required for oxidation to carbon dioxide and water. It is useful for monitoring and control of waste water discharges.

The following items and materials are needed to perform the COD test:

Sample being tested (approximately 200 ml)
Reflux apparratus (condenser)
250 ml reflux flasks
Distilled water
Glass beads
Tweezers
Mercuric sulfate
Sulfuric acid
0.25 N dichromate solution
0.05 N FAS solution
Ferroin indicator solution
5 ml, 10 ml, 15 ml pipettes—a narrow tube in which fluid is drawn into by suction. Used for exact volume measurements.
Pipette bulb—used to suck liquid into pipette.
Metler balance—a weighing device.
50 ml buret—a long narrow volumetric tube with a stop cock to control flow.
Buret stand
Used acid bottle
Ceramic funnel

Begin the COD test proceedure by measuring 10 ml of the sample with the 10 ml pipette. Place the measured sample in the 250 ml reflux flask. Perform the same for distilled water. The distilled water sample is referred to as the blank (zero pollutional strength) and is needed to determine any testing errors. Add to the sample and blank 0.2 grams of mercuric sulfate and glass beads (use tweazers to hanlde glass beads). Measure 15 ml of sulphuric acid with the 15 ml pipette. Very slowly add 2.0 ml of the sulphuric acid to the sample. Then add the remaining 1.3 ml to the sample flask and mix thoroughly. Add 10ml of dichromate solution, and mix thoroughly. Now connect the flask to the condensor.

Repeat for the blank.

Turn on the condensor water and heat. Reflux for two hours.

Turn the heat off. Rinse the condensor by pouring ten to 20 ml of distilled water into the condensor top. Allow the rinse water to flow down the condenser tubing and empty into the flask. Disconnect the flask from the condensor. Dilute the sample and blank solutions with distilled water to approximately 70 ml. Cool the solution to room temperature.

Set up buret in a buret stand. Close buret stopcock and fill with 0.05 N FaS Solution. Record the beginning volume in ml from the buret. This should be done for every sample or bland tested.

Add2-3 drops of ferrion indicator solution to the blank and sample. The ferrion indicator will turn both solutions a blue-green color.

Place sample flask below the buret, so that 0.05 N FAS solution can be dripped into the flask. Slowly open and keep right hand on the stopcock. Holding the flask in your left hand' drip the 0.05 N FAS solution into the flask. Keep the sample and FAS thoroughly mixed by swirling the flask with your right hand. Continue this until a slight color change is noticed. Then very slowly proceed dripping and mixing until as reddish brown color appears: The color changes quick, so be prepared to immediately stop dripping once the change occurs. Record the volume remaining in the burrette. Repeat for the blank.

Place the ceramic funnel in the mouth of an old acid bottle. Pour the spent solutions into the funnel to collect the glass beads.

After the spent sample has drained, close and label the bottle. The spent solution will be properly disposed of latter. Finally, clean and dry all glassware and equipment.

Cleaning and Maintaining the Mouse

Macro- and micro-edit these instructions following the guidelines in this chapter. The discussion is not yet on disk.

Planet Earth

Micro-edit this macro-edited manuscript following the guidelines in this chapter. Follow Associated Press guidelines and do not use the serial comma (described in Chapter 12).

Cleaning and Maintaining the Mouse

Most functions of the microcomputer are controlled by its "mouse." Rolling the mouse on the table moves the pointer on the video screen, and pressing its button when the pointer is positioned executes the function.

Because the rubber ball in the mouse is contained in a well inside the mouse but makes contact with the table as it rolls, dirt and grease from the table accumulate on the ball, in the well, and on the inside rollers. The mouse should be thoroughly cleaned every hundred hours of computer use to keep it in good working order.

The mouse is a small box connected to the computer by a cord and contains a hard rubber ball, rollers which translate the motion of the ball to sensors, the button, and a flat cylinder which screws on to retain the ball.

Disassemble the mouse by pressing the cylinder with the palm of the hand and rotating it a quarter turn counterclockwise. The cylinder and ball will now come out freely.

Three lint-free cloths and a cleaning solution are needed to clean the mouse. Any cleaning solution safe for rubber, plastic, and metal will do, but an easy alternative is ordinary rubbing alcohol. Completely soak one cloth in the cleaner and lay it flat on the table. Leave the other two dry.

Roll the ball vigorously on the wet cloth for thirty seconds and dry it with another cloth. Leave the ball on the dry cloth for a few minutes to dry completely, as it is slightly porous and will absorb some of the cleaning solution. Clean the cylinder and the inside of the mouse with the damp cloth while the ball is drying. Be sure to roll the inside rollers several times so that all sides are cleaned. Dry the well and cylinder with the third cloth, again being sure to roll the inside rollers. When the ball is dry, put it back in the well and reconnect the cylinder by applying pressure and rotating clockwise.

Keeping the table clean is the best way to maintain the mouse between cleanings. Lint-free rolling pads are also available. Besides keeping a cleaner surface than a table, they also provide better traction for the rubber ball. Regardless of the care taken, however, thoroughly cleaning the mouse periodically is essential to making it work well and lasting a long time.

Planet Earth

by Larry Juchartz

Today Earth Day 1990, a reaction is sorely needed to an Ann Arbor News article on Feb. 15 which identified the factors contributing to the thickened blanket of greenhouse gases hovering over us ("Adapting to a warmer world," by Steve Eisenberg). But, because it must adhere to the principle of objective journalism, the paper could only report the news, not react to it.

It won't be enough to sit home today, and read the special articles, watch special broadcasts and plant a special tree, if Monday will find us just going about business as usual. Everything we hear now should make us think, evaluate, and respond.

We'll hear about rising concentrations of carbon dioxide from burning fossil fuels and destruction of rainforests, about increased level of methane generated by termites, cattle droppings and rice paddies, and about chlorofluorocarbons (CFCs) and how they play a dual role in environmental damage by contributing to the blanket of greenhouse gasses while also eroding the protective layer of ozone covering the earth.

Many of these facts will be well known, since its become fashionable to be well-versed in envirospeak. In fact, mention of CFC's may do no more than remind us that these compounds are the main ingredient of our cooling systems—and isn't summer coming? Time to drag the air conditioner out of the garage.

Sadly, some of us will respond in just this way. But, others might realize how ironic it is that air conditioning is our only defense against the oppressive heat which may come in a greenhouse affected future.

On a summer morning, when the radio announces a pre-dawn temperature of 90 degrees, we'll leave our air conditioned homes and drive our air conditioned, CO_2-producing cars to our air conditioned offices. There we'll watch the sun broil the blacktop outside, and pray that next year will finally bring some relief, that it will bring some return to the normal summer days we knew as children.

And there lies the problem: never before have human beings been asked to change their entire way of living and to prevent something that may not even happen. Indeed, it is almost impossible to comprehend the exact predicament facing us. The greenhouse effect may turn out to be nothing more than a worst-case scientific scenario; but if we take a "wait-and-see" approach, and continue to live as we do, we will put into place an irreversible warming trend, if the scenario is in fact reality.

If only 100 years of industrial and technological "advancement" have placed us at the perceived brink of climatic disaster, then another 20 years will lock us into a pattern there is no escape from. This is what the scientists know. But because they have no historical precedent on which to accurately base predictions for the future, they break into opposing camps and quibble over inconsistent computer models rather than offer constructive guidelines for the sure salvation of the planet.

The summer of 1988 gave us a taste of our possible future. It brought drought and record-breaking heat to much of the North American continent. Yet the summer was only one degree hotter than the historical "normal" summer temperature. In a greenhouse world, temperatures are predicted to rise an average of five to seven times that. This means that instead of 10 consecutive 95-degree days in 1988, there could be 50 days in 2010.

According to the scientists who take a benign view to the issue this nightmare can be prevented, but planting hundreds of thousands of trees in North America and Europe to replace those being destroyed in South America. However this discounts the phenomenon of acid rain, which is already eroding the forests on these continents. Why would new trees be any less susceptible to acidity than their older and more firmly rooted counterparts?

The benign view scientists also point to the oceans as a possible "natural" antidote to a warming world, saying that these bodies should be able to absorb much of the excess heat. But the oceans, too, are sick with pollution and they would be experiencing even more trauma in the form of melting polar ice caps and rising sea levels. Can they really be counted on to preform exactly as expected?

The problem is: nothing is certain. And uncertainty is anathema to human nature. Without visible evidence, we are ill-equiped to do much more than argue, and perhaps worry, about a given crisis, even when that crisis is one of unprecedented magnitude. Unless we encounter that 50-day string of unbearable heat this summer, debate and not decision will continue to be our only response to the threat of being roasted alive.

Oren Lyons, a Dakota Indian chief, addressed the complex situation before us, when he spoke to the United Nation general assembly about ecology in 1978. A full decade before the terrible summer of heat and garbage returned the environment into our national consciousness, the chief counseled:

"This is a time to be strong, a time to think of the future, and to challenge the destruction of our grandchildren. It is a time to move away from the four-year cycle of living that this country goes through form one election to the next, and to think instead about the generations of people to come."

The greenhouse effect as an issue will have to transcend the infighting of political parties and ideologies if future generations are to survive. It must become a global grass-roots concern. We cannot look to our governments to pave the way, because governments are opposed—to radical change, especially when it effects economic growth. Since the changes needed to forestall global warming are definite social and economic steps backward, we have to step forward and lead our governments into taking swift and decisive action.

Many of our modern-day "necessities" must quickly come to be seen as destructive luxuries until science can discover safe alternatives for them. And until those discoveries come, life will be uncomfortable. Yet, it will be no less comfortable than it as for the generations preceeding us, who never knew electric light or running hot water or supersonic air travel, but still were able to make important contributions to science and culture. Today, as we acknowledge what we've done to the earth and reevaluate our place on it, our generation has only to sort through its own "contributions" and determine whether they actually improve life or threaten to terminate it.

Answer Key: COD Test Procedure

Three versions of the edited manuscript appear below: the macro-edited original (with questions for the author), the complexity edit, and the correctness edit. Note the different types of changes made in each pass.

Macro-edit version

Questions for author

1. Is there a need for 5-ml pipettes?
2. Specify numbers of flasks, beads, tweezers, and pipettes?
3. Specify how to dispose of the spent solutions?
4. Indicate how to calculate the COD?
5. OK to drop the descriptions of equipment? Would the readers be familiar with them?

The editor has prepared questions for the author to clarify points, expand the discussion, and make it more appropriate for the audience. After the questions are answered, the editor copymarks the manuscript for the macro-edit.

Copymarking

Chemical Oxygen Demand

(COD) Test Procedure

The ~~chemical oxygen demand~~ (COD) test procedure is used to measure polutional strength of industrial and domestic wastewaters. This allows measurement of a waste in terms of the total quantity of oxygen required for oxidation to carbon dioxide and water. It is useful for monitoring and control of waste water discharges.

The following items and materials are needed to perform the COD test:

Indent list

Sample being tested (approximately 200 ml)
Reflux apparatus ~~(condenser)~~
250 ml reflux flasks
Distilled water
Glass beads
Tweezers
Mercuric sulfate
Sulfuric acid
0.25 N dichromate solution
0.05 N FAS solution
Ferroin indicator solution
~~5 ml~~ 10 ml, 15 ml pipettes ~~a narrow tube in which fluid is drawn into by suction. Used for exact volume measurements.~~
Pipette bulb ~~used to suck liquid into pipette~~
Metler balance ~~a weighing device~~
50 ml buret ~~a long narrow volumetric tube with a stop cock to control flow~~
Buret stand
Used acid bottle
Ceramic funnel

¶ When a wastewater sample is tested, the same procedures should be used for a

Begin the COD test proceedure by measuring 10 ml of the sample with the 10 ml pipette. Place the measured sample in the 250 ml reflux flask. Perform the same for distilled water. ~~The~~ distilled water sample ~~is~~ referred to as the blank (zero pollutional strength), and is needed to determine any testing errors. Add to the sample and blank 0.2 grams of mercuric sulfate and glass beads (use tweazers to hanlde glass beads). Measure 15 ml of sulphuric acid with the 15 ml pipette. Very slowly add 2.0 ml of the sulphuric acid to the sample. Then add the remaining 1.3 ml to the sample flask and mix thoroughly. Add 10ml of the dichromate solution, and mix thoroughly. Now connect the flask to the condensor.

Repeat for the blank.

Turn on the condensor water and heat. Reflux for two hours.

Turn the heat off. Rinse the condensor by pouring ten to 20 ml of distilled water into the condensor top. Allow the rinse water to flow down the condenser tubing and empty into the flask. Disconnect the flask from the condensor. Dilute the sample and blank solutions with distilled water to approximately 70 ml. Cool the solution to room temperature.

Set up buret in a buret stand. Close buret stopcock and fill with 0.05 N FaS Solution. Record the beginning volume in ml from the buret. This should be done for every sample or bland tested.

Add2-3 drops of ferrion indicator solution to the blank and sample. The ferrion indicator will turn both solutions a blue-green color.

Place sample flask below the buret, so that 0.05 N FAS solution can be dripped into the flask. Slowly open and keep right hand on the stopcock. Holding the flask in your left hand' drip the 0.05 N FAS solution into the flask. Keep the sample and FAS thoroughly mixed by swirling the flask with your right hand. Continue this until a slight color change is noticed. Then very slowly proceed dripping and mixing until as reddish brown color appears: The color changes quick, so be prepared to immediately stop dripping once the change occurs. Record the volume remaining in the burrette. Repeat for the blank.

Place the ceramic funnel in the mouth of an old acid bottle. Pour the spent solutions into the funnel to collect the glass beads.

After the spent sample has drained, close and label the bottle. The spent solution will be properly disposed of later. Finally, clean and dry all glassware and equipment.

(add) Calculate the milligrams of oxygen per liter required to oxidize the organic wastes in the sample according to the formula below:

$$\text{COD} = \frac{(A-B) \times M \times 8000}{\text{mL sample}}$$

where A = mL FAS used for the blank; B = mL FAS used for the sample; M = molarity of FAS

Discussion As the instructions were not yet on disk, the editor has minimized changes in the macro-edit to simplify the typist's work. Unnecessary description in the list of materials has been omitted, and the need to run the blank test has been clarified. One paragraph has been split and two sentences dropped. The spelling of the technical term "ferroin" has been corrected. The directions for calculating the COD have been added in response to the questions for the author. The 5-ml pipettes and descriptions of equipment have been dropped from the list of materials and equipment, also following the questions. However, the author wanted no changes in response to questions 2 and 3, stating there was no need for that information, given the audience.

Obviously, there is much more to be done in editing these instructions. However, the macro-edit is designed to check organization (which is fine here) and content, without cluttering the text to be typed, and the amount done here is sufficient.

Complexity-edit version

Changes from the macro-edit have been incorporated by typists and a clean version printed for complexity editing.

Chemical Oxygen Demand (COD) Test Procedure

The COD test procedure is used to measure pollutional strength of industrial and domestic wastewaters. ~~This allows measurement of a waste in terms of the total quantity of~~ It determines how much oxygen is required ~~for oxidation~~ to oxidize organic matter to carbon dioxide and water. It is useful for monitoring and control of waste water discharges. The blank should be tested at the same time the sample is tested or immediately after.

The following items and materials are needed to perform the COD test:

Sample being tested (approximately 200 ml)
Reflux apparatus (2)
250 ml reflux flasks
Distilled water
Glass beads
Tweezers
Mercuric sulfate
Sulfuric acid
0.25 N dichromate solution
0.05 N FAS solution
Ferroin indicator solution

10 ml, 15 ml pipettes
Pipette bulb
Metler balance
50 ml buret
Buret stand
Used acid bottle
Ceramic funnel

When a wastewater sample is tested, the same procedures should be used for a distilled water sample referred to as the blank (zero pollutional strength), needed to determine any testing errors. Begin the COD test procedure by measuring 10 ml of the sample with the 10 ml pipette. Place the measured sample in the 250 ml reflux flask. Perform the same for distilled water. Add to the sample 0.2 grams of mercuric sulfate and glass beads (use twezers to hanlde glass beads to the flask). Measure 15 ml of sulphuric acid with the 15 ml pipette. Very slowly add 2.0 ml of the sulphuric acid to the sample. Then add the remaining 1.3 ml to the sample flask and mix thoroughly. Add 10 ml of the dichromate solution, and mix thoroughly. Now connect the flask to the condensor.

Repeat for the blank.

Turn on the condensor water and heat reflux for two hours.

Turn the heat off. Rinse the condensor by pouring ten to 20 ml of distilled water into the condensor top. Allow the rinse water to flow down the condenser tubing and empty into the flask. Disconnect the flask from the condenser. Dilute the sample and blank solutions with distilled water to approximately 70 ml. Cool the solution to room temperature.

Set up buret in a buret stand. Close the buret stopcock and fill with 0.05 N FaS Solution. Record the beginning volume in ml from the buret. This should be done for every sample or blanc tested.

Add2-3 drops of ferroin indicator solution to the blank and sample. The ferroin indicator will turn both solutions a blue-green color.

Place sample flask below the buret, so that 0.05 N FAS solution can be dripped into the flask. Slowly open and keep right hand on the stopcock. Holding the flask in your left hand drip the 0.05 N FAS solution into the flask. Keep the sample and FAS thoroughly mixed by swirling the flask with your right hand. Continue this until a slight color change is noticed. Then very slowly proceed dripping and mixing until as reddish brown color appears. The color changes quick, so be prepared to immediately stop dripping once the change occurs. Record the volume remaining in the burette.

Place the ceramic funnel in the mouth of ~~an old~~ a used acid bottle. Pour the spent solutions into the funnel to collect the glass beads.

After the ~~spent~~ sample has drained, close and label the bottle. Dispose of the spent solution ~~will be~~ properly ~~disposed of latter~~. Finally, clean and dry all glassware and equipment.

Calculate the milligrams of oxygen per liter required to oxidize the organic wastes in the sample according to the formula below:

$$COD = \frac{(A - B) \times M \times 8000}{mL\ sample}$$

where A = mL FAS used for the blank

B = mL FAS used for the sample

M = molarity of FAS

Discussion. The complexity editing has clarified the use of the blank in the second paragraph by adding one sentence and moving another from later in the instructions. Otherwise, the organization of the manuscript was good, so little else has been done. In several places, sentence-level editing has focused on simplifying statements and making them more straightforward.

Correctness-edit version

The changes from the complexity edit have been processed by typists and a clean version printed for correctness editing.

Chemical Oxygen Demand (COD) Test Procedure

The COD test procedure is used to measure pollutional strength of industrial and domestic wastewaters. It determines how much oxygen is required to oxidize organic matter to carbon dioxide and water. It is useful for monitoring and controlling ~~of~~ wastewater discharges.

When a wastewater sample is tested, the same procedures should be used for a distilled water sample referred to as the blank (zero pollutional strength), needed to determine any testing errors. The blank should be tested at the same time the sample is tested, or immediately after.

The following equipment and materials are needed to perform the COD test:

Sample being tested (approximately 200 ml)
Reflux apparatus (2)
250 ml reflux flasks
Distilled water
Glass beads
Tweezers
Mercuric sulfate
Sulfuric acid
0.25 N dichromate solution
0.05 N FAS solution
Ferroin indicator solution
10 ml and 15 ml pipettes
Pipette bulb
Metler balance
50 ml buret
Buret stand
Used acid bottle
Ceramic funnel

Begin the COD test procedure by measuring 10 ml of the sample with the 10 ml pipette. Place the sample in the 250 ml reflux flask.

Add to the sample 0.2 grams of mercuric sulfate. Use tweezers to add several glass beads to the flask. Measure 15 ml of sulfuric acid with the 15 ml pipette. Very slowly add 2.0 ml of the sulfuric acid to the sample. Then add the remaining 13 ml to the sample flask and mix thoroughly. Add 10 ml of the dichromate solution and mix thoroughly. Now connect the flask to the condenser.

Turn on the condenser water and reflux both the sample and blank for two hours.

Turn the heat off. Rinse the condenser by pouring 10 to 20 ml of distilled water into the condenser top. Allow the rinse water to flow down the condenser tubing and empty into the flask. Disconnect the flask from the condenser. Dilute the sample and blank solutions with distilled water to approximately 70 ml. Cool the solution to room temperature.

Set up a buret in a buret stand. Close the buret stopcock and fill the buret with 0.05 N FaS solution. Record the beginning volume in ml from the buret. This should be done for every sample or blank tested.

Add 2-3 drops of ferroin indicator solution to the blank and sample. The ferroin indicator will turn both solutions a blue-green color.

Place sample flask below the buret, so that 0.05 N FAS solution can be dripped into the flask. Keeping your right hand on the stopcock, slowly open the stopcock. Holding the flask in your left hand, drip the 0.05 N FAS solution into the flask. Keep the sample and FAS thoroughly mixed by swirling the flask with your ~~righ~~ left, hand. Continue this until the color changes slightly. Then very slowly keep dripping and mixing until a reddish brown color appears. The color changes quick~~,~~ly, so be prepared to stop immediately dripping once the change occurs. Record the volume remaining in the bur-rette.

Place the ceramic funnel in the mouth of a used acid bottle. Pour the spent solutions into the funnel to collect the glass beads. After the sample has drained, close and label the bottle. Dispose of the spent solution properly. Finally, clean and dry all glassware and equipment.

Calculate the milligrams of oxygen per liter required to oxidize the organic wastes in the sample according to the formula below:

$$\text{COD} = \frac{(A - B) \times M \times 8000}{\text{mL sample}}$$

where A = mL FAS used for the blank
B = mL FAS used for the sample
M = molarity of FAS

Discussion. Few changes have been made in the correctness edit; however, the abbreviation *ml* has been changed to *mL* to fit the standard, and problems in spelling, grammar, and punctuation have been corrected. Note the change from "right" to "left" hand in the paragraph beginning "Place sample flask." Such obvious errors aren't always obvious to an editor until he or she has cleaned up a manuscript.

Cleaning and Maintaining the Mouse

The version below is the final edited and proofread version of the instructions. Compare it to the original, find the changes, and determine why they were made.

The COD Test Procedure instructions were written by Greg Swanson and submitted, correctly edited, as a technical writing course assignment (© 1986 Greg Swanson; used with permission).

Cleaning and Maintaining the Mouse

Most functions of the computer are controlled by its "mouse." Rolling the mouse on the table moves the pointer on the video screen, and pressing its button executes the function marked by the pointer.

The mouse is a small box connected to the computer by a cord. It contains a hard rubber ball, rollers that translate the motion of the ball to sensors, a button, and a flat cylinder that holds the ball in the mouse.

Because the rubber ball makes contact with the table as it rolls, dirt and grease from the table accumulate on the ball, in the well it sits in, and on the inside rollers. The mouse should be cleaned thoroughly every 100 hours of computer use to keep it in good working order.

Materials Needed

You will need three lint-free cloths and half a cup of cleaning solution to clean the mouse. Any cleaning solution safe for rubber, plastic, and metal will do. Ordinary rubbing alcohol works well.

Cleaning Procedure

Soak one cloth in the cleaner and lay it flat on the table. Keep the other two cloths dry. Disassemble the mouse by pressing the cylinder with your palm and rotating the cylinder a quarter turn counterclockwise. The cylinder and ball will now come out freely.

Roll the ball vigorously on the wet cloth for thirty seconds and dry it with another cloth. Leave the ball on the dry cloth for a few minutes to dry completely, as the ball is slightly porous and will absorb some of the cleaning solution.

Clean the cylinder and the inside of the mouse with the damp cloth while the ball is drying. Be sure to roll the inside rollers several times so that all sides are cleaned. Dry the well and cylinder with the third cloth, again being sure to roll the inside rollers.

When the ball is dry, put it back in the well and reconnect the cylinder by pressing it in place and rotating it clockwise.

Keeping the Mouse Clean

The best way to maintain the mouse between cleanings is to keep the table clean. Also, lint-free rolling pads can provide a cleaner surface than a table and better traction for the rubber ball. Regardless of the care you take to keep the mouse clean, however, you should clean it thoroughly every 100 hours of computer use to make it work well and last a long time.

The instructions for cleaning and maintaining the mouse were written by Tod Fringke and submitted, correctly edited, as a technical writing course assignment (© 1987, Tod Fringke; used with permission).

In revising these instructions, the editor reversed the order of paragraphs 2 and 3 so that the general description of the mouse comes before a discussion of why the mouse needs regular cleaning. Paragraph 5 has been moved to come before the discussion of the cleaning procedure in paragraph 4.

Headings have been added to help users first assemble the materials needed to clean the mouse and then follow the procedures for cleaning. The final heading separates general maintenance instructions from the cleaning procedures.

Sentences have been tightened in paragraphs 1, 2, 4, and 7, and the first sentence in paragraph 3 has been split to first clarify what "mouse" refers to and then describe its components. Other sentences have been revised for clarity and economy. The first sentence in paragraph 7 has been made periodic to emphasize the best way to maintain the mouse between cleanings.

The cleaning procedures in paragraph 6 have been broken into three separate paragraphs to control the pace of the discussion and make the directions easier to follow. Paragraph 7 has been separated from the cleaning procedures with a heading.

Answer Key: Planet Earth

The changes from the micro-edit (complexity and correctness edits) have been incorporated in the version below. Compare the two versions and determine why the changes were made.

Planet Earth

by Larry Juchartz

An Ann Arbor News article on Feb. 15 identified the factors contributing to the thickened blanket of greenhouse gases hovering over us ("Adapting to a warmer world," by Steve Eisenberg). But because it must adhere to the principle of objective journalism, the paper could only report the news, not react to it.

Today, Earth Day 1990, a reaction is sorely needed. It won't be enough to sit home today and read the special articles, watch special broadcasts and plant a special tree, if Monday will find us just going about business as usual. Everything we hear now should make us think, evaluate and respond.

We'll hear about rising concentrations of carbon dioxide from burning fossil fuels and destruction of rainforests. We'll hear about increased levels of methane generated by termites, cattle droppings and rice paddies, and about chlorofluorocarbons (CFCs) and how they play a dual role in environmental damage by contributing to the blanket of greenhouse gases while also eroding the protective layer of ozone covering the earth.

Many of these facts will be well known, since it's become fashionable to be well versed in envirospeak. In fact, mention of CFCs may do no more than remind us that these compounds are the main ingredient of our cooling systems—and isn't summer coming? Time to drag the air conditioner out of the garage.

Sadly, some of us will respond in just this way. But others might realize how ironic it is that air conditioning is our only defense against the oppressive heat that may come in a greenhouse-affected future.

On a summer morning when the radio announces a pre-dawn temperature of 90 degrees, we'll leave our air-conditioned homes and drive our air-conditioned, CO_2-producing cars to our air-conditioned offices. There we'll watch the sun broil the blacktop outside and pray that next year will finally bring some relief, some return to the "normal" summer days we knew as children.

And there lies the problem.

Never before have human beings been asked to change their entire way of living to prevent something that may not even happen. Indeed, it is almost impossible to comprehend the exact predicament facing us. The greenhouse effect may turn out to be nothing more than a worst-case scientific scenario, but if we take a "wait-and-see" approach and continue to live as we do, we will put into place an irreversible warming trend if the scenario is in fact reality.

If only 100 years of industrial and technological "advancement" have placed us at the perceived brink of climatic disaster, then another 20 years will lock us into a pattern from which there is no escape. This is what the scientists know. But because they have no historical precedent on which to base predictions for the future accurately, they break into opposing camps and quibble over inconsistent computer models rather than offer constructive guidelines for the sure salvation of the planet.

The summer of 1988 gave us a taste of our possible future by bringing drought and record-breaking heat to much of the North American continent. Yet that summer was only one degree hotter than the historical "normal" summer temperature. In a greenhouse world, temperatures are predicted to rise an average of five to seven times that. Instead of 10 consecutive 95-degree days in 1988, there could be 50 such days in 2010.

This nightmare can be prevented, according to the scientists who take a benign view to the issue, by planting hundreds of thousands of trees in North America and Europe to replace those being destroyed in South America. However, this discounts the phenomenon of acid rain, which is already eroding the forests on these continents. Why would new trees be any less susceptible to acidity than their older and more firmly rooted counterparts?

The benign-view scientists also point to the oceans as a possible "natural" antidote to a warming world, saying that these bodies should be able to absorb much of the excess heat. But the oceans, too, are sick with pollution, and they would be experiencing even more trauma in the form of melting polar ice caps and rising sea levels. Can they really be counted on to perform exactly as expected?

The problem is, nothing is certain, and uncertainty is anathema to human nature. Without visible evidence, we are ill-equipped to do much more than argue and perhaps worry about a given crisis, even when that crisis is one of unprecedented magnitude. Unless we encounter that 50-day string of unbearable heat *this summer,* debate and not decision will continue to be our only response to the threat of being roasted alive.

Oren Lyons, a Lakota Indian chief, addressed the complex situation before us when he spoke to the United Nations General Assembly about ecology in 1978. A full decade before the terrible summer of heat and garbage returned the environment into our national consciousness, the chief counseled:

"This is a time to be strong, a time to think of the future and to challenge the destruction of our grandchildren. It is a time to move away from the four-year cycle of living that this country goes through from one election to the next, and to think instead about the generations of people to come."

If future generations are to survive, the greenhouse effect as an issue will have to transcend the infighting of political parties and ideologies. It must become a global grass roots concern. We cannot look to our governments to pave the way, because governments are opposed to radical change, especially when it affects economic growth. Since the changes needed to forestall global warming are definite steps backward, both socially and economically, we have to step forward and lead our governments into taking swift and decisive action.

Many of our modern-day "necessities" must quickly come to be seen as destructive luxuries until science can discover safe alternatives for them. Until those discoveries come, life will be uncomfortable. Yet it will be no less comfortable than it was for the generations preceding us, who never knew electric light or running hot water or supersonic air travel but still were able to make important contributions to science and culture. Today, as we acknowledge what we've done to the earth and reevaluate our place on it, our generation has only to sort through its own "contributions" and determine whether they actually improve life—or threaten to terminate it.

ASSIGNMENT

The editing project described below is adapted from an article entitled "An Editing Project for Teaching Technical Editing," which appeared in *Technical Communication.* This assignment was designed for use in a classroom situation, to help students develop their ability to edit technical material. If you are not enrolled in a technical editing course, the assignment will be just as beneficial if you have an experienced editor review it.

Instructions

For this editing assignment, you will edit a manuscript to be submitted to a technical journal. You will solicit a manuscript and edit it to the standards of that journal, conferring with the author during the editing. This assignment will acquaint you with publication standards, develop your editing skills, introduce you to working with writers preparing technical documents for pub-

The editorial "Planet Earth" was published in the *Ann Arbor News,* Vol. 155, No. 112 (April 22, 1990), p. B7 (© 1990, Larry Juchartz; used with permission).

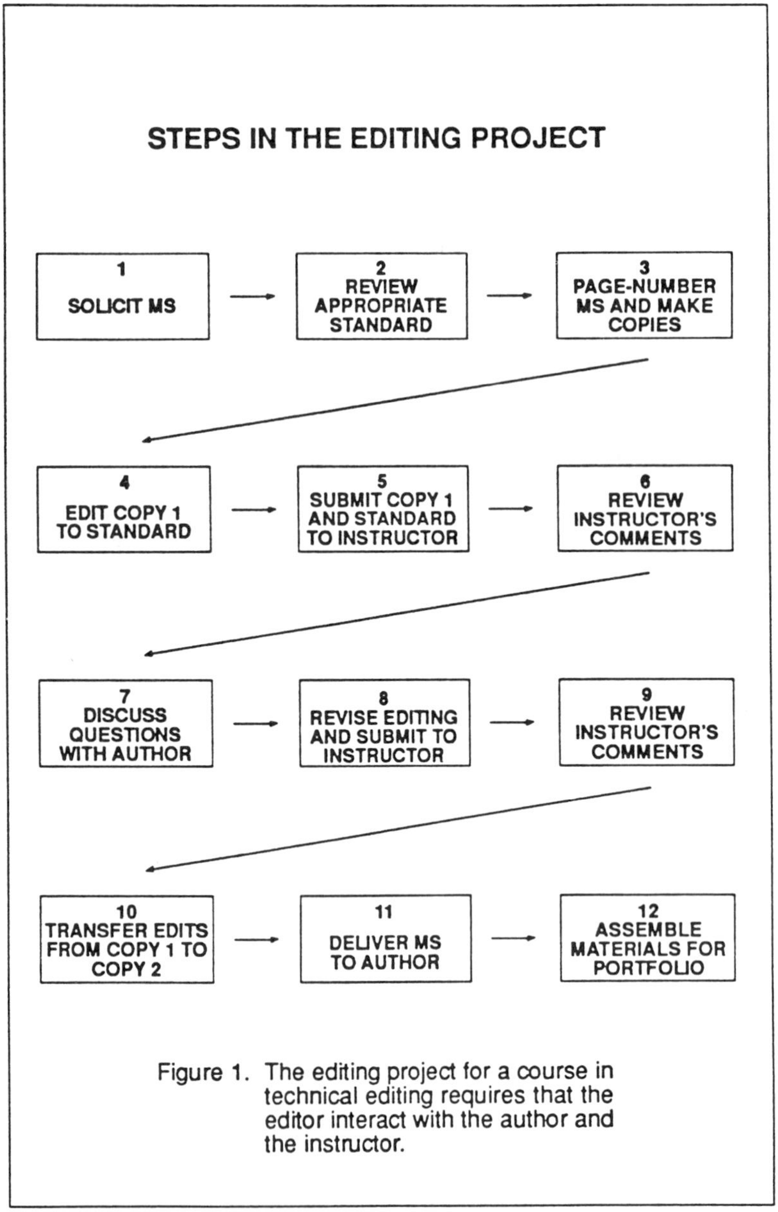

Figure 4.6. Steps in the editing process.

lication, help you gain confidence with the science and technology of your field, and provide an important document for your portfolio. The steps in the assignment are shown in Figure 4.6.

In a college or university setting, finding a technical manuscript to edit is easy because science and technology faculty are expected to publish frequently. Working with a technical professional will give you valuable experience for professional work in editing. Working with a faculty author has added advantages: Most faculty have published before, and they are committed to helping students. You may approach a professional in your community for a manuscript, but it is often more difficult to arrange writer–editor conferences with professionals in business or government.

Step 1: Solicit MS

Solicit a manuscript (MS) designed to report results of research in a technical journal. The more technical the manuscript, the more this assignment will teach you about technical editing and help you create a good editing sample for your portfolio. If you have any question about the suitability of the manuscript, consult your instructor.

Tell the author about the assignment and the course. Be sure to mention that your instructor will review your editing. Stress that only you and your editing instructor will examine the article, to ensure confidentiality. Remember that the author is doing you a favor by providing a manuscript and don't take too much of his or her valuable time.

Determine whether the text is on disk. Also, find out which journal the author plans to submit the MS to and when. Agree with the author on when you can pick up the complete MS (including all graphics and captions) and when he or she needs it back for review. This deadline will determine your schedule for the entire project. As you plan the project, remember that the worst mistake an editor can make is to miss a deadline.

Step 2: Review Appropriate Standard

Examine the target journal's instructions to authors or the recommended style guide. Examine the instructions carefully before you begin to edit and follow them exactly.

Determine whether the journal wants the article submitted with all graphics placed at the end of the text (in the order in which they are called out in the text) or merged with the text (placed after the page on which the graphic is called out). Make sure all graphics are numbered and labeled carefully, so they can be matched with their captions.

Step 3: Page Number MS and Make Copies

Page number the manuscript in pencil. Make three *clean* copies of the complete package. Plan to keep the author's original unmarked (but page numbered) and do all your editing on copy 1. When you have edited the manuscript and your revisions have been reviewed by your instructor, transfer your changes neatly to copy 2. Save copy 3 for the "before" sample in your portfolio. Copy 2 is the "after" sample, which reveals your ability to edit technical material.

Step 4: Edit Copy 1 to Standard

Do not mark on the originals of the tables and figures. Instead, write your editing marks on photocopies or vellum overlays. Follow the guidelines for editing text (this chapter) and graphics (Chapter 5). Use a red pen.

If you have questions about the technical content of the article, try to solve them through research. For this project, you will probably know more about the technical content of the article than your instructor. If you still have questions after research, address them to the author, as you will later in the professional situations that this project is designed to resemble.

Step 5: Submit Copy 1 and Standard to Instructor

Submit to your instructor your questions for the author, the standard, and copy 1 with the macro-edits you can make before the author answers your questions. Also submit to the instructor any questions or suggestions about the organization of the article. Establish with your instructor a schedule for review.

Step 6: Review Instructor's Comments

Review the instructor's comments and (if necessary) discuss any edits, questions, or suggestions.

Step 7: Discuss Questions with Author

Make an appointment with the author to discuss any questions or suggestions you have about the organization of the article. Take a copy of your suggestions in case the author requests one. Avoid letting the author see your edits until you and the instructor are finished with the manuscript. Most authors will want to see your edits, but if you have them with you, he or she might be more interested in seeing what you have done than in answering your questions. Do not get into matters of style in this conference; concentrate on getting the information you need to finish editing the manuscript.

Step 8: Revise Editing and Submit to Instructor

Following the author conference, revise your edits in copy 1 with black or blue ink and resubmit copy 1 to the instructor.

Step 9: Review Instructor's Comments

Review the instructor's comments and revise your edits appropriately. Discuss any questions.

Step 10: Transfer Edits from Copy 1 to Copy 2

Copy 1 will contain your edits; your instructor's markings, comments, and questions; erasures or line-outs; short questions for the author; white-out splotches; and perhaps other smudges. Once you have decided exactly how to copymark the MS, transfer the necessary editing marks neatly to copy 2 and make clean copies for your portfolio.

In a memo to the author, present a few suggestions that might help him or her prepare similar articles. Remember, however, the suggestions in Chapter 4 about the author conference. Make a clean copy of the suggestions for your portfolio.

Step 11: Deliver MS to Author

Deliver copy 2 and the original MS to the author. If the author wishes, discuss your changes. Address major changes first; don't mention changes in spelling, punctuation, or grammar. If the author is interested in suggestions for writing such articles, give him or her your memo of suggestions. Don't force your advice on the writer.

Step 12: Assemble Materials for Portfolio

Assemble in a file for your portfolio:

- A copy of the target journal's instructions to authors and reference formats
- Copy 3 of the original (unmarked)
- Copy 1 (your working copy)
- A copy of copy 2 (returned to the author)
- A copy of any questions to the author about the MS
- A copy of your suggestions to the author
- A 250-word discussion of what you learned from the assignment

The 250-word discussion is not to be shown to prospective employers; rather, review it before each interview and use it to support your claims that you have learned what technical editing involves and can edit technical material.

When the article appears in the journal, make a good copy for your portfolio. Examine the final article carefully to see what the journal's copyeditor did with your editing, and be prepared to explain those changes in interviews.

CHAPTER FIVE

Editing Graphics

Graphics are becoming increasingly important in technical documents. They are used in project reports, proposals, instructions, parts manuals, brochures, fact sheets, repair manuals, and other documents—sometimes even in correspondence. They are important to materials for oral presentations as well, as discussed in Chapter 8. Graphics enable writers to present many kinds of technical information more clearly and emphatically than they could with words.

Several different terms are commonly used for graphics. They are sometimes referred to generally as *illustrations*. However, some textbooks and professional guides divide graphics into two types—tables and illustrations—and illustrations are often called *figures*. To avoid confusion, this text follows common practice and refers to graphics as *tables* and *figures* (illustrations). Tables are usually prepared by typists or typesetters, and figures are usually prepared by graphic artists (but sometimes by writers and editors—see below).

Tables use vertical columns and horizontal rows to present information in a matrix format; figures do not. Figures include line graphs, bar graphs, circle (pie) graphs, scatter charts, drawings, photographs, cartoons, flow charts, organization charts, and schedules.

To explore the editing of graphics for technical documents, this chapter

- Reviews the importance of graphics in printed and online documents
- Discusses the editor's role in creating graphics
- Presents general guidelines for all graphics
- Examines tables and how to edit them
- Examines specific types of figures and how to edit them
- Describes how to choose the right graphic
- Discusses the layout of text and graphics

From your studies or work experience in technical writing, you will be familiar with many of the general principles outlined in the first part of this chapter. Now you should examine these guidelines for creating graphics from a different standpoint: that of the *editor* of graphics created by someone else.

THE IMPORTANCE OF GRAPHICS

Graphics can be used for information that would be difficult to present verbally, such as the information in photographs or in large bodies of statistics. The information in Figure 5.1 would take several sentences (pages?) to present in text, and the result would not be nearly as impressive. Test results, cost data, assembly or operating instructions, and other kinds of information can often be presented more simply, clearly, and dynamically in graphic form than in words.

Graphics can help editors avoid unnecessarily long or repetitive presentations of information in text. They can highlight facts for readers who might skim a document quickly (as people look first at the pictures in a magazine

Figure 5.1. The visual impact of a photograph. (Source: U.S. Department of the Interior, Bureau of Reclamation.)

or newspaper) and for readers who are accustomed to finding information summarized in tables and figures. Graphics also balance the appearance of a document, mixing text and graphics to avoid solid pages of text (which intimidate many readers) or an overabundance of illustrations (which makes documents look like comic books).

As graphics become more essential to printed documents, so too does color. Developments in printing have allowed newspapers like *USA Today* to produce color graphics at low cost; readers are particularly drawn to color graphics. In the past, technical journals used color only when absolutely necessary—to show, for example, stains in electron microscope photographs. However, more scientific journals are now using color, even merely to dress up graphics, as *Scientific American* often does. Some technical journals print color only when the author agrees to offset the increased publication costs. For example, *Nature* charges authors £500 (about $875) a page for full-color figures.

As readers of technical documents become accustomed to fancier and more abundant graphical presentations in popular journalism, they will come to expect higher concentrations of graphics in technical documents as well.

For less-adept readers, graphics have particular value in presenting technical information. If reading skills continue to decline, newspapers, magazines, and ultimately business and government publications will rely more heavily on graphics to enhance comprehension. The tendency to depend on graphics is not new. In the Middle Ages, when literacy was rare, shopkeepers hung signs in front of their shops with a painting of what they sold. Today, packages of food often include a pictorial representation of the contents. A photograph or drawing can make food look more appealing than can words on a package. Also, marketing specialists know that more than 10 percent of the people in the United States are illiterate, but they still buy food. Although technical editors do not prepare materials for an illiterate audience (think how difficult that would be!), they should recognize the growing tendency of all readers to rely on visual presentations.

THE EDITOR'S ROLE IN CREATING GRAPHICS

Technical editors often serve five major roles in creating graphics for reports, brochures, and other technical documents. They contribute to the design of the graphics, often establishing the standards to be followed (such as line thicknesses in graphs, and type sizes and fonts in tables (see, for example, Figure 5.2). Editors plan and coordinate the production of graphics, as described in Chapter 2. Also, they examine the relationship of the document's graphics and text to ensure that the proper graphics have been chosen to present the information. They edit the *roughs* (sketches) for clarity and conformance to standards. Finally, editors proofread the camera-ready graphics.

Although newsletter editors and editors in small companies and government offices may regularly use computer graphics programs such as Mac-

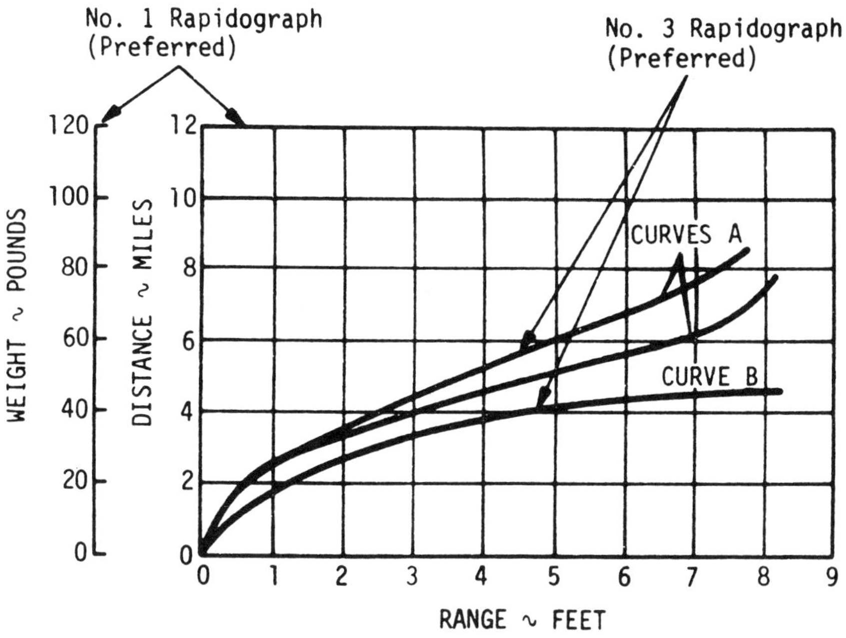

Figure 5.2. Sample instructions on line weights for graphics. (Source: Martin Marietta Corporation, *Martin Marietta Style Manual,* p. 168. Martin Marietta Corporation, © 1982, used with permission.)

Draw or Freehand, many technical editors do not create graphics. Often, technical writers make sketches that are edited by the editor and then submitted to graphic artists for preparation. When graphics are to be reprinted from other company or agency documents, the editor locates the original artwork in the archives or finds the artists who produced the graphics and gets *stats* (photostats) or a print of the computer file. As more technical graphics are produced with computers, it can be easy to get a print of the original graphic. However, sometimes it is very difficult to find the original graphic in the system or on one of many disks. Therefore, many editors file stats of graphics that are likely to be used in other documents.

Whether the graphics for a technical document are created by technical writers on PCs or on line, or by graphic artists on computer-aided equipment or by hand ("on the boards"), editors need to know how graphics are prepared so they can contribute to the process.

The editor's role will vary, depending on whether the graphics are produced by the writers or by graphic artists. Concepts for a document's graphics should be developed by the writers and sketched at the same time as the first draft of the text is written. Editors may edit the text and graphics lightly before the draft is reviewed, as described in Chapter 6. Pending revisions from review, the text and graphics are edited to standards, and the graphics are submitted to the appropriate production groups (tables to typesetters, figures to graphic artists).

Often, as in the astronomy and technology example in Chapter 4, a graphic that is *firm* (not likely to be changed or dropped in review) will be submitted to the artist for production before the first draft is reviewed so the preliminary drawing can appear in the review draft. The artist's drawing is proofread against the original by the editor or a proofreader. The editor incorporates the graphics into the text for printing the review draft, usually by having each graphic appear on a separate page after the page that refers to it.

Editors should receive sketches for a document's graphics from the writers or create their own roughs from information provided by the writers. If the editors are not generating the graphics, they should compile the information for the graphic artists, adding any necessary *callouts* (labels) to the roughs. The graphics should be reviewed for accuracy, clarity, conformance to standards for abbreviations and capitalization, and so forth.

Writers and editors should examine the roughs carefully. Editing the roughs is important no matter who will create the final graphics. Writers often overlook standards when they create graphics, but artists sometimes do too, especially when they work simultaneously on graphics for several documents prepared according to different standards. Once the graphics are final, the editors deliver the text and graphics to the layout staff.

Companies and agencies tend to rely on certain types of graphics that are especially appropriate for the work they do. Power companies, for example, customarily use tables to provide information about power generation, consumption, and costs. Auto parts manufacturers use drawings (especially exploded views) to identify parts of assemblies for automobiles. Medical research organizations rely on electron microscope photographs. The best preparation for editing graphics, however, is to learn about all types of graphics.

Students should recall the graphics they have seen in lab reports, in textbooks in their major courses, and in materials their professors have used in lectures. Faculty can usually provide additional samples of technical documents indicating the types of graphics used most commonly in their specialty. Students and technical professionals should consult many other sources, including those listed in the bibliography, and not rely solely on the information in any one book.

In addition to examining sample reports and publications in your field, another good way to develop a sense of how to create effective graphics is to hear experts discuss why particular graphics are or are not effective. You might study graphic design texts on your own to learn about page and screen design and about graphic arts equipment and processes such as stripping and airbrushing. The best way to develop a sense of what makes graphics effective is to study art and design and examine how proportion, perspective, color, balance, and focus can be combined with illustration techniques.

Another option is to look at graphics in other company or agency documents to see how similar material has been presented. Consult other editors, writers, and especially graphic artists if you are not sure how to present a particular type of material.

GENERAL GUIDELINES FOR GRAPHICS

Although information displays can vary widely in their designs, some general guidelines apply to all technical and business graphics.

Customize each graphic for your audience. To make every graphic appropriate in form and content for the intended audience, editors must know who that audience is (see the suggested questions about audience in Chapter 4). Also, different types of graphics are appropriate for different audiences, as discussed in Chapter 4 and throughout this chapter.

Choose the appropriate type of graphic for the material. When the document is planned, the writers and the project manager decide which information is especially significant and appropriate for graphic presentation.

Once writers identify the material to be portrayed in the graphic, editors can help them decide on the most appropriate type of graphic for it. For example, data dependent on a number of variables is best presented as a formal table. If the data adds up to 100 percent, consider a circle chart or a bar graph, especially for a lay audience. If the subject is an assembly, try a photograph, exploded-view drawing, or cutaway drawing. If your data involves a dependent and an independent variable, a line graph may work best. To identify company staff, think about photographs or an organization chart. When editors discuss with writers the possible graphics for a document, they should also estimate the cost of creating these graphics to ensure that the document's budget is sufficient.

Editors should suggest to artists the graphics they think would be effective. Experienced technical artists are experts in presenting information graphically, and editors should get in the habit of asking them for their suggestions.

Follow the appropriate standard when editing graphics. Editors should understand the company or agency standards for graphics (size limits, formats, line thicknesses, etc.). A sample graphics standard is shown in Figure 5.3. If writers create the graphics for a document, editors should be able to provide them with details about graphics standards. A company or agency style guide should be available from the publications department. If artists will be drawing the graphics, editors should review the standards before they edit the roughs to make sure that the graphics will conform to the standards. For example, if no illustration can be more than 6 inches high and 3 inches wide in the document, there is no sense in requesting a complicated wiring diagram for the document.

Editors preparing an article for publication in a journal should follow explicitly that journal's instructions for authors, as discussed in Chapter 8. The journal may instruct contributors to follow a style guide like the Council of

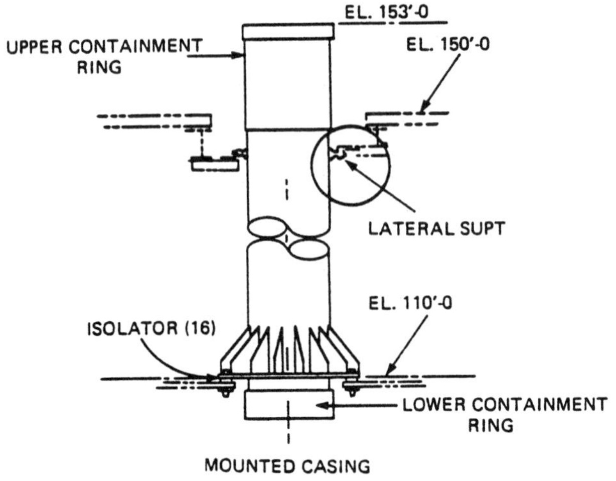

Fig. 4.2. Mounted casing and lateral support detail.

3. **Draw the different parts of an object in proportion to one another, unless it is indicated that certain parts are enlarged.**
4. **When a sequence of drawings is used to illustrate a process, arrange them from left to right or from top to bottom; they may be numbered for added clarity.**
5. **Label parts in the drawing so that text references to them are clear.**
6. **Place a zero before all decimal points that are not preceded by another number (e.g., 0.12).**

Figure 5.3. Typical drawing guidelines. (Source: Martin Marietta Energy Systems, *Document Preparation Guide,* page 4–15. Martin Marietta Energy Systems, © 1986, used with permission.)

Biology Editors' *CBE Style Manual* or the Government Printing Office *Style Manual.* If no style guide is preferred, follow the guidelines in this text, but always adapt your work to fit company or agency policy.

Keep each graphic simple and legible. Each graphic should illustrate a point, and it should be uncomplicated and straightforward. A reader may look at a graphic for no more than a few seconds, so limit the amount of information in any graphic and make sure its message is readily apparent. Edit the table title or figure caption until it presents the focus of the graphic clearly.

Graphics should be simple rather than dressed up with different typefaces, fancy borders, and so forth. Used selectively, however, border rules, patterns, and other devices can help display information more clearly.

For example, consider the shading in Figure 5.4. The shaded area represents a region of high heat flow from underground formations. A line ringing the area would not mark it as clearly; the line might be interpreted as a road or some other feature. Also, the shading draws the reader's attention to the area. This shading can be easily accomplished with options in computer graphics programs. By hand, the illustrator can use dry-transfer line, grid, or textured patterns such as Zipatone. The pattern is cut from a sheet to fill the desired area, then applied by rubbing the back of the sheet to transfer it to the graphic. Labels can be superimposed on such patterns, and alphabets and number sets are available in scores of typefaces. Line patterns, symbols, and dingbats are also available in a great variety of sizes and weights.

Acknowledge borrowed graphics. If editors use a graphic published in another report or printed document, they must cite the source directly below the graphic, as in Figure 5.4. Also, if the source is not one of their in-house documents or a U.S. government publication, editors should formally request the publisher's permission to reprint the graphic. For federal govern-

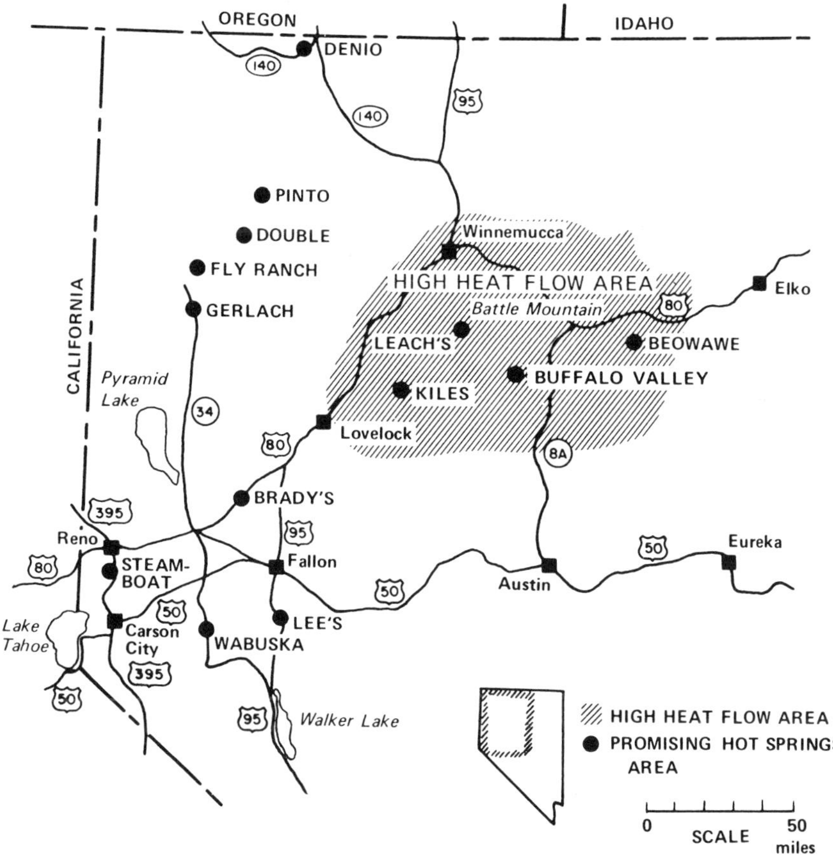

Figure 5.4. Shading to identify an area in a graphic. (Source: Eaton, William W., *Geothermal Energy* [Washington, DC: U.S. Energy Research and Development Administration, Office of Public Affairs, 1975], p. 7.

ment publications and *all* borrowings, use a "source tag" (see Figure 5.4) to identify the source fully, even if it is stated in a list of works cited.

Place each graphic near what it illustrates. In technical documents, graphics supplement the discussion in the text; they do not replace it. Graphics should appear close to the discussion they illustrate, referred to in the text in a *callout* sentence that directs the reader's attention to the figure. Some style guides suggest a parenthetical reference, as in: "The test data (Figure 3) indicate that the valve will operate effectively at temperatures up to 160 degrees C." Other standards prefer the callout to be worked into the sentence more, as in: "The test data shown in Figure 3 indicate that the valve will operate effectively at temperatures up to 160 degrees C." Editors should be consistent and not mix the styles. Also, the discussion in the text should refer to a graphic by its figure or table number, rather than saying "as shown in the figure below," because the sequence of the graphics may be changed when the document is revised or laid out. In a document organized into chapters, figure and table numbers include the chapter number: "Table 2.1." If the graphics are separate from the text in a manuscript, editors should also indicate where the graphics are to be placed in the text with a set-off comment such as

[[INSERT FIGURE 5.4 HERE]]

When graphics are mixed with text on a page, they are called *dropped-in*. Most technical documents use dropped-in art, as this text does. In some documents, and in review drafts of most documents, graphics appear on the page after their callout when they won't fit on the same page.

Indicate necessary reductions or enlargements. If a graphic must be reduced or enlarged to fit a page layout, a scaling wheel is used to determine the percentage of reduction or enlargement needed. A scaling wheel (or *proportional scale*), shown in Figure 5.5, has two disks joined in the center so that the inner disk, which indicates the size of the original, can be rotated to align with the desired size of the reduction or enlargement. The arrow marking "percentage of original size" then indicates the percentage to specify to the printer or camera operator. A number less than 100 percent indicates that a reduction is needed, more than 100 percent, an enlargement. The wheel setting in Figure 5.5 illustrates that a graphic 9 inches tall must be reduced to 55 percent of its vertical height to fit 5 inches of vertical page space. To request this reduction, the editor would send the graphic to the camera room or printer with the instruction "Shoot at 55%."

Scaling a graphic to fit the available space in a layout requires attention to both the height and width of the graphic. When a graphic 9 inches tall and 7 inches wide is shot at 55 percent to reduce the height to 5 inches, the graphic will become only 3.8 inches wide, which might leave considerable white space on each side. This space may be effective in focusing attention on the graphic, but it might create an unbalanced layout for the page. Reducing or enlarging graphics in a document can also result in varying type sizes in their labels. Editors should try to avoid this so the graphics appear

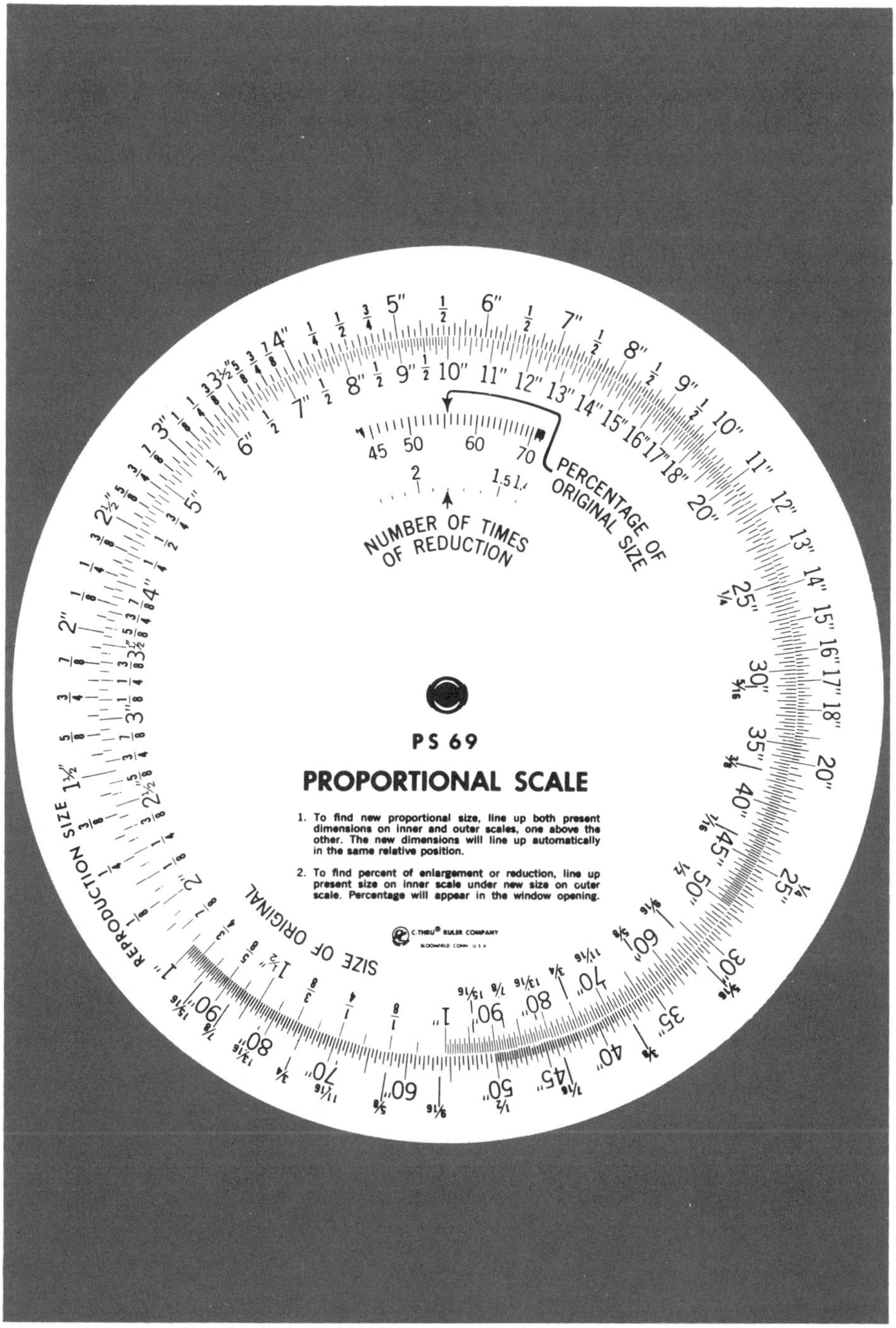

Figure 5.5. A scaling wheel for measuring reductions and enlargements. (Source: C-THRU Company, © 1985, used with permission.)

consistent; however, some variation may be permitted to avoid the expense of relabeling the graphics.

Some graphics are too wide for a page but cannot be reduced because their labels would become too small to read. Editors should avoid the temptation to risk the legibility of a graphic by reducing it to make it fit the page. Instead of reducing the graphic too much, you could place it on a *turn page.* For instance, a table can be turned sideways so that its leftmost edge is placed at the bottom of the page. With this arrangement, when the document is turned 90 degrees in a clockwise direction the table and its title (or the figure and its caption) can be read properly. Figure 5.48, later in this chapter, illustrates a turn page.

Editors should avoid having too many turn pages in a document. If a document requires a large number of turn pages, consider printing the entire document on turn pages and binding at the top to create a flip chart, rather than binding at the left side.

Some graphics are too big for the finished size of the document and must be printed on a *fold page* larger than the regular page size of the document. Fold pages are awkward and should be avoided if possible. They often require special printing and hand collating, and they complicate binding. If large graphics must be used, consider splitting them on two facing pages or placing them on fold pages in an appendix at the end of the document.

Label each graphic carefully. For each graphic, make sure all axes, curves, wedges, sections, parts, and so forth are labeled clearly. Many standards specify type no smaller than 8 point (after reduction), preferably in a sans-serif typeface such as Helvetica (illustrated in Chapter 4). In a printed document, the labels in all graphics should be one size, preferably boldface all caps. Labels should be placed near what they identify; if necessary, *leader lines* can connect labels to what they identify, as in Figure 5.2. Editors and artists should not automatically label every part of a graphic. Some common parts don't need labels, such as screws or washers in an assembly drawing. Labels are explanatory, and editors should ask which explanations the readers will need.

Include a concise caption. Tables have *titles* and figures have *captions.* Both are often called captions. Many style manuals suggest that figures be numbered with arabic numerals and that the number and caption be centered beneath the figure. Many standards once suggested that tables be numbered with uppercase Roman numerals, but today's trend is toward arabic numerals for table numbers as well as figures. Also, most standards require that the table number and title be placed above the table. These conventions are followed in this book.

Captions should focus clearly on the main point or subject of the graphic. Most style guides prefer a concise phrase caption that identifies clearly the point of the graphic. That is the format illustrated throughout this text. Other caption formats can be used, however. Complete-sentence captions are sometimes used, as in Figure 5.57. In some journals and other documents, extended captions involving discussion are used, as shown in Figure 5.6. Unless

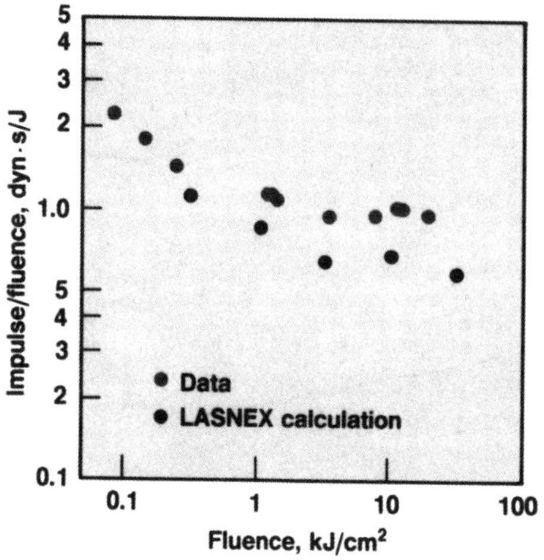

Fig. 1

Nova data on impulse coupling for an aluminum sample in vacuum compared to results from the LASNEX code. The spot diameter was 1.66 cm for shots 1 to 5 (five data points on the left), 0.77 cm for shots 6 to 8, and 0.5 cm for shots 9 to 11 (three data points on the right). For shots performed at similar fluences, the agreement among data points is good. Additional postshot analysis of all targets will determine the mass loss and is required before these new data can be unambiguously compared to results from the LASNEX code.

Figure 5.6. An extended caption, for discussing a graphic. (Source: "Laser Lethality: Experiments with the Nova and Janus Lasers," Lawrence Livermore National Laboratory, *Energy and Technology Review.* [June–July, 1986], p. 30.)

a specialized type of caption is called for, editors should use a concise phrase caption.

Use neither too few nor too many graphics. In any technical document, too much text without graphics makes reading monotonous for some readers, and too many graphics with little text can make a document seem oversimplified. The more your readers know about the subject, the more accustomed they are to relying on the discussion in the text. However, for readers who are unfamiliar with your subject or who don't read well, more graphics can help them grasp the information and become interested in the discussion. Lay and middle-level audiences rely on and expect simple graphics for information, and expert audiences expect graphics too.

Balance the arrangement of text and graphics. Throughout a document and in graphics for oral presentations, editors should strive for an effective balance of text and graphics, as discussed later in this chapter.

Recognize the advantages and disadvantages of desktop publishing and on-line graphics programs. Developments in computer graphics, especially for PCs, have made it much easier for technical writers to develop their own graphics for technical documents with graphics software. Periodicals such as *MacWorld, MacUser, Byte,* and *PC Week* discuss graphics and desktop publishing software regularly. Graphics programs allow shading patterns, slanted line marking, and different typefaces and type sizes, among other features, to label and highlight a graphic. For many technical documents, graphics produced on a PC and printed on a 300 dot per inch (DPI) laser printer are of acceptable quality for reproduction.

Having writers and editors produce the graphics for a document is an excellent way to eliminate the cost of having graphics prepared by graphic artists. However, producing effective graphics on a computer requires learning a graphics program, and that takes time. An expensive laser printer might be required to produce graphics that will reproduce well. Also, experts sometimes must be consulted to determine which desktop publishing and computer graphics programs to use and how.

Making graphics effective requires keeping them simple and straightforward. Most software programs have so much capability that it can be difficult for some writers (and editors) to discipline themselves and not add that extra cross-hatching, different typeface, or shading to suggest three dimensions. Many computer graphics programs tend to make simple graphics unnecessarily elaborate (for example, by shading the sections of a circle graph). Some ignore conventions of technical and business graphics. For example, many programs center the largest wedge of a circle graph at the top of the chart instead of beginning the left edge of the wedge at the top with the remainder following clockwise, as many standards specify.

The growing capabilities of desktop and on-line graphics programs have led some publications departments to reduce their graphic art staff and rely on writers to prepare their own graphics. This cost-containment strategy is dangerous, however, because professional graphic artists know much more about constructing effective technical graphics than do most technical writers and even editors. Graphic artists are trained in the theory and practice of presenting information visually. They can not only develop good concepts for particular types of information but also suggest how weak graphics could be revised to make them better.

The more complicated the equipment used to create graphics for technical documents, the greater the chance that equipment will fail or that sufficiently knowledgeable operators will not be available for creating or revising the graphics. When equipment is complex or heavily used, chances are greater that the editor relying on it will encounter delays in meeting the document production schedule.

Mark revisions to graphics clearly and simply. Editors should follow the copyediting and proofreading symbols used in their company or agency style guide. Changes should be indicated clearly on the rough art, or on a photocopy or tissue paper overlay of the final graphic. Explanatory notes should be circled so that they will not be considered part of the graphic and added to it.

Budget enough time and money to produce the graphics. Whether graphics are produced by writers or by artists, they take longer to create, revise, and proofread than text. To control art costs, editors should not spend substantial time editing a graphic—and an artist shouldn't begin work on it—until it is sure to be used in the document. Graphics can be very expensive to produce, even moreso when the schedule for document production requires artists to work overtime.

Review the graphics in their final form. Whether or not editors proofread the graphics during the production stages, they should review them before the document is laid out. Editors often have the final responsibility for their accuracy, clarity, and conformity to standards, as well as for the effectiveness of the layout and printing of the document. They should thus review the graphics in their final form in the print masters, especially to make certain that the right caption has been applied to each graphic.

Invite criticism of the graphics in documents you edit. Editors should solicit the opinions of experienced writers, other editors, and artists regarding their designs for graphics and their editing of them. Only by having experts tell you which graphics are well done and why, and how others could be improved, can you learn more about creating good graphics. Editors should seek out fellow employees who will critique their work honestly and fairly. Nothing is learned from someone who only says "Good" or "These could be better."

These general guidelines apply to both types of graphics—tables and figures—and editors should keep them in mind when they edit both. Different types of graphics, however, are effective for presenting different types of information, so editors should understand how to use and edit tables and various types of figures.

TABLES

Tables are used to organize data into a matrix of columns and rows so readers can examine the data conveniently and see relationships in it. Tables usually present numerical data, but tables with word content are also common, especially in less technical material for lay audiences. Tables present data that involves two sets of variables.

Formal Tables

Figure 5.7 presents a table of information about average precipitation by month for six cities. Each data point in the table depends on the city and the month—that is, on two variables. The data allows the reader to see patterns such as the higher amounts of January rainfall on the West Coast and the heavy summer and fall rainfall in Miami.

Figure 5.7 could contain many more cities (in the table on which it is based, data is presented for 70 cities). It could continue for pages, but then it should go in an appendix to the document rather than in the body, unless the document is a reference book with many such tables.

Tables are far more common in technical documents than they are in pub-

Table 5-2. Average Precipitation per Month, 1951--1980

	J	F	M	A	M	J	J	A	S	O	N	D
Dallas	1.7	1.9	2.4	3.6	4.3	2.6	2.0	1.8	3.3	2.5	1.8	1.7
Detroit	1.9	1.7	2.5	3.2	2.8	3.4	3.1	3.2	2.3	2.1	2.3	2.5
Miami	2.1	2.1	1.9	3.1	6.5	9.2	6.0	7.0	8.1	7.1	2.7	1.9
New York	3.2	3.1	4.2	3.8	3.8	3.2	3.8	4.0	3.7	3.4	4.1	3.8
San Francisco	4.7	3.2	2.6	1.5	0.3	0.1	*	0.1	0.2	1.1	2.4	3.6
Seattle	6.0	4.2	3.6	2.4	1.6	1.4	0.7	1.3	2.0	3.4	5.6	6.3

(* less than 0.05)

Figure 5.7. A typical data table. (Based on Table 350, "Normal Monthly and Annual Precipitation—Selected Cities," in *Statistical Abstract of the United States 1988,* 108th ed. [Washington, DC: U.S. Department of Commerce, 1987], p. 202. Hereafter cited as *Statistical Abstract 1988;* other editions cited by year.)

lications for a lay audience. Tables are especially useful for presenting numerical data, and one significant difference between technical and general documents is the amount and level of data presented to support conclusions. Some documents, especially research studies that have produced a mass of experimental findings, must show large bodies of data, and tables are often the best way to display them.

Figure 5.8 presents demographic data on which studies or interpretations might be based. This information could have been written out in sentence and paragraph form, as the data in most tables could, but a table is far more economical. The statistics are grouped in sets based on logical divisions of the information, and formatting has been used to divide the sets.

Boldface type can be used to emphasize parts of a table, and footnotes can provide necessary explanation. Footnotes are then part of the table, making up the *legend.* Editors should use legends if they must to present needed clarification, but if many notes are required, the explanation should be presented in the accompanying text or worked into the row or column headings.

In the table body, footnote references should be made with lowercase letters or symbols, to avoid readers' interpreting them as exponents.

No. 759. Persons Below Poverty Level and Below 125 Percent of Poverty Level, by Race of Householder, and Family Status: 1959 to 1982

[Persons as of **March** of **following year.** For explanation of poverty level, see text, p. 429]

RACE OF HOUSEHOLDER, AND FAMILY STATUS	BELOW POVERTY LEVEL					BELOW 125 PERCENT OF POVERTY LEVEL				
	1959	1969	1979 [1]	1981	1982	1959	1969	1979 [1]	1981	1982
NUMBER (mil.)										
All persons [2]	**39.5**	**24.1**	**26.1**	**31.8**	**34.4**	**54.9**	**34.7**	**36.6**	**43.7**	**46.5**
In families	34.6	19.2	20.0	24.9	27.3	49.3	28.6	28.1	34.2	37.0
Householder	8.3	5.0	5.5	6.9	7.5	11.8	7.4	7.8	9.6	10.3
Related children under 18 years	17.2	9.5	10.0	12.1	13.1	24.3	13.9	13.4	15.7	16.8
Children 5-17 years	(NA)	7.1	7.1	8.4	9.0	(NA)	3.3	(NA)	11.0	11.6
Other family members	9.0	4.7	4.5	5.9	6.7	13.2	7.3	7.0	8.9	9.8
Unrelated individuals	4.9	5.0	5.7	6.5	6.5	5.6	6.1	8.0	9.0	8.9
White	**28.5**	**16.7**	**17.2**	**21.6**	**23.5**	**41.8**	**24.5**	**25.2**	**31.0**	**33.1**
In families	24.4	12.6	12.5	16.1	18.0	37.2	19.5	18.4	23.4	25.3
Householder	6.2	3.6	3.6	4.7	5.1	9.2	5.4	5.3	6.8	7.3
Related children under 18 years	11.4	5.7	5.9	7.4	8.3	17.5	8.7	8.2	10.2	11.1
Other family members	6.9	3.4	3.0	4.0	4.6	10.5	5.4	4.8	6.3	7.0
Unrelated individuals	4.0	4.0	4.5	5.1	5.0	4.7	5.0	6.5	7.2	7.2
Black [3]	**11.0**	**7.1**	**8.1**	**9.2**	**9.7**	**13.1**	**9.5**	**10.3**	**11.4**	**11.9**
In families	10.1	6.2	6.8	7.8	8.4	12.1	8.5	8.8	9.7	10.3
Householder	2.1	1.4	1.7	2.0	2.2	2.6	1.9	2.2	2.5	2.7
Related children under 18 years	5.8	3.7	3.7	4.2	4.4	6.8	4.9	4.7	4.9	5.2
Other family members	2.2	1.2	1.3	1.6	1.8	2.7	1.8	1.9	2.2	2.5
Unrelated individuals	.9	.9	1.2	1.3	1.2	1.0	1.0	1.4	1.6	1.4
In families with female householder, no husband present [2, 4]	**10.4**	**10.4**	**13.5**	**15.7**	**16.3**	**11.8**	**13.0**	**17.4**	**19.8**	**20.3**
In families	7.0	6.9	9.4	11.1	11.7	8.0	8.6	11.6	13.3	13.9
Householder	1.9	1.8	2.6	3.3	3.4	2.2	2.3	3.3	4.0	4.1
Related children under 18 years	4.1	4.2	5.6	6.3	6.7	4.6	5.2	6.7	7.3	7.6
Other family members	1.0	.8	1.1	1.5	1.6	1.2	1.1	1.6	2.0	2.1
Unrelated individuals	3.4	3.5	3.8	4.3	4.1	3.8	4.3	5.4	6.0	5.8
In all other families [2, 5]	**29.1**	**13.7**	**12.6**	**16.1**	**18.1**	**43.1**	**21.7**	**19.2**	**24.0**	**26.2**
In families	27.5	12.3	10.6	13.8	15.6	41.3	19.9	16.5	20.9	23.1
Householder	6.4	3.2	2.8	3.6	4.1	9.6	5.0	4.5	5.6	6.1
Related children under 18 years	13.1	5.3	4.4	5.8	6.4	19.7	8.8	6.6	8.4	9.2
Other family members	8.1	3.9	3.4	4.4	5.1	12.1	6.1	5.4	6.9	7.8
Unrelated individuals	1.6	1.4	2.0	2.2	2.3	1.8	1.8	2.7	3.0	3.1
PERCENT OF POPULATION										
All persons [2]	**22.4**	**12.1**	**11.7**	**14.0**	**15.0**	**31.1**	**17.4**	**16.4**	**19.3**	**20.3**
In families	20.8	10.4	10.2	12.5	13.6	29.7	15.4	14.4	17.2	18.4
Householder	18.5	9.7	9.2	11.2	12.2	26.2	14.3	13.1	15.7	16.7
Related children under 18 years	26.9	13.8	16.0	19.5	21.3	37.9	20.2	21.3	25.5	27.4
Children 5-17 years	(NA)	13.6	15.3	18.7	20.5	(NA)	6.3	(NA)	24.5	26.3
Other family members	15.9	7.2	6.1	7.8	8.7	23.3	11.3	9.5	11.7	12.7
Unrelated individuals	46.1	34.0	21.9	23.4	23.1	52.7	41.8	30.7	32.6	31.9
White	**18.1**	**9.5**	**9.0**	**11.1**	**12.0**	**26.7**	**14.0**	**13.1**	**15.9**	**16.9**
In families	16.5	7.8	7.4	9.5	10.6	25.2	12.0	10.9	13.8	14.8
Householder	15.2	7.7	6.9	8.8	9.6	22.6	11.7	10.2	12.8	13.6
Related children under 18 years	20.6	9.7	11.4	14.7	16.5	31.6	14.9	15.9	20.2	22.0
Other family members	13.3	5.8	4.7	6.1	6.9	20.4	9.2	7.5	9.6	10.4
Unrelated individuals	44.1	32.1	19.7	21.2	20.7	50.8	40.0	28.6	30.3	29.7
Black [3]	**56.2**	**32.2**	**31.0**	**34.2**	**35.6**	**66.8**	**43.2**	**39.9**	**42.4**	**43.8**
In families	56.0	30.9	30.0	33.2	34.9	67.0	42.2	39.0	41.2	43.2
Householder	50.4	27.9	27.8	30.8	33.0	61.0	37.9	36.2	38.8	41.1
Related children under 18 years	66.7	39.6	40.8	44.9	47.3	78.0	52.7	51.1	53.1	56.1
Other family members	42.5	20.0	18.2	21.2	22.2	53.4	29.5	26.1	29.0	30.1
Unrelated individuals	57.4	46.7	37.3	39.6	40.3	64.0	53.9	45.4	49.2	47.4
In families with female householder, no husband present [2, 4]	**50.2**	**38.4**	**32.0**	**35.2**	**36.2**	**57.1**	**47.8**	**41.3**	**44.2**	**44.9**
In families	40.4	38.2	34.9	38.7	40.6	56.2	48.0	43.2	46.5	48.1
Householder	42.6	32.7	30.4	34.6	36.3	49.7	41.5	38.2	42.4	43.7
Related children under 18 years	72.2	54.4	48.6	52.3	56.0	79.4	66.2	58.0	60.7	64.0
Other family members	24.0	17.5	16.9	21.0	21.2	29.8	25.0	23.9	28.0	28.1
Unrelated individuals	52.1	38.7	26.0	27.7	26.6	59.1	47.5	37.0	39.0	37.9
In all other families [2, 5]	**18.7**	**8.0**	**7.0**	**8.8**	**9.8**	**27.7**	**12.6**	**10.6**	**13.1**	**14.2**
In families	18.2	7.4	6.3	8.1	9.1	27.3	11.9	9.8	12.3	13.5
Householder	15.8	6.9	5.5	7.0	7.9	28.6	10.9	8.8	10.8	11.8
Related children under 18 years	22.4	8.6	8.5	11.6	13.0	33.8	14.4	13.0	16.9	18.6
Other family members	15.3	6.4	5.1	6.5	7.3	22.8	10.2	8.1	10.1	11.1
Unrelated individuals	36.8	26.2	16.9	18.1	18.8	42.8	32.4	22.9	24.6	24.5

NA Not available. [1] Population controls based on 1980 census; see text, p. 428. [2] Beginning 1969, includes races not shown separately. Beginning 1979, includes members of unrelated subfamilies not shown separately. For earlier years, unrelated subfamily members are included in the "In families" category. [3] For 1959, Black and other races. [4] For persons in families, sex of family householder; for unrelated individuals, sex of individual. [5] Includes male unrelated individuals.

Source: U.S. Bureau of the Census, *Current Population Reports*, series P-60, No. 144.

Figure 5.8. Table ruling to facilitate interpretation. (Source: *Statistical Abstract 1985*, p. 455.)

Figure 5.9 demonstrates the use of words in a table. Figure 5.9 uses good spacing of rows and columns to clarify its information, but notice how the table seems at first to exclude many people.

The data in a table should be organized into logical rows and columns. Figure 5.10 illustrates the restructuring of a table to make the division into variables more logical.

Figures 5.7 through 5.10 are *formal* tables, matrices with a number of columns and rows. Formal tables are more complex than informal tables, which are discussed below. Editors should follow their style guide about *ruling* and *boxing* formal tables. Tables may be ruled with vertical and/or horizontal lines to separate the columns and rows, and they may be boxed with lines enclosing them. Rules are useful in tight tables but should be used only if readers will have trouble reading the table without them. Indicate clearly in circled comments the ruling needed for a table. Boxing is useful for a dropped-in table, especially if text is wrapped around it. Boxing makes a table (or a figure) stand out more as a separate unit.

Each formal table should have a table number and a concise title indicating the subject of the table. The title of Figure 5.7, "Average Precipitation per Month, 1951–1980," clearly states the table's content. By convention, table titles are set in initial capitals, with the table number and title centered above the table, as shown in Figures 5.7, 5.8, and 5.10. This format is very common, so editors should follow it unless the company or agency style guide requires otherwise. The title of the table in Figure 5.9 is handled differently, and the table is not numbered, as is common in documents with few graphics and in brochures and advertisements. In this book, tables are given figure numbers and captions because they are samples for illustration purposes only.

SPF SELECTION GUIDE

Skin Type	Pigmentation	Sunburn/Tanning History	Minimum Recommended SPF per Hours of Exposure (midday summer sun)			
			4 hrs.	3 hrs.	2 hrs.	1 hr.
I	Very fair skin; freckling; blond, red or brown hair	Always burns easily, never tans	SPF 30	SPF 24	SPF 20	SPF 15
II	Fair skin; blond, red, or brown hair	Always burns easily, tans minimally	SPF 20	SPF 15	SPF 15	SPF 8
III	Brown hair and eyes, darker skin (light brown)	Burns moderately, tans gradually & uniformly	SPF 15	SPF 15	SPF 8	SPF 6
IV	Light brown skin; dark hair & eyes (moderate brown)	Burns minimally; always tans well	SPF 15	SPF 8	SPF 6	SPF 4
V	Brown skin; dark hair and eyes	Rarely burns, tans profusely (dark brown)	Data unavailable: Recommended same as Skin Type IV			
VI	Brown-black skin; dark hair and eyes	Never burns, deeply pigmented (black)	Data unavailable: Recommended same as Skin Type IV			

Figure 5.9. A common application of word tables: consumer information. (Source: Johnson & Johnson, "Which SPF Number Is Right For You?" Johnson & Johnson Consumer Products, Inc., © 1987, used with permission.)

TABLE 1.—Resulting Hydraulic Heads for Transient One-Dimensional Linear Example

Data	End of Time Step 1	End of Time Step 2
Node 1, m (ft)	33.27 (109.08)	33.27 (109.07)
Node 2, m (ft)	36.44 (119.47)	36.39 (119.32)
Node 3, m (ft)	39.58 (129.76)	39.54 (129.65)

FIG. 7(*a*).—Example of Table *Incorrectly* Set Up

TABLE 1.—Resulting Hydraulic Heads for Transient One-Dimensional Linear Example

Data (1)	Node 1, m (ft) (2)	Node 2, m (ft) (3)	Node 3, m (ft) (4)
End of time step 1	33.27 (109.08)	36.44 (119.32)	39.58 (129.76)
End of time step 2	33.27 (109.07)	36.39 (119.32)	39.54 (129.65)

FIG. 7(*b*).—Same Table with Columns Set Up Vertically and Correctly Labeled and Numbered

Figure 5.10. Proper table orientation. (Source: American Society of Civil Engineers, *ASCE Authors' Guide to Journals, Books, and Reference Publications,* p. 11. American Society of Civil Engineers, © 1983, used with permission.)

Although tables are an excellent way to present a mass of data, they can contain only so much before they become too complex to interpret or too hard to read. Resist the temptation to reduce a table's type size to cram in more information or to make the table fit available layout space. Technical staff often want to include every bit of data that supports their interpretations, and consequently they sometimes create overly large, complex tables.

Editing a Formal Table

Before editing a table, review the specifications for the document to determine page (or screen) size and type face and size, and check the dummy layout of the document to determine the space available for the table. If possible, try to postpone editing each table until its data is complete so that the size of the table can be estimated. NA (not available) or ND (no data) may be acceptable in the final table if data cannot be obtained. However, editors often discover in manuscripts the phrase "(To come)" when the data for the table is not complete; the writer should then be queried.

Editors should decide whether to use vertical or horizontal *(landscape)* table format, based on logical divisions of the variables in the data. Check the row and column headings and subheadings for logical divisions of the categories and for grammatical parallelism if it is feasible. Rarely do technical writers create illogical headings, but revisions or typing errors sometimes create problems.

Editors should check the data in the table against the original to make sure it was transcribed properly. This is especially important in one-to-one proofreading of a table typed on disk for the first time. Also, mark the table for ruling, boxing, type sizes and faces, and unusual symbols such as Greek letters, using the company or agency style sheet. Type face and size should be consistent in tables throughout the document (as far as the document budget permits). Sans-serif typefaces are standard in tables, and boldface is commonly used for column and row headings. A smaller type in the same face might be chosen to make large tables fit the document page or screen size, but the data in the table should be no smaller than 6 point to ensure legibility.

Editors should limit what Edward Tufte calls "nondata ink," or printing that does not help the graphic communicate. Print that does not contain data, such as shading or patterning the column and row headings, should not be used unless it significantly improves the table's organization. An attempt to limit nondata ink is one reason why some organizations do not box tables and rule only complex ones.

Justifying text in a table is more difficult than in a printed page, as each line has less space for the typesetter to work with. Consequently, entries are often unjustified. Numbers are ordered vertically on the decimal or ones column, and zeros should be placed in front of decimal fractions. If the numbers in a table are no higher than four digits, no commas should be used after the thousands; if five digits or higher, commas should be used after the thousand in all numbers of four or more digits. In word tables, grammatical parallelism should be used for a balanced and more readable presentation.

Editors should proofread tables carefully or assign the job to experienced proofreaders. The more data in a table, the greater the chance that something was typed incorrectly. An error in one digit can affect an important total, and an omission such as the one in Figure 5.9 could lead readers to question the thoroughness of the company's research.

Informal Tables

Two- and three-column lists in a body of text are often called *informal tables*. Informal tables differ from formal tables in a number of ways. They are not set off as significantly from the text and are often embedded within it. They are not called out in the text with a table number, and they have no title. They are rarely ruled or boxed. Informal tables have only a few columns, often without headings. The characteristics of formal and informal tables are compared in the informal table below:

Formal Tables	Informal Tables
Table number	None
Table title	None
3+ columns and rows	2 or 3 columns, >3 rows
Ruled	None
Boxed	None

When two bodies of information to be compared or related can be stated concisely in a two-column list, writers can set up an informal table for the information. Informal tables can also be used to break long paragraphs of text, making it easier for readers to process. Also, informal tables are used in many reference lists and conversion charts, such as:

States and their ZIP codes

English measurements and metric equivalents

Degrees Fahrenheit and degrees Centigrade

Cities and their area codes

Elements and their chemical symbols

(The list above is not an informal table; it does not match each item in the left column with an item in another column.)

Informal tables are often used for glossaries and lists of symbols and abbreviations as in Figure 5.11. The information in this informal table could be written out in sentences, but the two-column list is a more efficient way to present it. The items in the left-hand column in an informal table should be listed according to some rationale.

Writers should be careful to set off only those lists that are long enough to warrant being displayed as informal tables. Information such as the U.S. population statistics below (based on U.S. Bureau of the Census statistics) should not be set up as an informal table, but written into a sentence.

Females in the U.S.	124,928,000
Males in the U.S.	118,987,000

TO CONVERT	TO	MULTIPLY BY
angstrom	meter	1×10^{-10}
BTU	joule	1.05587×10^{3}
cubic foot	meter^{3}	2.83169×10^{-2}
foot	meter	3.048×10^{-1}
maxwell	weber	1×10^{-8}
quart	liter	9.4635295×10^{-1}
rutherford	becquerel	1×10^{6}
square foot	meter^{2}	9.290×10^{-2}

Figure 5.11. Sample abbreviations in informal table format.

FIGURES

Line Graphs

Line graphs are very common graphics in technical documents. Writers use them to describe the relationship between two bodies of information, one of which depends on the other. For example, a line graph could chart plant growth over time. The independent variable, time, is plotted on the *x* (horizontal) axis. Plant growth depends on time, so it is the dependent variable graphed on the vertical *(y)* axis.

Other independent variables could be correlated to plant growth, such as amounts of water, fertilizer, and sunlight. Plant growth is usually dependent on all three, so in graphs with any of these three as the independent variable, plant growth is still the dependent variable graphed on the *y* axis.

Figure 5.12 places temperature on the *y* axis and time of day on the *x* axis to suggest that air temperature depends on time of day for any particular location. The information in Figure 5.12 could be presented in sentences: At 2 A.M., the temperature was 14 degrees. At 4 A.M., it was 13 degrees. And so on, for ten more sentences. No one would want to read twelve such sentences, and the pattern of temperature change can be seen much more easily in the line graph.

In each line graph, the *x* and *y* axes must be properly labeled, and the units of measurement must be indicated. The *x* and *y* axes in Figure 5.13

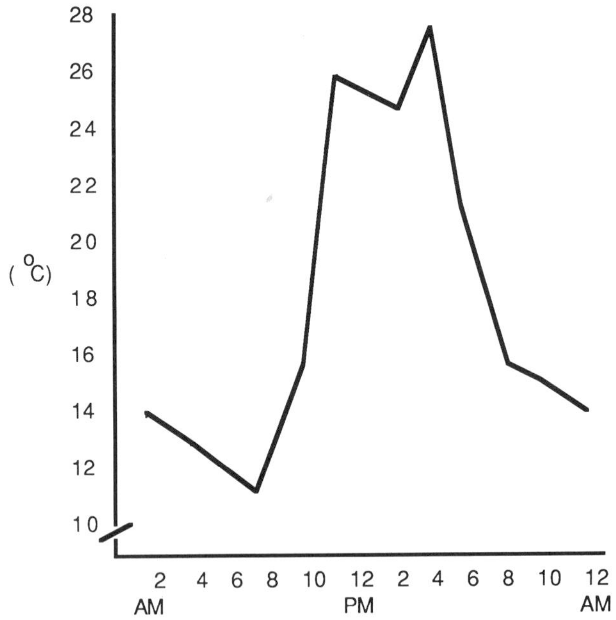

Figure 5.12. Graphing of the dependent variable on the vertical (y) axis.

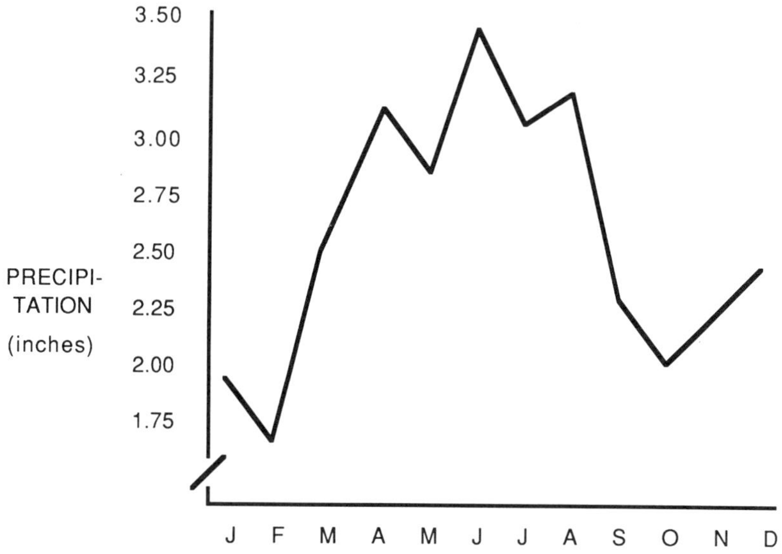

Figure 5.13. A line graph based on a table (Figure 5.7).

indicate other characteristics of time graphs. A *time plot* graph (in which time is the independent variable) can measure time in hours, months, or any other segment of time. Also, in any line graph the axes need not begin at 0. In Figure 5.13, there is no reason to have a *y* coordinate for 0 or 0.5 inches of precipitation. Some editors might omit the coordinate for 1.0 inch, as there is no measurement below 1.5 inches, but it is easier for readers if each axis begins with a whole number. A significant "break" in an axis should be clear, as in the dot chart in Figure 5.14.

Line graphs should be used only when there are enough data points present that a line is the proper way to show them. A continuous line suggests data for the intervals between coordinates marked on the axes. When there are a limited number of data points, joining them by a line is misleading. Consider how the graph of temperatures in Figure 5.12 would look if the data intervals were 10 minutes instead of 2 hours.

Tic marks are often placed along the axes of graphs to indicate the data reference points, especially when there is not room to label all the variables, as in Figure 5.15.

When writers or artists have more than one line in a line graph (Fig. 5.15) or juxtapose related graphs for comparison (Fig. 5.16), editors must be careful that the graph does not become cluttered and difficult to interpret.

Charting different sets of data in a line graph can result in problems of scale that can make the graph confusing. In Figure 5.16, the decision to juxtapose different sets of data in a two-part line graph has led to misrepresentation. Notice how the rates for aggravated assault and larceny/theft might seem equal and the burglary rates less than the robbery rates until the reader sees the different scale at the right of the graphic. Editors should avoid such possible misinterpretation by not aligning graphs with dissimilar scales.

Several solutions are possible for the problem in Figure 5.16. The editor could rearrange the data, setting the higher property crime rate section above the violent crime rate section. However, this does not solve the problem of scale. The burglary rates would appear twice as high as the violent crime rates, although they are actually about six times as great. Extending the vertical scale for property crimes (using the same interval for the violent crime rates) would require spreading the property crime results, and the graph would be five or six times as tall as it is now. Reducing the graph to fit an 8-inch image area in a document would require new axis and line labels for legibility. Also, the middle of the graph would be empty, with no data plotted between 340 and 1,200—there would be no data for 20 intervals. This empty space could be shrunk by changing the vertical scale so that each interval represents 100 crimes rather than 40. However, this would make the murder rate line overlap the *x* axis and seem to be zero.

Editors could fix this graph in one of two ways, depending on the audience. If the readers have technical backgrounds and are accustomed to inter-

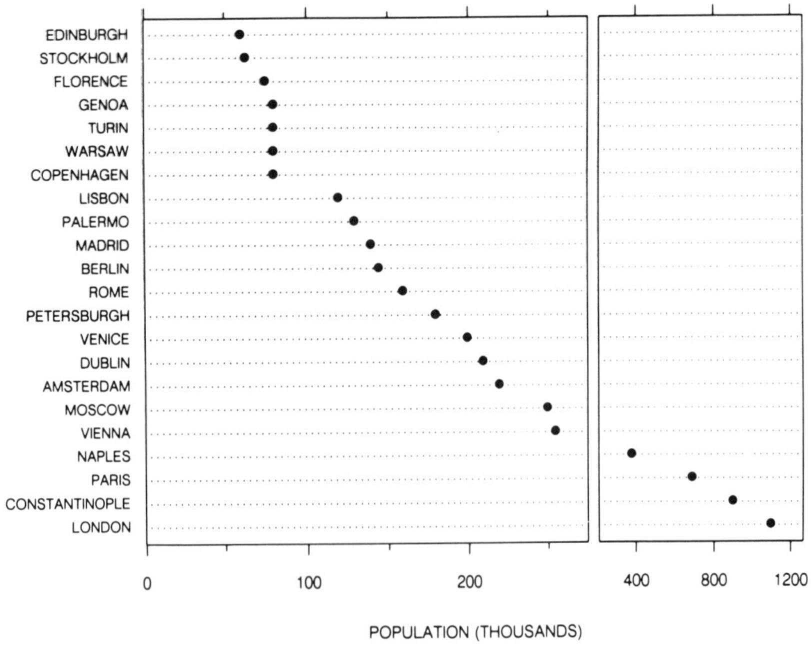

Figure 5.14. A clear scale break. (Source: Cleveland, William S., *The Elements of Graphing Data* [Wadsworth, 1986], p. 90. Bell Telephone Laboratories, © 1985, used with permission.)

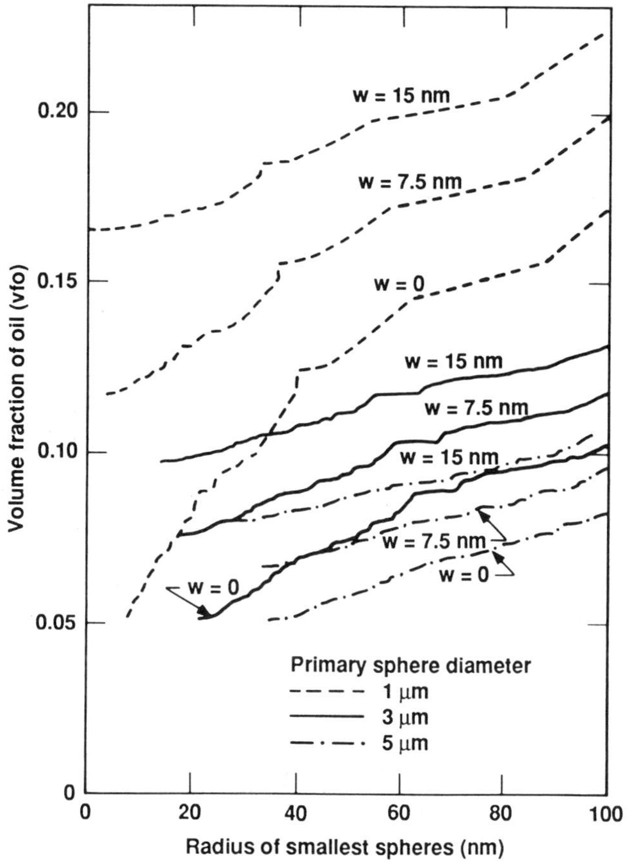

Figure 5. **Volume fraction of oil as a function of primary sphere size, size of smallest spheres, and interfacial thickness *w*.**

Figure 5.15. Labeling in a multiple line graph. (Source: Chen, C., Cook, R. C., Haendler, B. L., et al., *Low-Density Hydrocarbon Foams for Laser Fusion Targets: Progress Report—1986* [Livermore, Calif.: Lawrence Livermore National Laboratory, 1987], p. 11.)

preting data graphics, a semilogarithmic base 10 scale could be used. The same unit length on the *y* axis could represent 1 to 10 crimes (per 100,000 inhabitants), 10 to 100, 100 to 1,000, and so on. Plotting to a semilog base 10 scale would result in Figure 5.17.

For technical readers, such a graph would be preferable to the original, which they might interpret correctly only on a second examination. If readers are not familiar with semilog charts, the simplest solution is to break

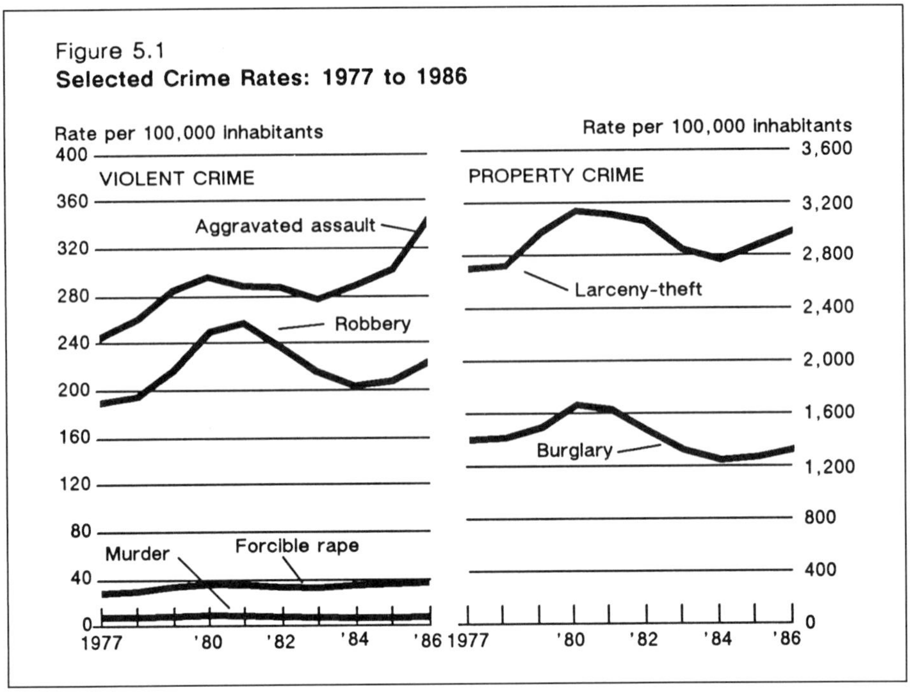

Figure 5.16. A misleading multiple line graph. (Source: *Statistical Abstract 1988,* p. 154.)

Figure 5.16 into two separate line graphs with different vertical scales and a break in the *y* axis for the graph of property crime rates, as the *x* axis was broken in Figure 5.14. However, breaks in the scale can be misleading, as many lay and middle-level readers pay far more attention to the data plots than the axes. The two figures could be stacked on a page as separate figures, but they should not appear side by side. This way, readers are far less likely to be led astray.

One valuable use of the line graph is to give an overview of more detailed information presented in another graphic, especially a table. Figure 5.18, a shaded line graph showing the surplus or deficit of the federal budget from 1935 to 1987, is based on Figure 5.19.

A multiple-line graph can be an effective way to present related bodies of data, but editors should be careful to set up such a graphic only if the lines do not overlap excessively or differ so much in the dependent variable that the graph becomes spread out, as would Figure 5.17 on crime rates if the vertical scale were not logarithmic.

The areas bounded by the lines in a multiple-line graph can be shaded or textured with patterns, as seen in Figure 5.20. Shaded line graphs can look impressive to a lay audience, especially when produced with computer-assisted drawing equipment. But notice how the patterns in Figure 5.20 could draw readers' attention to the areas between (rather than below) the lines in the graph. Shading can make a graphic difficult to interpret quickly and easily.

To make line graphs present bodies of information clearly and efficiently, editors should make sure they are labeled appropriately, scaled properly, and kept simple and straightforward. Line graphs are excellent for showing

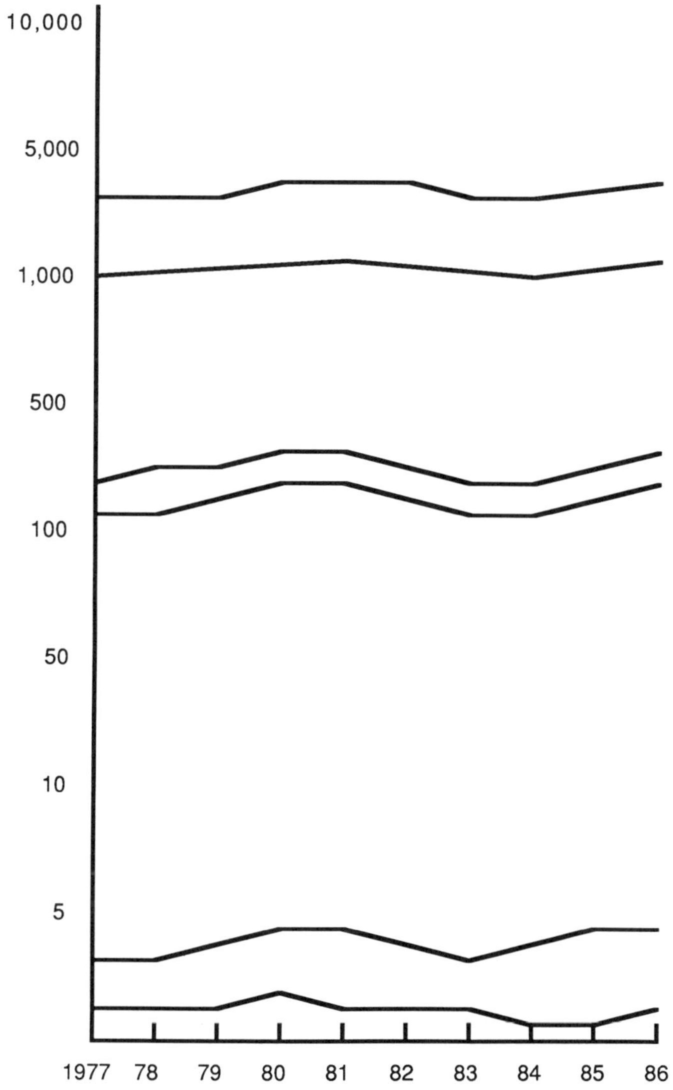

Figure 5.17. Semilog scale for crime rates in Figure 5.16.

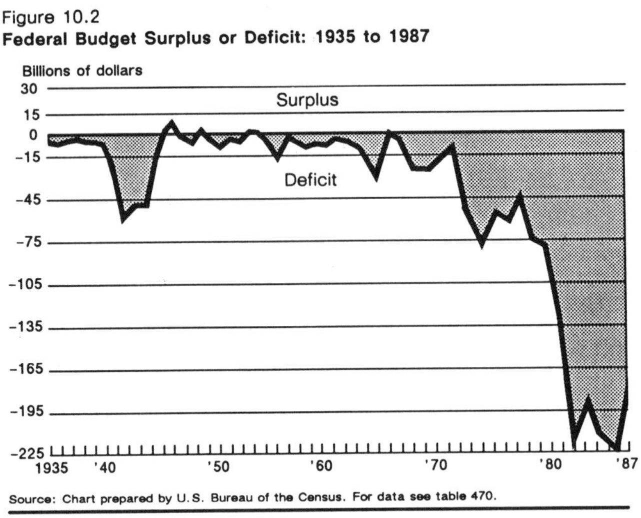

Figure 5.18. Use of a shaded line graph to emphasize data in a table (cf. Figure 5.19). (Source: *Statistical Abstract 1988,* p. 290.)

YEAR	RECEIPTS			OUTLAYS					SURPLUS OR DEFICIT (—)		
	Total [1]	Federal funds	Trust funds	Total [1]	Federal funds	Trust funds	Human resources	National defense	Total [1]	Federal funds	Trust funds
1945	45.2	41.9	3.6	92.7	94.8	−.6	1.9	83.0	−47.6	−53.0	4.3
1950	39.4	35.3	3.5	42.6	38.4	5.1	14.2	13.7	−3.1	−3.1	−1.6
1955	65.5	58.2	3.1	68.4	62.3	3.0	14.9	42.7	−3.0	−4.2	.1
1960	92.5	75.6	7.8	92.2	74.9	8.1	26.2	48.1	.3	.8	−.3
1961	94.4	75.2	9.4	97.7	79.4	9.0	29.8	49.6	−3.3	−4.2	.4
1962	99.7	79.7	9.9	106.8	86.5	8.9	31.6	52.3	−7.1	−6.8	1.0
1963	106.6	84.0	10.8	111.3	90.6	8.1	33.5	53.4	−4.8	−6.6	2.7
1964	112.6	87.5	11.2	118.5	96.1	9.2	35.3	54.8	−5.9	−8.6	2.0
1965	116.8	90.9	11.5	118.2	94.9	9.2	36.6	50.6	−1.4	−3.9	2.3
1966	130.8	101.4	12.9	134.5	106.6	10.8	43.3	58.1	−3.7	−5.2	2.1
1967	148.8	111.8	16.5	157.5	127.5	13.4	51.3	71.4	−8.6	−15.7	3.1
1968	153.0	114.7	17.6	178.1	143.1	16.9	59.4	81.9	−25.2	−28.4	.6
1969	186.9	143.3	20.1	183.6	148.2	15.7	66.4	82.5	3.2	−4.9	4.4
1970	192.8	143.2	22.3	195.6	156.3	17.8	75.3	81.7	−2.8	−13.2	4.5
1971	187.1	133.8	26.0	210.2	163.7	22.2	91.9	78.9	−23.0	−29.9	3.8
1972	207.3	148.8	28.4	230.7	178.1	25.5	107.2	79.2	−23.4	−29.3	2.9
1973	230.8	161.4	41.2	245.7	187.0	30.9	119.5	76.7	−14.9	−25.7	10.3
1974	263.2	181.2	46.1	269.4	201.4	33.9	135.8	79.3	−6.1	−20.1	12.2
1975	279.1	187.5	51.0	332.3	248.2	45.6	173.2	86.5	−53.2	−60.7	5.4
1976	298.1	201.1	61.8	371.8	277.2	56.2	203.6	89.6	−73.7	−76.1	5.6
1976 [6]	81.2	54.1	13.3	96.0	66.9	13.8	52.1	22.3	−14.7	−12.8	−.5
1977	355.6	241.3	70.3	409.2	304.5	56.9	221.9	97.2	−53.6	−63.1	13.4
1978	399.6	270.5	76.9	458.7	342.4	59.9	242.3	104.5	−59.2	−71.9	17.0
1979	463.3	316.4	86.0	503.5	374.9	65.7	267.6	116.3	−40.2	−58.5	20.3
1980	517.1	350.9	94.7	590.9	433.5	84.8	313.4	134.0	−73.8	−82.6	9.9
1981	599.3	410.4	106.0	678.2	496.2	94.2	362.0	157.5	−78.9	−85.8	11.8
1982	617.8	409.3	122.1	745.7	543.4	107.9	388.7	185.3	−127.9	−134.2	14.2
1983	600.6	382.3	147.3	808.3	613.2	124.4	426.0	209.9	−207.8	−230.8	22.9
1984	666.5	419.6	158.1	851.8	637.8	125.4	432.0	227.4	−185.3	−218.2	32.6
1985	734.1	459.5	197.5	946.3	725.9	152.7	471.8	252.7	−212.3	−266.4	44.8
1986	769.1	473.5	206.9	989.8	756.5	161.4	481.6	273.4	−220.7	−283.0	45.5
1987, est	842.4	526.1	215.1	1,015.6	769.1	164.8	500.5	282.2	−173.2	−243.0	50.3

Figure 5.19. A formal table that does not emphasize trends. (Source: *Statistical Abstract 1988,* p. 291.)

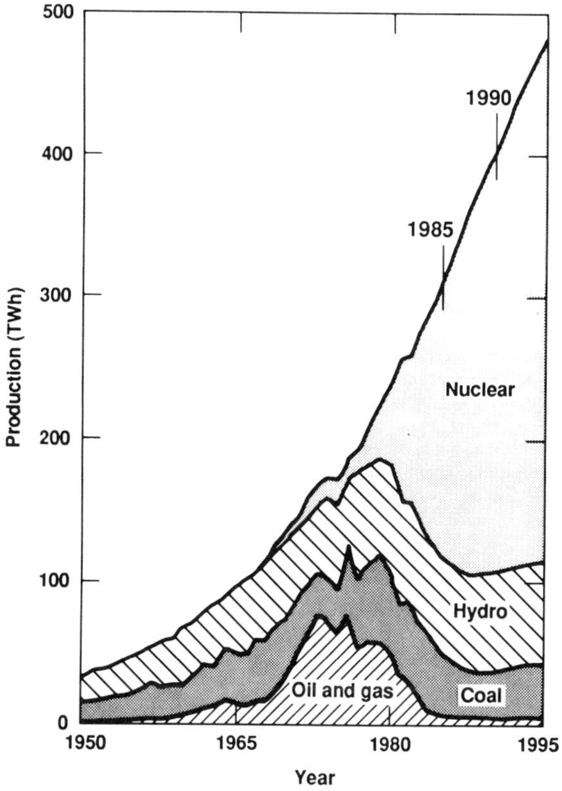

Figure 5.20. Distracting shading in a multiple line graph. (Source: Borg. I. Y., *Present and Future Nuclear Power Generation as a Reflection of Individual Countries' Resources and Objectives* [Livermore, Calif: Lawrence Livermore National Laboratory, 1987], p. 12.)

trends; the trends should be clearly visible and honestly presented in the scale. Note, for example, how Figure 5.21 suggests that profits have varied considerably during the six years shown. Such manipulation of the vertical scale misrepresents the minor fluctuations in profits the corporation has experienced. Careful readers will notice such manipulation, which an editor should support only for a good reason (and with a supervisor's blessing). The actual range of expenditure variation in Figure 5.21 is only 2.7%. The scale on the vertical axis should be tightened, and the line representing expenditure should be represented as nearly horizontal. Clearly, there would be little reason to use in a report or presentation a line graph showing a horizontal line. The editor should suggest that such a graph would be gratuitous (and misleading if the scale were manipulated).

Editing line graphs well requires attention to simplicity, clarity, and em-

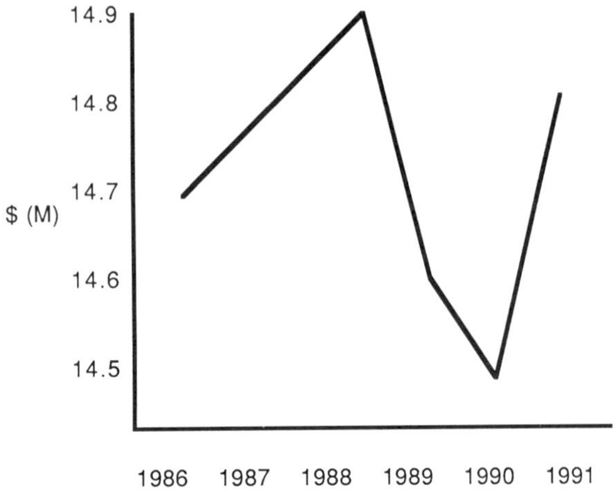

Figure 5.21. Manipulation of vertical scale in a graph of profits.

phasis. To achieve these goals, label axes and lines carefully, use proper scale, avoid dressing up graphs, and limit the amount of information presented.

To check your understanding of the principles of effective line graphs, complete the following exercise.

The line graph in Figure 5.22 has been submitted to you (the editor) to be created for a report. You can't submit it to the artist the way it is, but why not? Decide how best to prepare the graph for submission to the artist.

Copyediting this graphic would waste the editor's and the artist's time. Writing out instructions to place the months along the *x* axis and to place the dollar amounts (with additions for missing intervals) along the *y* axis would not address the question of where to put the data points. So the easiest "fix" for this graphic might be to resketch it as in Figure 5.23.

Scatter Graphs

Scatter graphs are frequently used to plot experimental data. They resemble line graphs in having an independent and a dependent variable, but there is often no line connecting the data points, or the line represents manipulation of the data. Figures 5.24 and 5.25 illustrate the use of the scatter chart.

When editing a scatter graph, editors should check the plotting of the data points against the data reported in the table or text the scatter graph is based on to make sure the graph is thorough, complete, and accurate. Also, if the data has been manipulated by a regression (least-squares) analysis, the editor should confirm the results.

Joining the data points plotted on a scatter graph rarely results in a straight

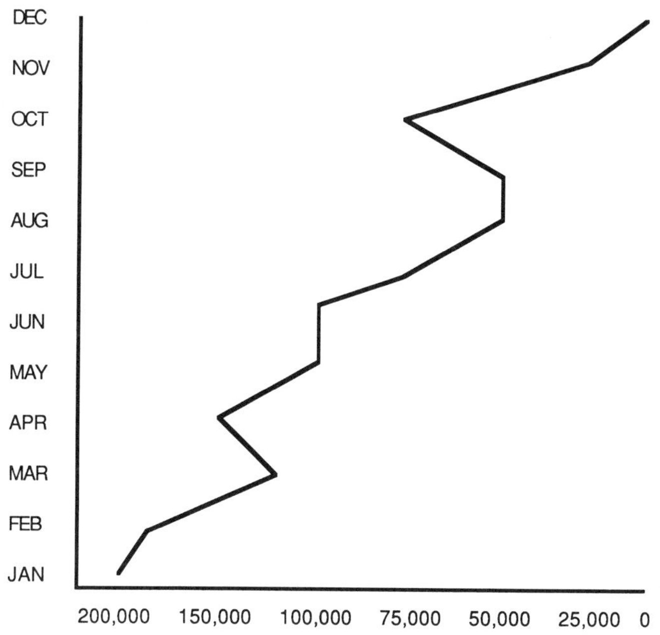

Figure 5.22. Improper graphing of gross income.

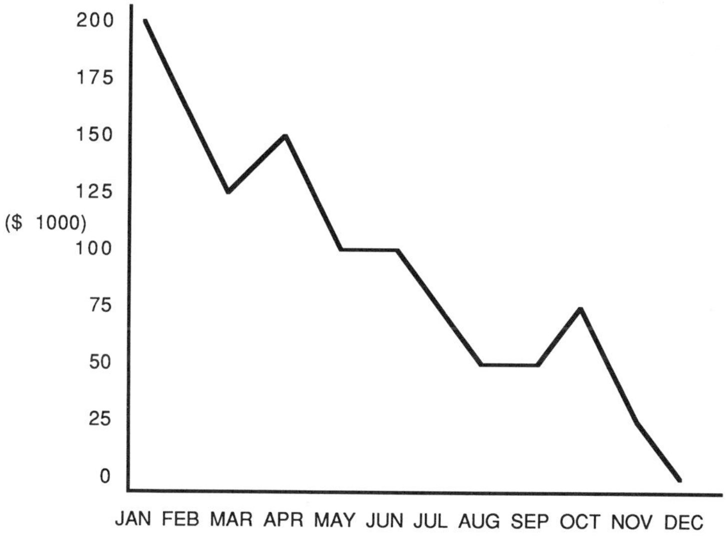

Figure 5.23. Proper graphing of gross income (Figure 5.22 revised).

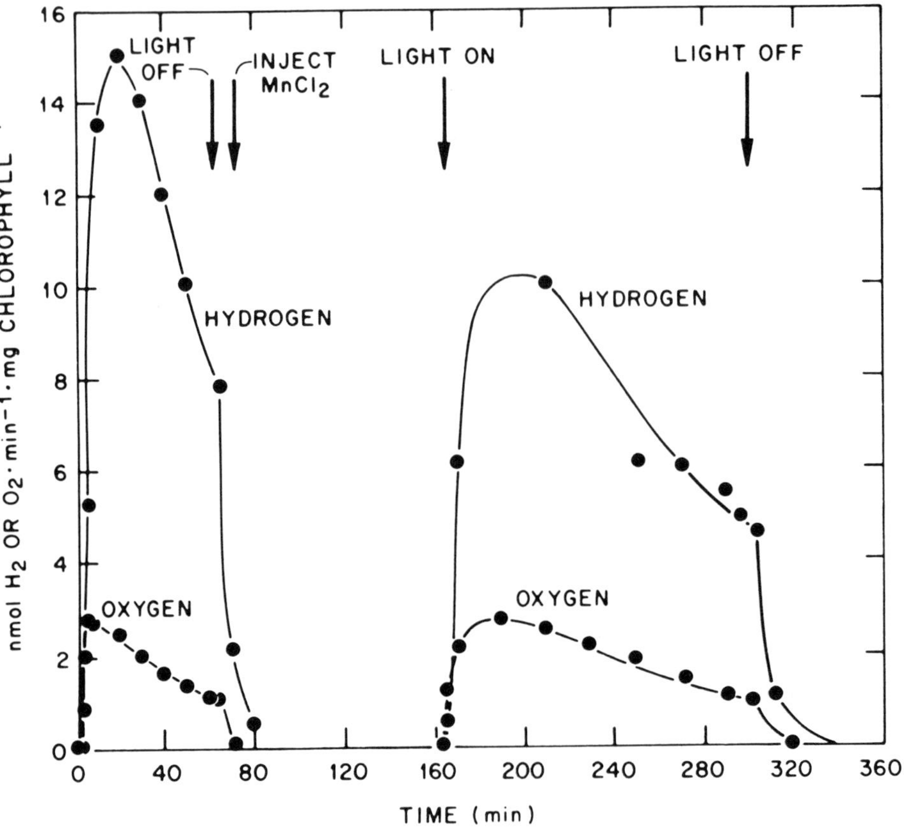

Figure 5.24. A scatter chart with data points connected. (Source: Woodward, Jonathan, et al., "Effect of exogenously added manganese chloride on the operational stability of the chloroplast-ferredoxin-hydrogenase system," *Enzyme Microbiology Technology*, Vol. 10 [February 1988], p. 124. Used with permission.)

line or a regular curve. However, the line that is the "best fit" of the data—that best shows the pattern in the data—can be determined through a linear or multiple regression analysis.

Surface Graphs

Surface graphs represent a recent development in the use of computer modeling of data. Surface graphs allow a three-dimensional representation of data. They are especially useful for measurements of intensity across an area, as shown in Figure 5.26. Editing surface graphs is more complex than line or scatter graphs because a third dimension must be checked against the source data.

Bar Graphs

Bar graphs are similar to line graphs. The dependent variable is usually charted on the *y* axis, and both axes must be labeled properly. However, in a bar graph, the bars represent the units of the variable, and no line connects the data. Figure 5.27 illustrates present (1986) and proposed nuclear capacity of several nations.

As Figure 5.27 reveals, bar graphs are useful for comparing data that is not time dependent as many line graphs are. For additional comparison within data sets in a bar graph, the bars can be divided (see Fig. 5.27).

Again, bar graphs should be kept simple. Dividing the bars too minutely

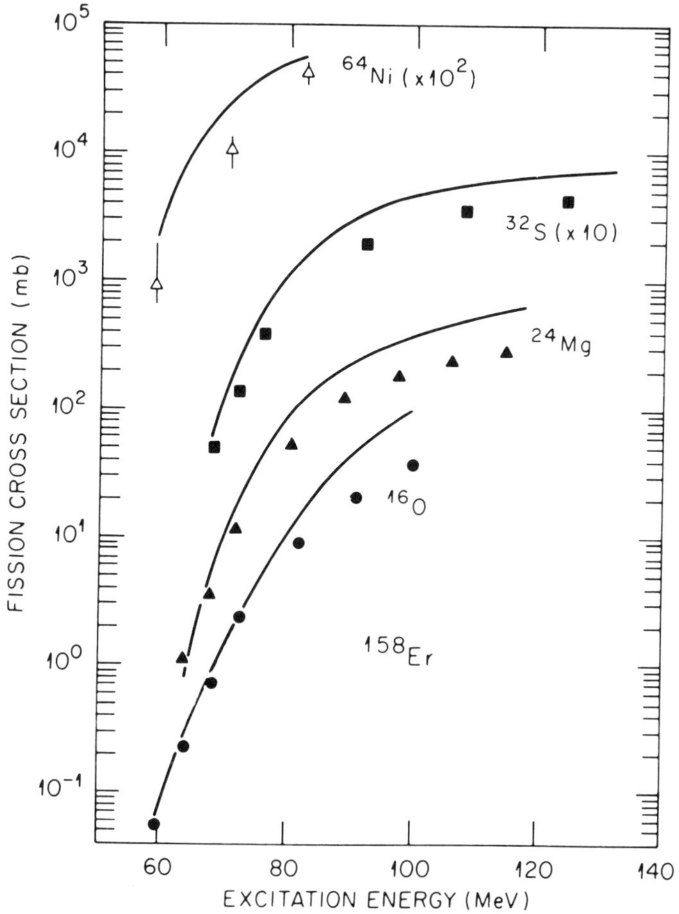

Figure 5.25. A scatter chart with curves based on analysis of data. (Source: Plasil, Frank, "Macroscopic Nuclear Physics," Oak Ridge National Laboratory *Review* Vol. 19, No. 3 [1986], p. 214.)

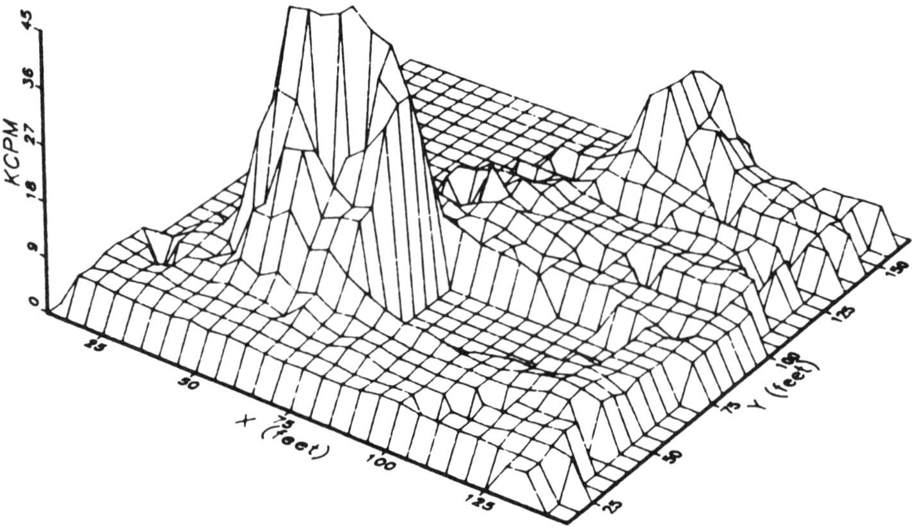

Figure 5.26. A surface graph of location and radiation levels of mill tailings. (Source: Croff, Allen G., "Radioactive Waste Management R&D," Oak Ridge National Laboratory *Review* Vol. 21, No. 4 [1988], 53.)

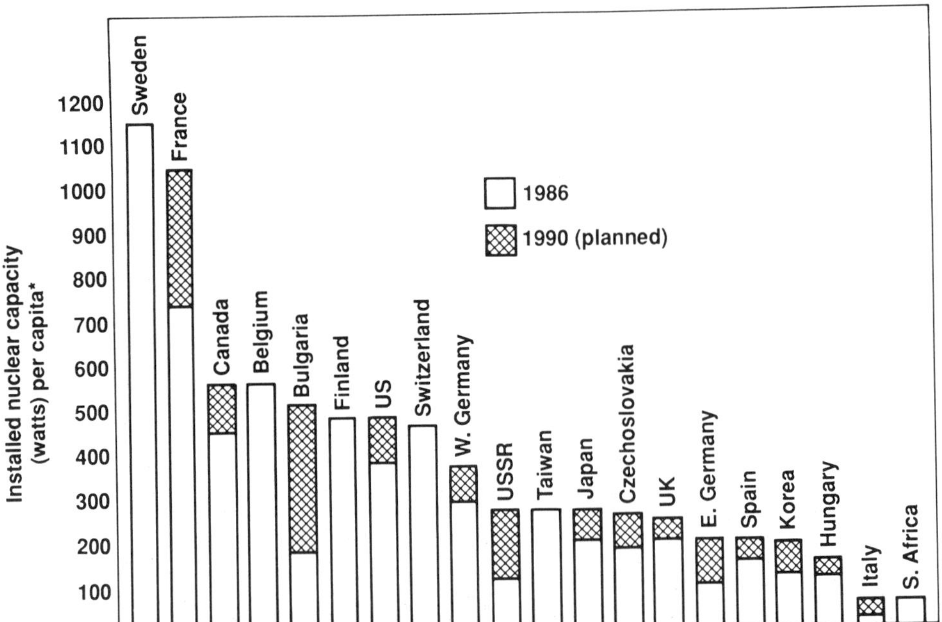

Figure 5.27. Divided bars for multiple data sets. (Source: Borg, I. Y., *Present and Future Nuclear Power Generation as a Reflection of Individual Countries' Resources and Objectives* [Livermore, Calif.: Lawrence Livermore National Laboratory, 1987], p. 5.)

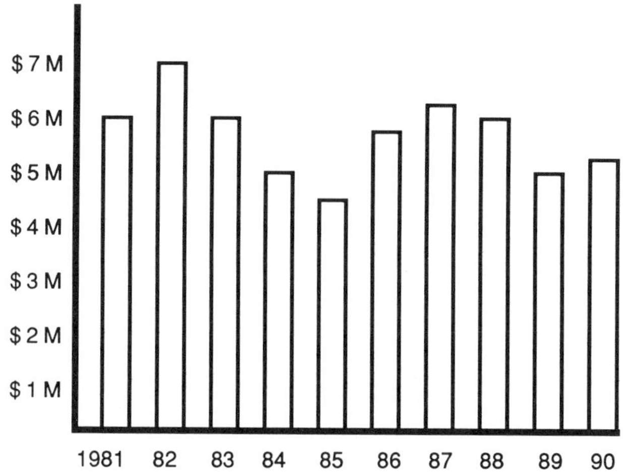

Figure 5.28. Jump in a design for a graphic.

or having too many bars can make the graph too complex. Also, care must be taken with bar width and spacing to avoid distracting optical effects in which the design seems to waver or "jump" as in Figure 5.28.

Line graphs and bar graphs are sometimes combined, as in Figure 5.29. Like line graphs, bar graphs can be used to highlight information in tables.

Circle Graphs

Editors frequently use circle graphs (pie charts) in documents for lay audiences. Circle graphs look like a pie cut into slices, each slice *(wedge)* representing a different set of data. They are frequently used to illustrate the percentage composition of a body of information, as in Figure 5.30.

Many style guides require that the largest wedge of a circle graph begin at the top and run clockwise, followed by the next largest wedge (see Fig. 5.30). As readers turn the pages of a bound document, they tend to notice first what is in the upper right corner, so the upper right quarter of the page or graphic should be the focal point.

Each wedge of a circle graph should be labeled horizontally inside the wedge if there is room or as close to it as possible. Circle graphs should be kept simple; editors should avoid the temptation to overcomplicate them with shading patterns as in Figure 5.31.

In Figure 5.31, the largest wedge does not begin at the top (twelve o'clock), followed by the next largest, as many readers would expect, because it is an "Other" category. So it should be placed in the top left of the chart as the last wedge or category. The "Toys, sports supplies, and equipment" wedge should begin at twelve o'clock, followed by the "Radio and TV . . ." wedge. Also, the shading in the wedges of Figure 5.31 and the attempt to give the

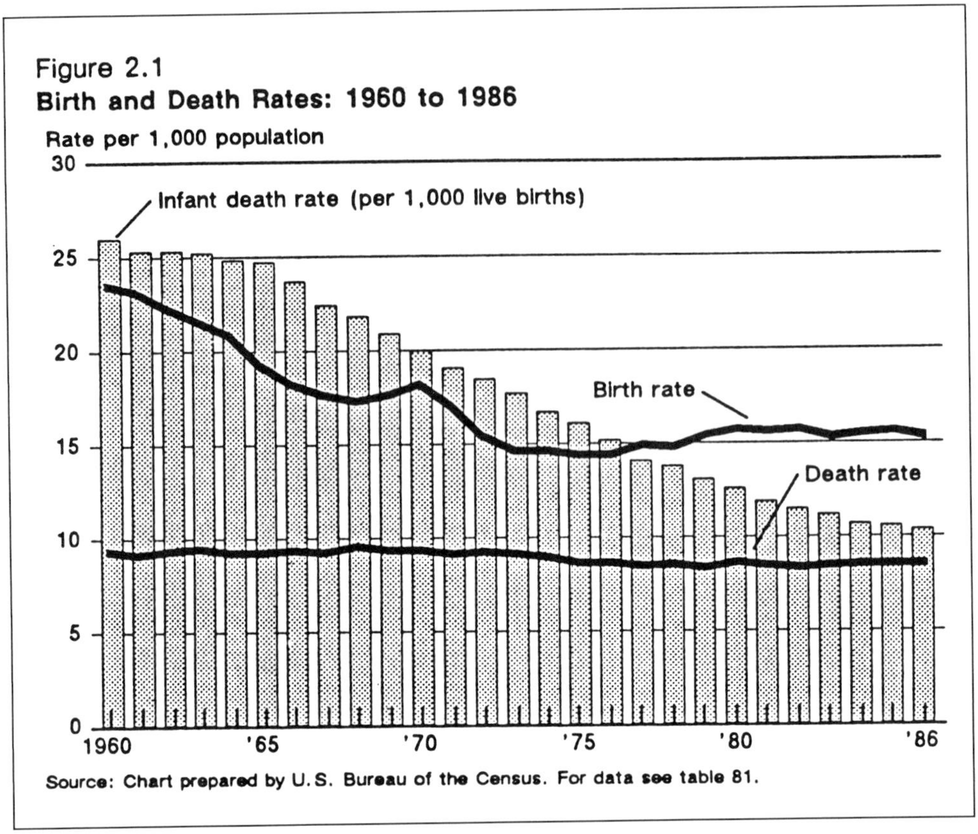

Figure 5.29. A combined line and bar graph for related data sets. (Source: *Statistical Abstract 1988,* p. 56.)

chart a three-dimensional look are unnecessary. The percentage labels inside the wedges and the labels outside the wedges are all the chart needs to communicate its information clearly.

This graph illustrates another weakness: Any circle graph with "Other" as the largest wedge is poorly designed. Editors should suggest reconstructing the graphic to break down the "Other" wedge or to at least indicate what "Other" includes.

The information in most circle graphs could be presented in the text or in an informal table. However, circle graphs are effective with readers who skim documents and tend to pay attention to the graphics. They are common in annual reports and other documents for nonexpert readers. They are easy to construct with computer graphics packages and easy for the reader to interpret *if* editors limit the number of wedges, arrange the wedges properly, and avoid excessive decorative shading.

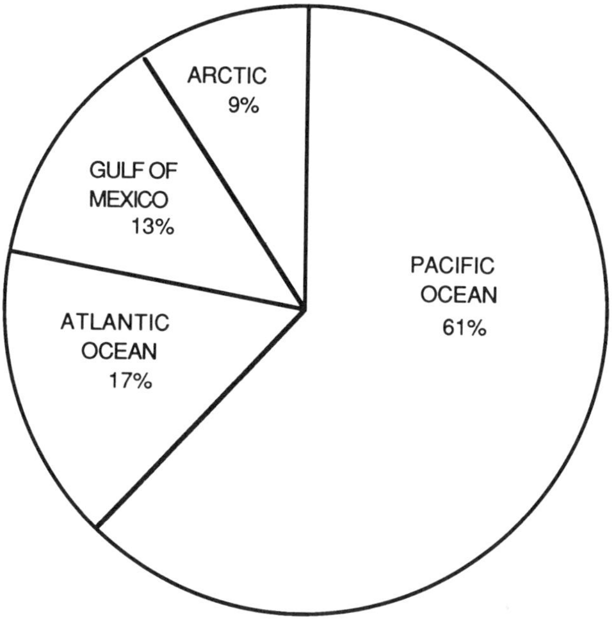

Figure 5.30. A typical circle graph, illustrating U.S. coastline frontage.

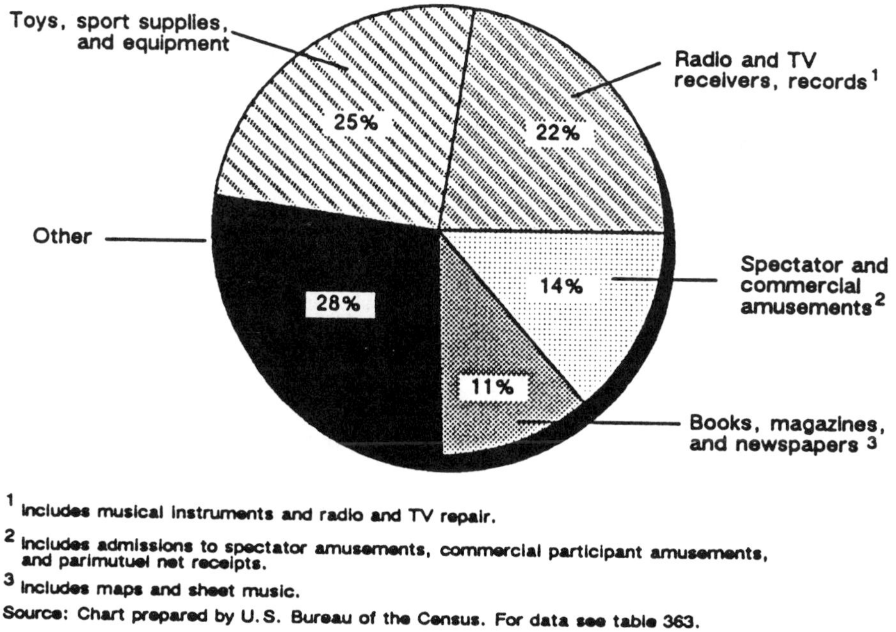

Figure 5.31. A circle graph with layout and shading problems. (Source: *Statistical Abstract 1988*, p. 208.)

As with line and bar graphs, circle graphs can be used to show related sets of information by juxtaposing related graphs.

Circle graphs, like bar graphs, simple line graphs, and informal tables, are more commonly used with lay or middle-level audiences than with expert audiences. Formal tables and complex line graphs are more commonly used with expert audiences.

Drawings

Drawings are common in technical documents because they can communicate a great deal of information quickly and clearly, if they are kept simple and labeled sufficiently. They allow the reader to picture a simplified version of what is being described.

Simple drawings can be prepared easily with graphics packages for PCs. Technical writers and editors can use graphics programs to create bar, line, and circle graphs and some simple drawings. For more complicated drawings, writers and editors often sketch a concept and submit it to a graphic artist with any reference information needed. For example, editors often provide the actual object to be drawn (such as a part for an automobile) or a photograph of it. Editors often consult with artists to determine the best type of drawing for what is to be illustrated.

Technical drawings, on the other hand, are often created by engineers, architects, or graphic artists rather than technical writers. Problems of scale, perspective, and technique sometimes require training in technical illustration to create *projections.*

To produce good drawings for a written document or an oral presentation, editors should keep several guidelines in mind:

- Drawings should show only the essentials.
- Technical experts as well as editors should review drawings carefully for accuracy.
- Drawings should have parts labeled clearly, in callouts no smaller than 6-point bold type.
- Leader lines should be used only where necessary to indicate what the callouts identify.
- Different techniques can emphasize particular parts of a drawing.

Exploded-view drawings

An exploded-view drawing such as Figure 5.32 can give readers a good sense of what the parts of an assembly look like, what they are called, and how they fit together. Exploded views are very common in parts lists, maintenance and repair manuals, and assembly instructions.

If possible, editors revising or proofing an exploded-view drawing should examine the disassembled item in question to ensure that all parts of the

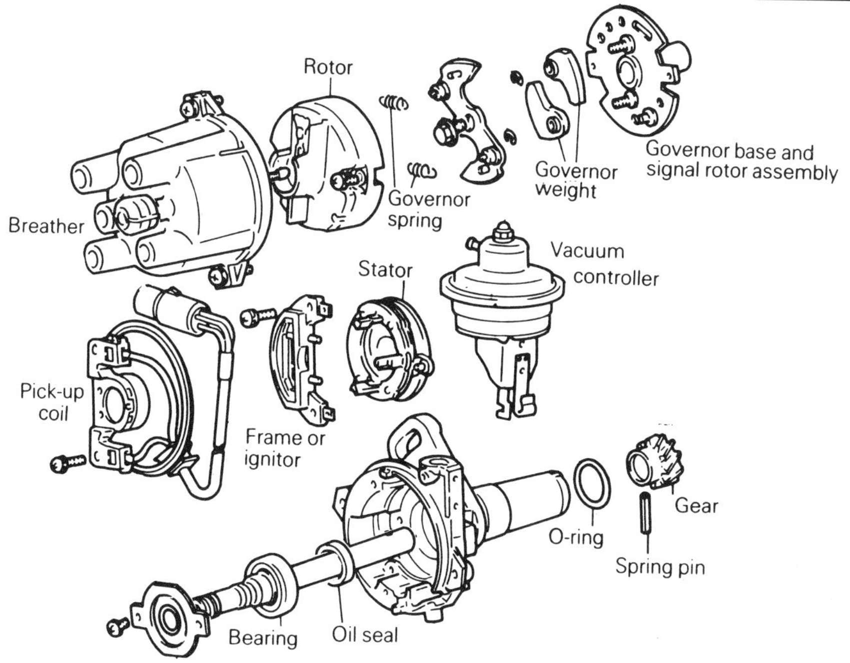

Figure 5.32. An exploded view with callouts. (Source: Chilton Book Company, © 1987, used with permission.)

item are presented to scale and in the proper physical relationship to each other. Also, be especially careful that labels for parts in the drawing have not been switched. (Often these labels are also confusingly called callouts.) Note in Figure 5.32 that all of the parts are not labeled. Often, screws, bolts, nuts, and other hardware are not; however, all important components should be labeled.

In Figure 5.33, the key (or *legend*) beneath the drawing provides the part names. If parts in an exploded view have long names that would take up too much room in the drawing, numbers can be used to denote them.

Occasionally a component in an assembly drawing needs to be shown in greater detail than the rest of the assembly. For example, in Figure 5.34, the liquid injector has been *popped out.* Popping out part of an assembly is a good way to provide large-scale information (to facilitate repair, for example) without drawing the entire assembly to that scale. The graphic can focus on the subassembly or component to be illustrated but still show how that part fits into the whole. Be sure to identify what parts are to be labeled and how.

Exploded-view drawings are useful in many kinds of technical documents. Like most drawings, they require skill and time to create. Editors should ask graphic artists for estimates of drawing time, based on comparable examples (see Chapter 10). Also, editors should keep their own records of how long it took the artist(s) to complete particular drawings for estimating future projects.

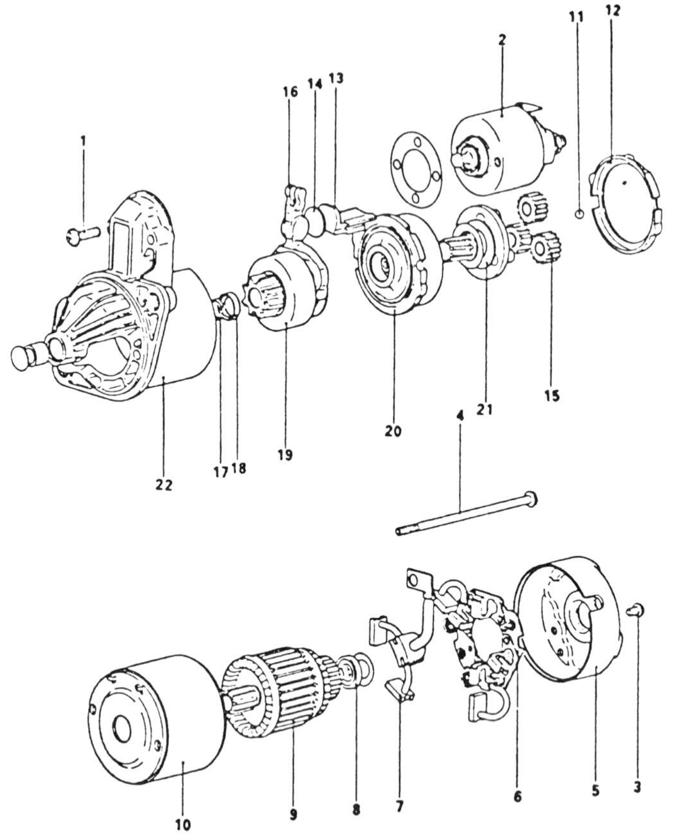

1. Screw
2. Magnetic switch
3. Screw
4. Screw
5. Rear bracket
6. Brush holder
7. Brush
8. Rear bearing
9. Armature
10. Yoke assembly
11. Ball
12. Packing A
13. Packing B
14. Plate
15. Planetary gear
16. Lever
17. Snapring
18. Stopring
19. Overrunning clutch
20. Internal gear
21. Planetary gear holder
22. Front bracket

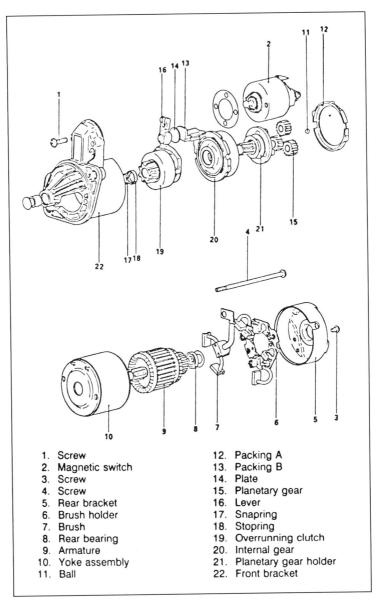

Figure 5.33. An exploded view with a legend. (Source: Chilton Book Company, © 1986, used with permission.)

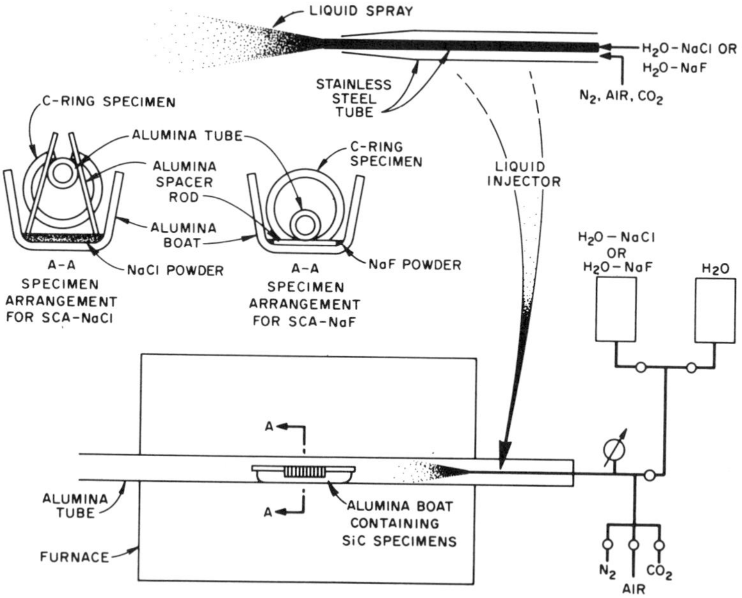

Figure 5.34. A popout to illustrate part of an assembly. (Source: Federer, J. I., *Corrosion of Materials by High-Temperature Industrial Combustion Environments—A Summary* [Oak Ridge, TN: Oak Ridge National Laboratory, 1986], p. 41.)

Cross sections

Cross sections (often called *sections*) show a view of an item as if it were cut along a center line at a right angle to the reader's line of sight. The cross section in Figure 5.35 shows shipping casks for spent nuclear fuel. A cross section may have dimensions and materials labeled. The aim is to present a two-dimensional view of an assembly (as here) or a volume (such as a casting or soil layers). As with exploded views, part of the section might be popped out and enlarged.

In a cross section, there is no attempt to achieve a sense of three dimensions, as in a cutaway drawing. Rather, the drawing presents an area, and shading is often used to differentiate parts of the area. Editing cross sections requires checking the scale carefully. Shading should also be checked against blueprints or other data because it is easy for an artist working from drawings rather than the assembly itself to shade an area improperly.

Cross sections are very common in architecture, construction, biological and physical sciences, and manufacturing. Figure 5.36, a geologic cross section of the rocks underlying the north rim of the Grand Canyon, illustrates one important advantage of drawings: They can present information that would be impossible to photograph. Note the indication of the scales in the lower right corner and the labels for layer thickness and geologic period at the right. This cross section presents several bodies of information:

- Strata names
- Strata thicknesses
- Strata colors
- Strata arrangement (layering order)
- Strata slope contours
- Section names (such as *Esplanade*)
- Paleontographical data
- Geological period
- Geological era

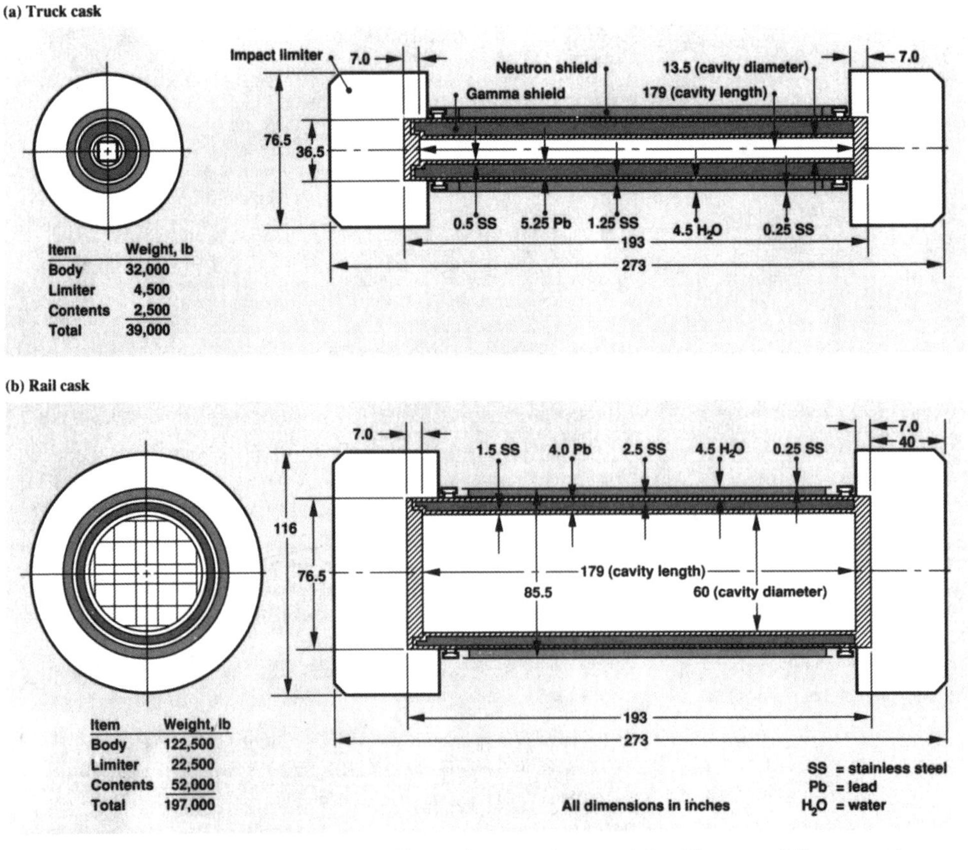

Figure 5.35. Cross-sections with dimensions and materials. (Source: "Transporting Spent Reactor Fuel Safely," in Lawrence Livermore National Laboratory, *Energy and Technology Review*. [June 1988], p. 4.)

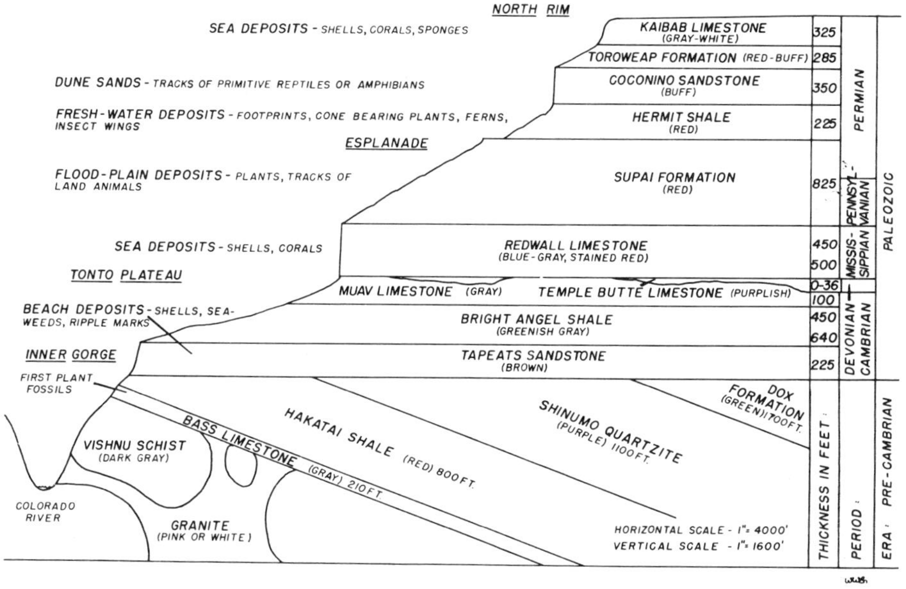

Figure 5.36. Geological cross-section, Grand Canyon. (Source: Grand Canyon Natural History Association, © 1970, used with permission.)

Figure 5.36 presents a great deal of information effectively. It is far too "busy" to serve as a visual aid for an oral presentation, but in a printed document it would work well.

Cutaway drawings

Cutaway drawings show the component parts of an assembly, making them similar to exploded views. However, in a cutaway the parts are shown assembled, but as if a chunk of the assembly had been cut away. Editors may choose to use labels and leader lines to identify some or all of the parts in a cutaway drawing, but they should keep the labeling simple. Cutaways are designed to show how all the parts fit together, so they are often more complicated than exploded views. Notice how the cutaway drawing in Figure 5.37 shows better than an exploded view would how the parts fit together. Likewise, cutaways present information that could be presented in a cross section of an assembly, but cutaways give a clearer sense of what the completed assembly looks like.

Projections and perspective drawings

Projections are pictorial representations of an object, assembly, or other subject. Drawings showing two dimensions are called *orthographic* projections. Figures 5.34 through 5.36 are examples; there is no attempt to represent the

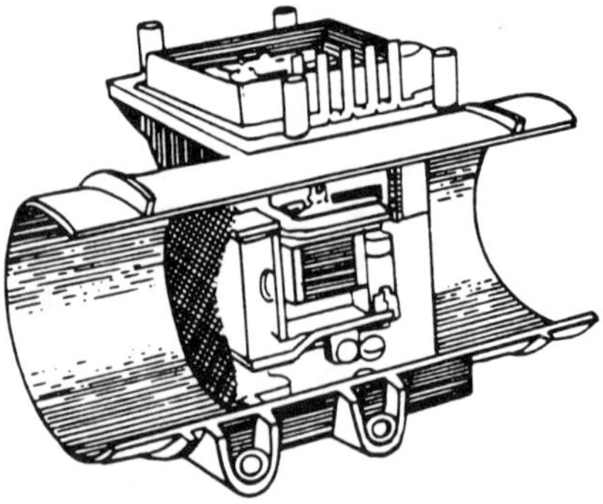

Figure 5.37. A typical cutaway drawing. (Source: Chilton Book Company, © 1987, used with permission.)

subject with a sense of three dimensions because the third dimension is unnecessary, as in cross sections.

Projections showing three dimensions are *axonometric* or *perspective* drawings. Axonometric projections show the object with lines drawn to scale; depending on the angles involved, the axonometric is called an *isometric, oblique,* or *cabinet* view. Axonometric projections represent the subject in three dimensions according to particular drawing principles. They are used in drawings for expert audiences such as engineers and architects rather than for a general audience.

Perspective drawings such as Figure 5.38 show the subject as it would appear to someone actually looking at it.

Wiring, schematic, and block diagrams

Wiring, schematic, and block diagrams are related: They describe electrical or electronic systems. A wiring diagram shows coded wires, connections, and so forth in an electrical system, emphasizing the wiring. A schematic diagram identifies the components of an electrical system through symbols and shows their connections. A block diagram shows the relationship of units or assemblies in the system without providing details about the components. All are drawn from engineering specifications, but not to scale. Instead, they show how things are connected, to enable construction or repair of the electrical system. Block diagrams are usually a less complex demonstration of how the components are connected.

Wiring and schematic diagrams are demanding to edit, because verifying their accuracy is a task for a technical expert. Block diagrams are less difficult, but arrows must be checked for transposition of direction. Rough and final art for wiring and block diagrams should be checked carefully against the original engineering drawings. If possible, the writer should check each wiring diagram against an actual assembly to verify wire colors, connections, etc. It is very easy for labels to be mistyped or transposed, frustrating users and perhaps causing the assembly to go up in smoke if wires are connected improperly.

In Figure 5.39, a wiring diagram for a Harley-Davidson motorcycle, note that the colors of the wire insulation are indicated to facilitate installation or repair. Editors should check such color labeling carefully because color labels are easily confused—for example, BLK is the standard abbreviation for black, but BK is also used. In the event of a discrepancy, editors ought to check the drawing against the actual assembly. Information about color coding is very useful in a wiring diagram but should be provided only if needed.

Editors should resist the temptation to overreduce wiring diagrams, block diagrams, and other graphics to fit the space on a page. Refer to the earlier discussion on oversized tables for potential reproduction and binding problems caused by excessive reduction or enlargement.

Figure 5.40 shows the conventional symbols used in schematics. Figure 5.41 shows and labels schematic symbols.

Block diagrams are not nearly as complicated as wiring diagrams, as is clear from Figures 5.42 and 5.43. They identify the components in an electronic assembly without providing data about them.

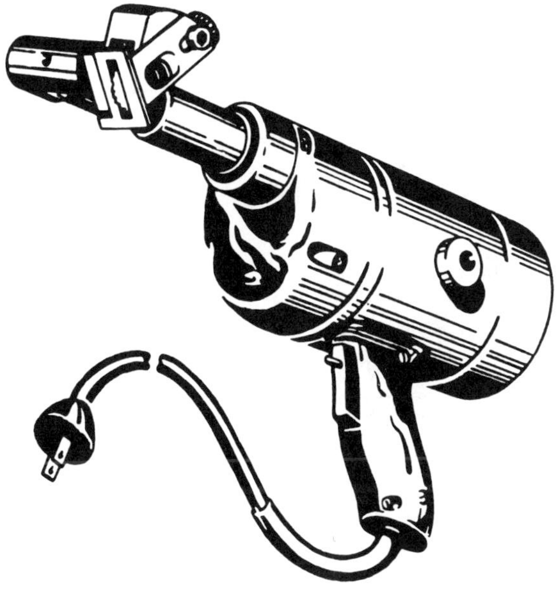

Figure 5.38. A persepctive drawing (of a Ketts Saw). (Source: U.S. Department of Transportation, Federal Aviation Administration, Flight Standards Service, *Airframe and Powerplant Mechanics Airframe Handbook,* p. 138.)

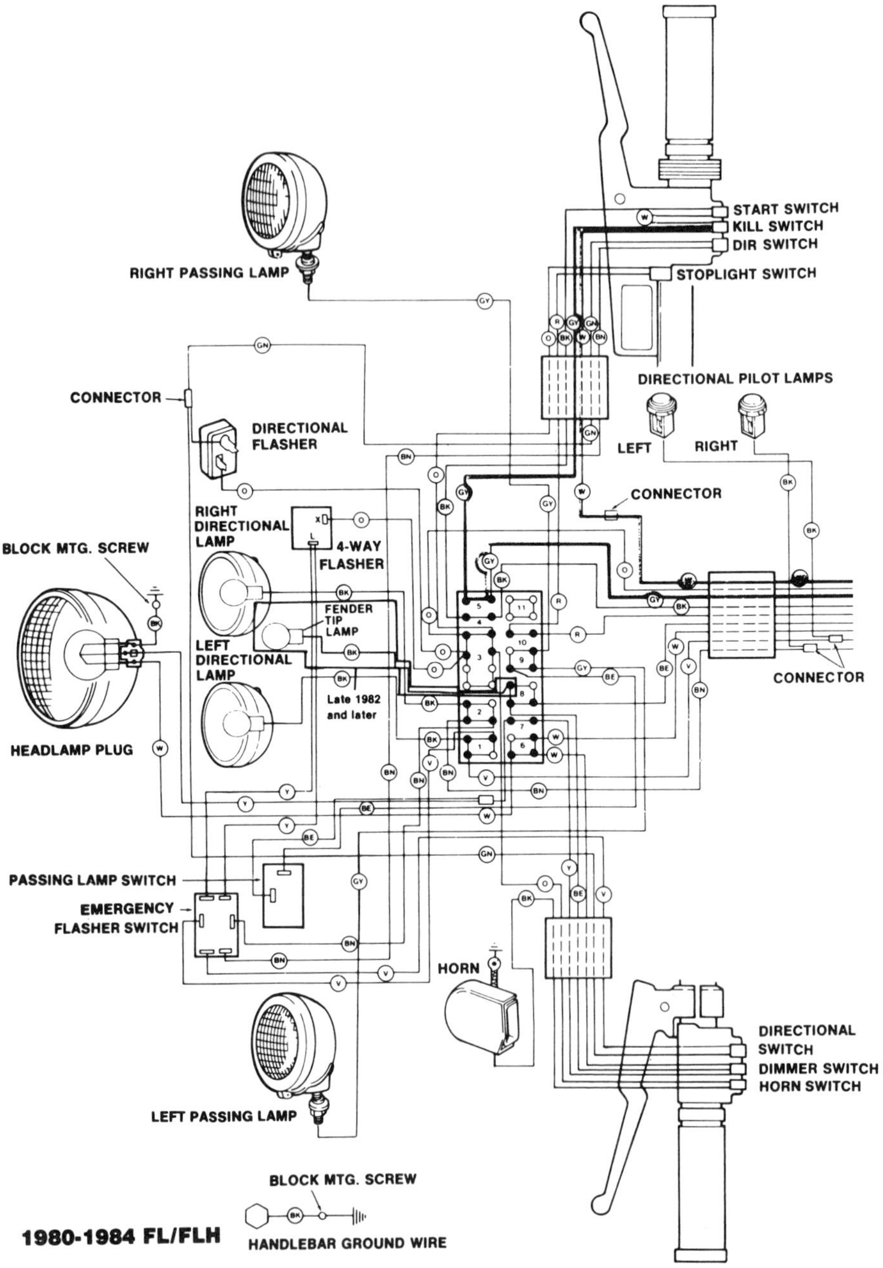

Figure 5.39. A combination wiring and block diagram. (Source: *Harley Davidson 74 & 80 Four-Speed V-Twins 1959–1984: Service, Repair, Maintenance,* p. 403. Intertec Publishing, © 1986, used with permission.)

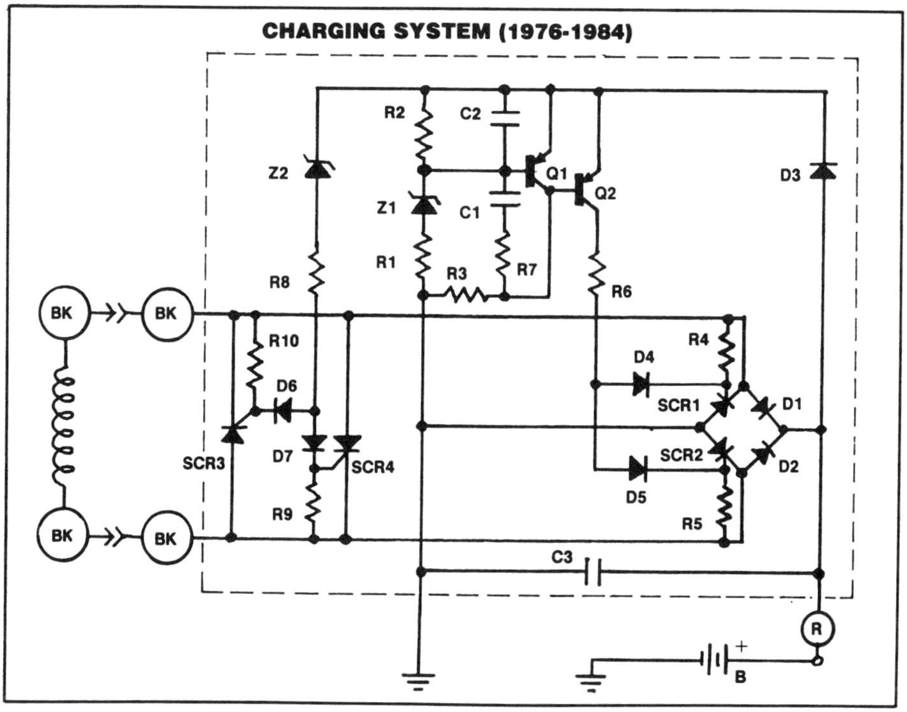

Figure 5.40. A schematic for a Harley-Davidson motorcycle. (Source: *Harley Davidson 74 & 80 Four-Speed V-Twins 1959–1984: Service, Repair, Maintenance,* p. 223. Intertec Publishing, © 1986, used with permission.)

The term *schematic* is also used to represent the arrangement of equipment in an assembly such as that in the laser in Figure 5.44. However, technical editors should reserve the term schematic for electrical diagrams to avoid confusion.

Cartoons

Cartoons are rarely used in technical documents designed to present information to knowledgeable readers, but they are useful for warnings, operating instructions, and humor. They are frequently used in oral presentations to experts to "break the ice," get attention, or embellish transparencies *(viewgraphs)* or slides. Cartoons are very useful for warnings and operating instructions when the readers are not skilled in the language of the document. The cartoon in Figure 5.45 could be used on a poster for a safety campaign, on gummed stickers placed on electrical hazards, or in instructions to maintenance or janitorial personnel.

Writers should make sure that cartoons will quickly and clearly serve their purpose and not offend anyone. Because humor is subjective, there is greater risk of offending with cartoons than with any other kind of drawing in a technical document, and editors must be sensitive to this risk. In Figure 5.45, for example, shading the worker's skin or depicting the worker as a woman might imply that some kinds of people are careless or ignorant. Editors must ensure that a cartoon's effectiveness is not jeopardized by poor taste or suggestions of racism, sexism, or ageism.

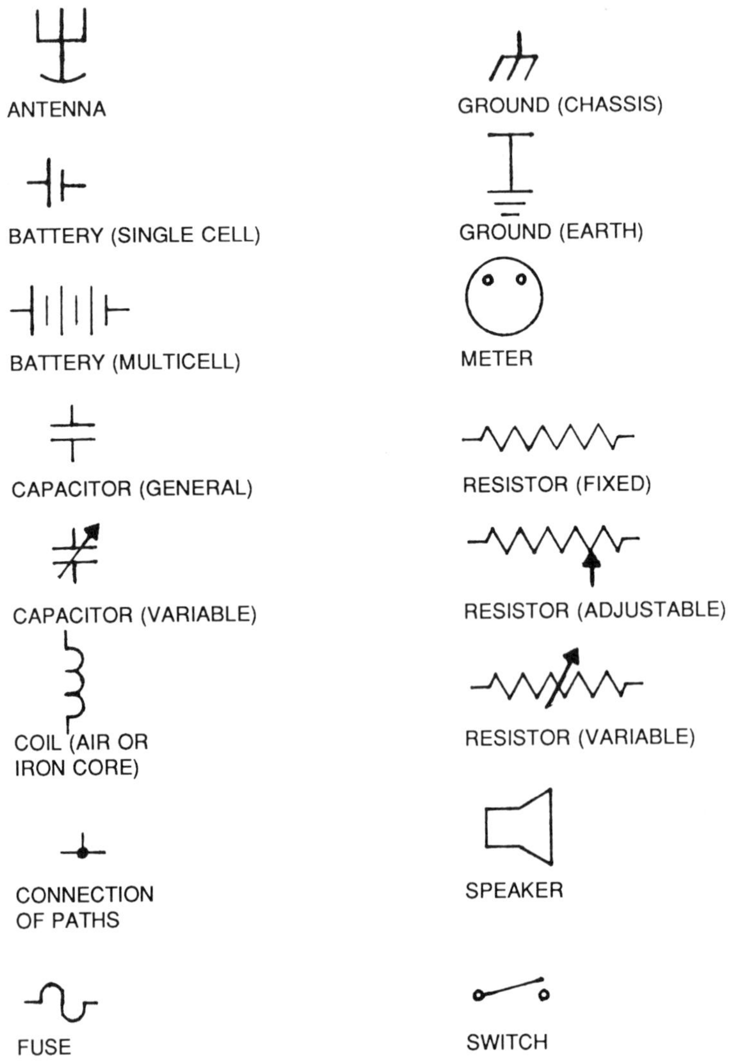

Figure 5.41. Standard symbols for schematic drawings.

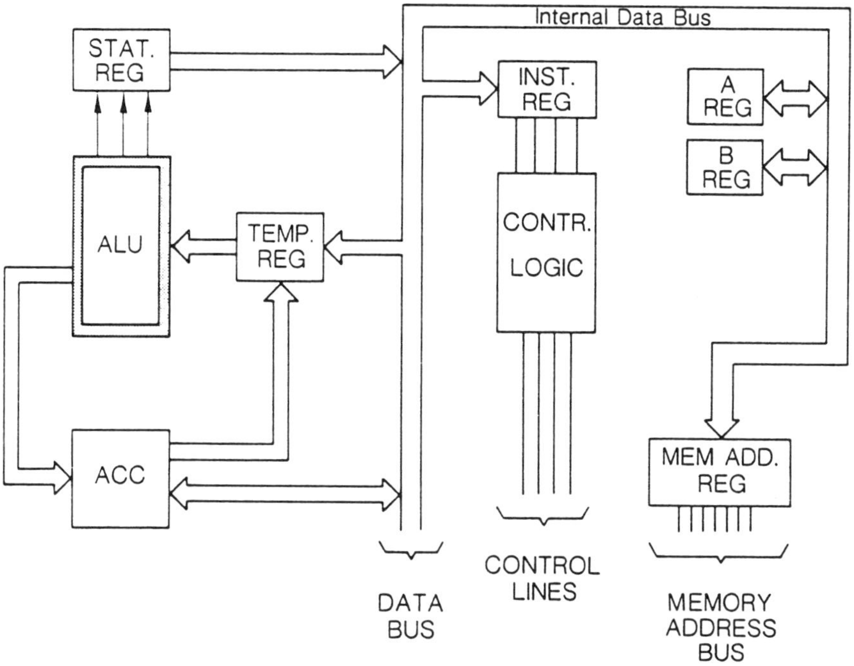

Figure 5.42. A block diagram illustrating flow directions. (Source: Baer, Charles J., and Ottaway, John R., *Electrical and Electronic Drawing* [McGraw-Hill, 1986], p. 207. Reproduced by permission of the Glencoe Division of Macmillan/McGraw-Hill School Publishing Company.)

Maps

Most readers of technical documents are familiar with maps, especially road maps. Good maps are a pleasure to use and can even be entertaining; some people collect them. Bad maps, however, can be a nightmare. Technical writers and editors working with maps should learn and follow the standards of the American Cartographic Association. Maps should have a clear indication of north (unless north is at the top of the map as the reader reads it). Also, most maps should contain the scale or a statement indicating that the map is not to scale. As with all graphics, maps should not be cluttered with unnecessary information.

Color coding of map regions must be done with care to ensure that the colors used for adjacent regions are different enough to enhance correct interpretation. Also, the color scheme should correspond to some logical pattern, as in the *USA Today* weather map. In that scheme, the regions of temperature are colored along a gradient from dark blue (coldest) to dark orange (warmest). This logical color scheme allows readers to see temperature patterns without relying on a numerical scale or a color-coded legend.

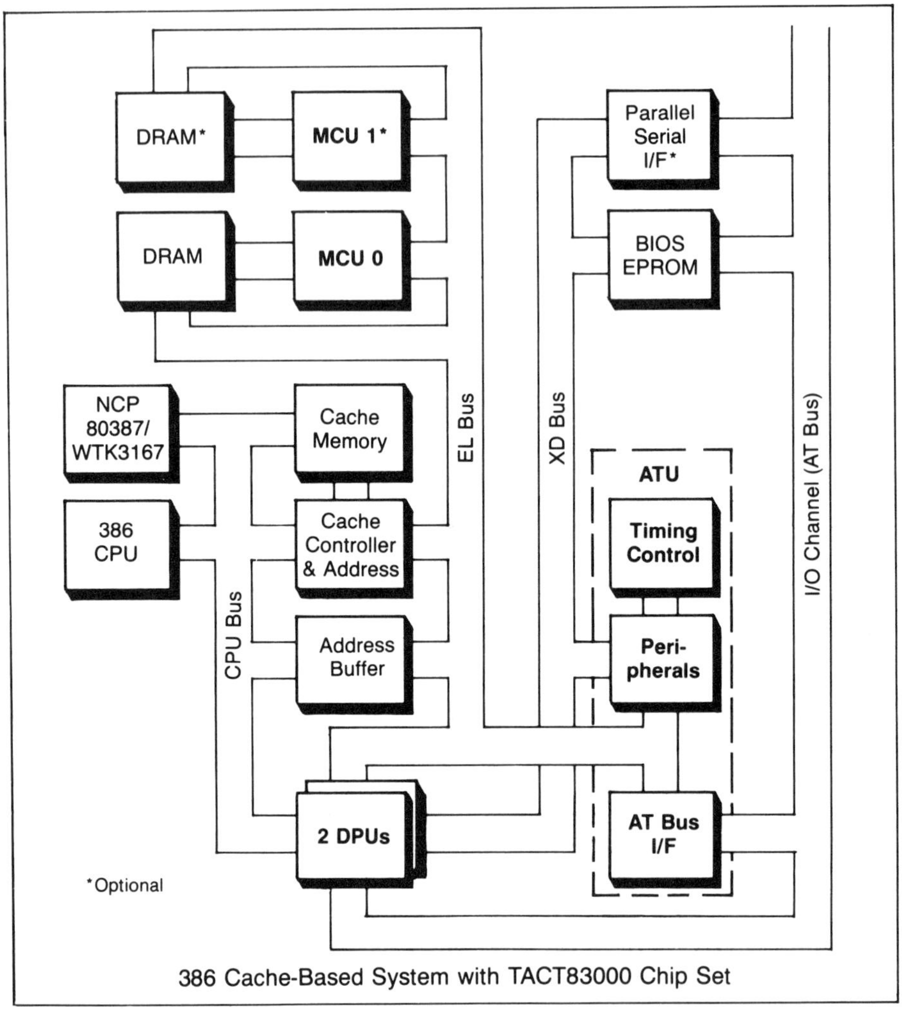

Figure 5.43. Shading to emphasize components in a block diagram. (Source: Texas Instruments, © 1990, used with permission.)

If maps use different type sizes, for example for the names of cities of different populations, editors must verify the type sizes carefully. Also, editors must check for consistent and accurate use of symbols.

Maps can present spatial information far more economically and clearly than it could be presented in text. Consider the number of sentences it would take to present the information in Figure 5.46. Also, consider the likelihood that however clear those sentences are, readers would have much more trouble grasping and remembering where those wells are if a map were not used.

The "Explanation" section of Figure 5.46, with the key presenting site

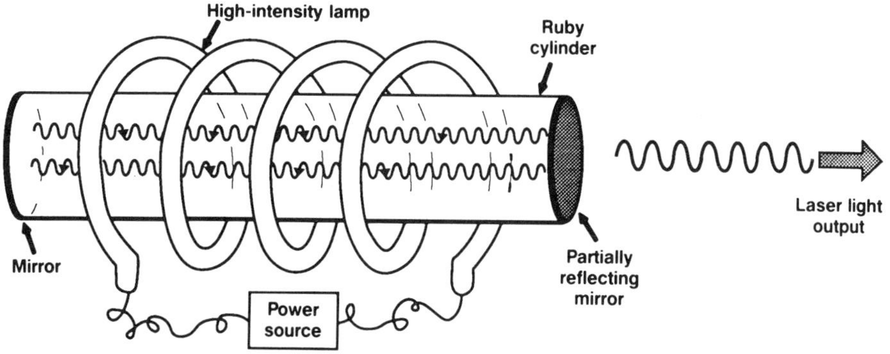

In this schematic of a simple ruby laser, the pumping radiation comes from a high-intensity white-light lamp. Stimulated light waves are reflected back and forth between parallel mirrors, building in intensity. A pulse of light finally flashes through the partially reflecting mirror. That pulse is coherent laser light.

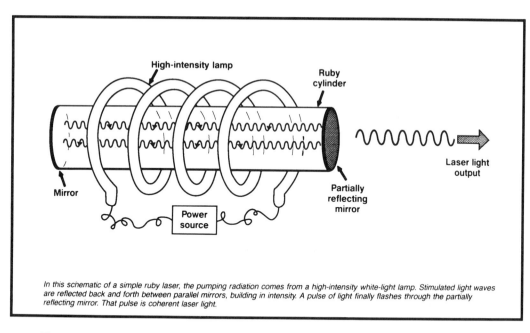

Figure 5.44. An engineering "schematic" (a type of block diagram). (Source: "How a Laser Works," in *Inertial Confinement Fusion* [Livermore, Calif.: Lawrence Livermore National Laboratory, 1988], p. 4.)

Figure 5.45. One use of cartoons: hazard warnings.

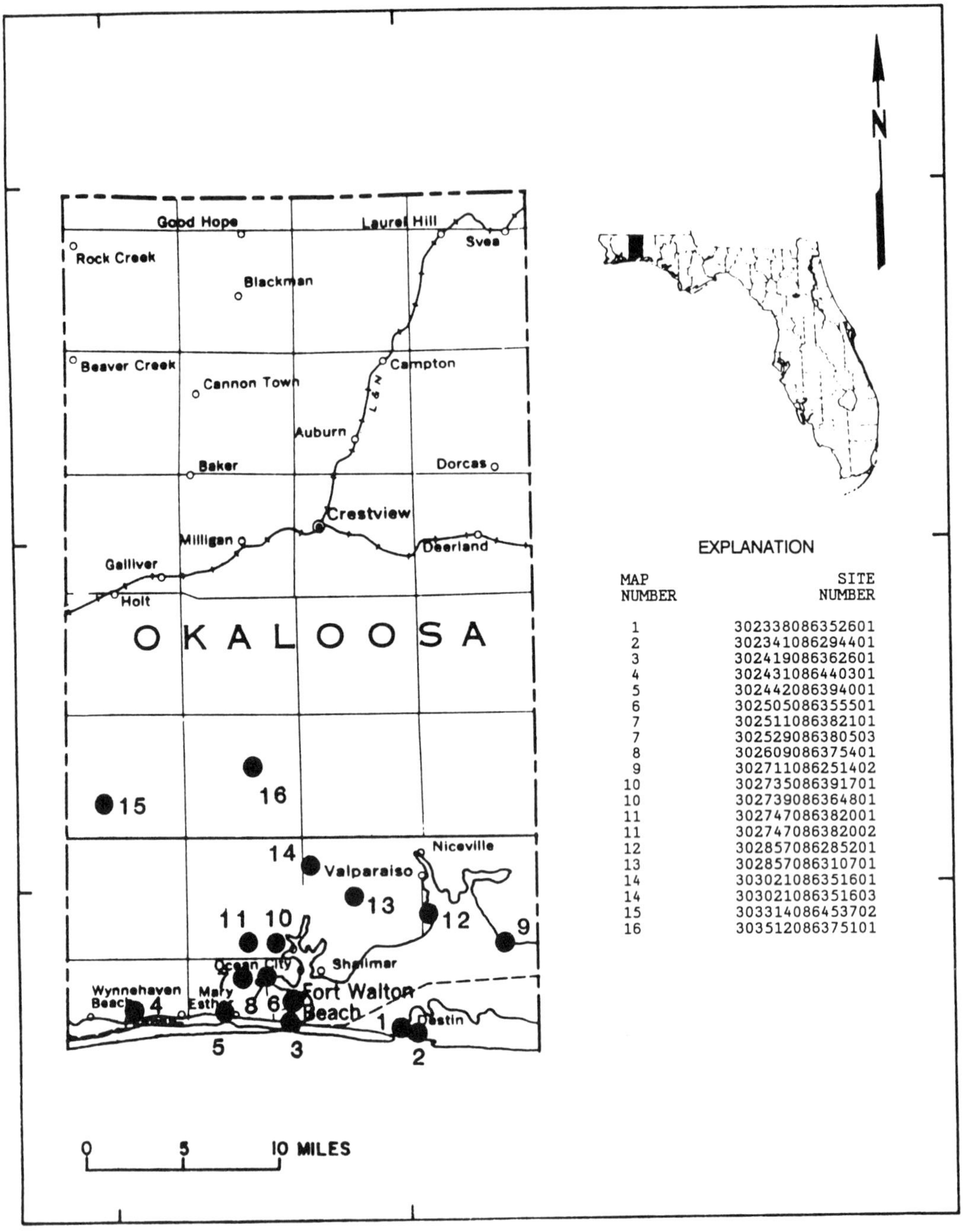

Figure 5.46. Use of an inset map to orient readers. (Source: *Water Resources Data Florida: Water Year 1984*, U. S. Geological Survey Water-Data Report FL-84-4, Vol. 4, p. 220.)

numbers for the map numbers, is called the *legend.* Writers should ensure that a legend is provided if needed. Also, points on a map should be labeled (or numbered, as here) according to a simple, clear pattern. Here, the numbers move from south to north and east to west, to correspond to population densities or some other demographic feature.

Maps depicting a possibly unrecognizable section of a region should include a reference map to orient the reader. In Figure 5.46, the small county map of Florida with Okaloosa County darkened orients the reader to its location within the state.

Writers creating maps must decide how much information to include, recognizing that the amount of detail must be controlled to keep the map readable. In Figure 5.46, main highways could have been included, but doing so would have cluttered the map unnecessarily.

Also, writers must decide where to include the information in a map. Some maps must be drawn to a small scale, creating areas that are difficult to label. Editors should place labels outside these areas, using leader lines to indicate what they identify. Also, take care to indicate all scales used on a map.

Maps are occasionally used in other graphics, but sometimes they serve little purpose. Figure 5.47 uses an unnecessary map to make the point that the statistics in the line graph are for the United States. Clearly, this graphic is misrepresentative. Note how the failure to orient the map properly (with north "up") makes it appear that the average age of people in the United States is not increasing significantly. The attempt to give the map a three-dimensional look has negated the message of the graphic, and the people in the graphic are merely clutter. Figure 5.48, based on Figure 5.47, without the map clutter and with proper orientation of the line graph, presents a clearer, more accurate statement of the statistics.

Figure 5.49, which shows patterns of relocation of retirees into Florida and California from 1975 to 1980, is not cluttered with unnecessary information. As the key indicates, the thickness of each arrow represents the number of retirees. However, it might not be clear at first to some readers that 4,000 to 7,999 retirees, not 1,000 to 3,999, moved to California from Washington. The number of retirees could have been presented by labeling the tail of each arrow, but that would have crowded the right third of the map.

Flow Charts

Flow charts are common in technical reports. Flow charts such as Figure 5.50 are especially valuable for identifying the steps in a process. Many procedures involve reiterative steps *(loops)* represented by the "No" arrow at the bottom left of Figure 5.50. If an activity or related activities must be repeated, the arrow carries the reader back to the beginning of the cycle. Editors must ensure that such loops are clearly marked and that readers will know easily when to repeat a step or sequence of steps.

Flow charts can present a great deal of information clearly if they are kept

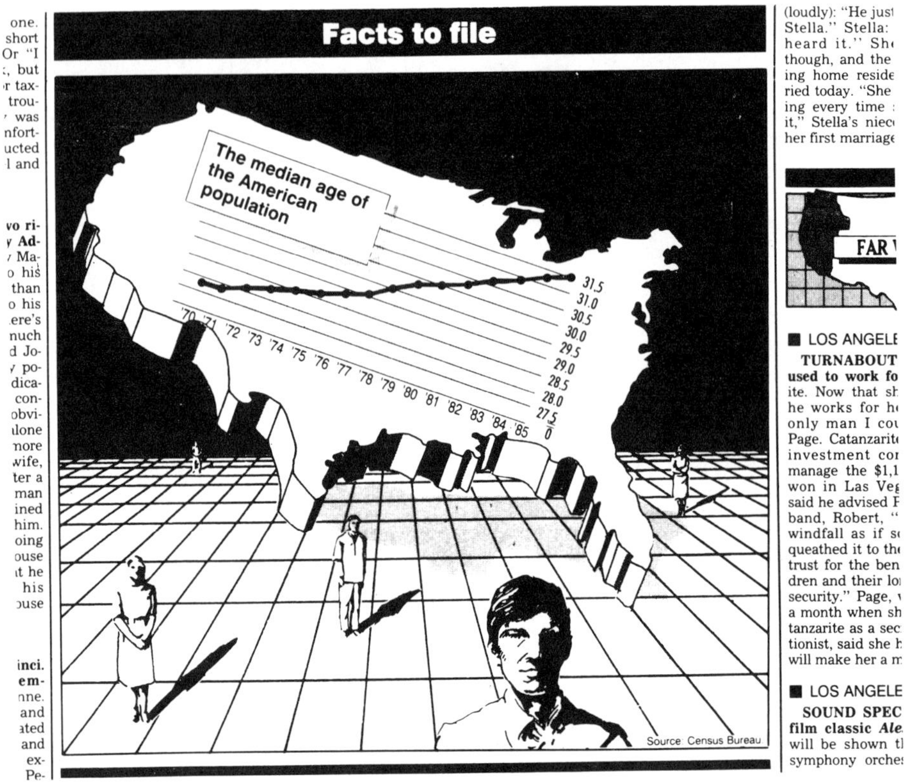

Figure 5.47. Improper orientation in a graphic. (Source: *Orlando Sentinel,* April 26, 1987, p. A-22, used with permission.)

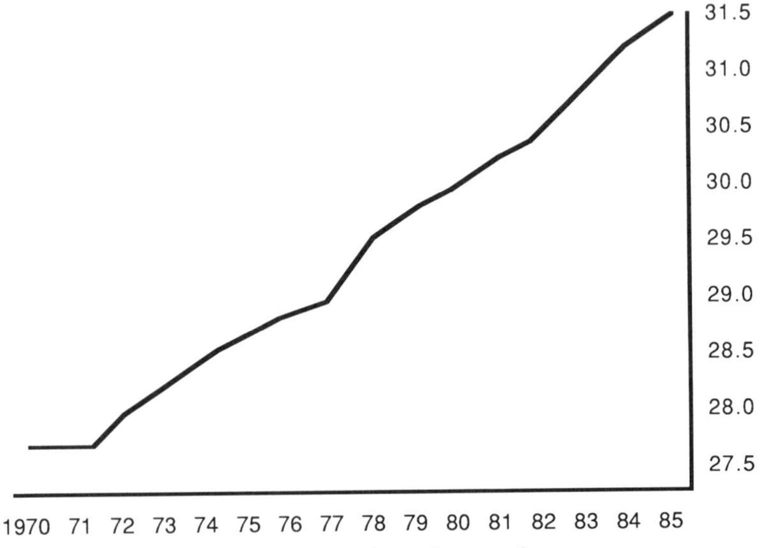

Figure 5.48. The graph of average age oriented properly.

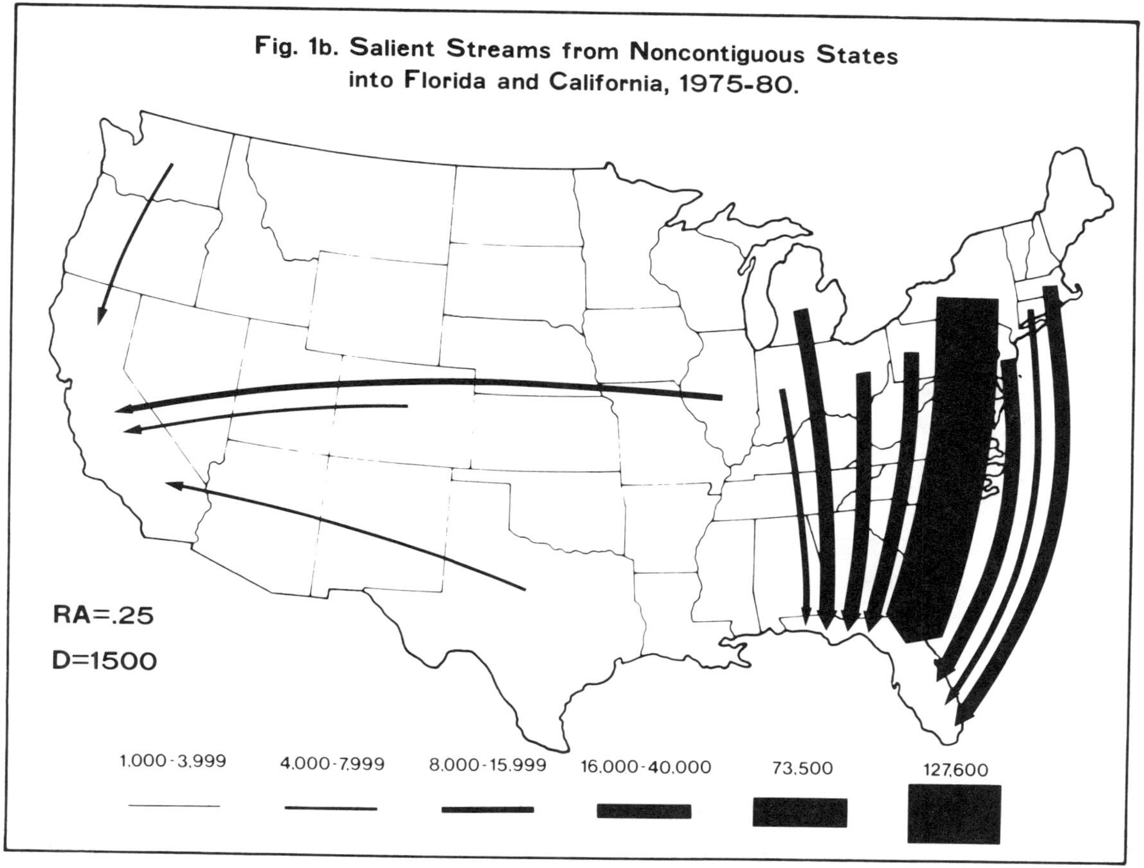

Figure 5.49. A map keyed to show relocation of retirees. (Source: Longino, Charles F., Jr., *State to State Migration Patterns of Older Americans for Two Decades* [Coral Gables, Fla.: Center for Social Research in Aging, University of Miami, 1986], p. 17, used with permission.)

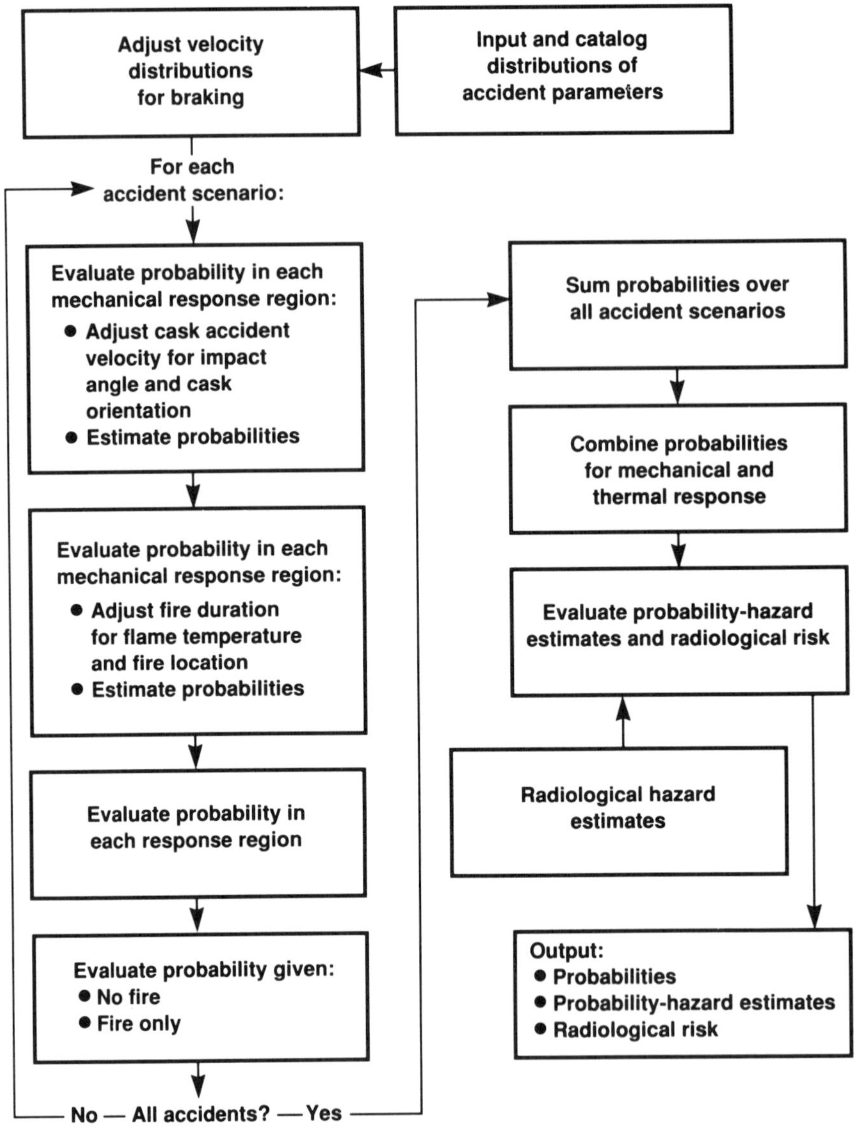

Figure 5.50. A flow chart for identifying tasks. (Source: "Transporting Spent Reactor Fuel Safely," in Lawrence Livermore National Laboratory, *Energy and Technology Review* [June 1988], p. 11.)

simple. Sometimes it is not possible to present every step in the process in one illustration, so artists might recommend an overview flow diagram with more detailed flow charts for individual steps. Editors should consult with the writers and artists to decide how much detail any flow chart should include. Editors should avoid the tendency to explain too much. Effective flow charts for lay and middle-level audiences should indicate beginning and end points (see Figure 5.51). Figure 5.52 is more complex, appropriate for expert and upper- to middle-level readers. Flow charts often involve actions to be taken by people, but they are also used to show the flow of electrical current or a substance through a system, as in Figure 5.52. Flow charts can also illustrate steps in a sequence that operates independently after the beginning, as in the test diagram shown in Figure 5.51.

Organization Charts

Organization charts are common in discussions of company or agency management and project staffing. They can illustrate well an organization's structure, either by listing names of managers and their staff or by listing departments or groups.

To be tactful, editors should use equally sized squares or rectangles for people or groups on the same level within the organization. Be careful to represent each person or group in the organization chart at the proper level, as many managers are concerned about other people's sense of their importance in the organization. Also, editors must check the lines of authority (and reporting) carefully, verifying who reports to whom. Figure 5.53 violates these guidelines. Editing Figure 5.53 to represent the organization more accurately could result in the chart in Figure 5.54.

When editors edit organization charts, they often must convince the writers and project managers not to try to include too many levels of people or groups. Many "org" charts contain too much information to be presented in one graphic and still be legible.

If executives in business and government continue the trend of reducing the number of middle managers in their organization, streamlining it so there are fewer levels of management between them and the supervisors and workers, organization charts will look different in the future. When levels of middle management are removed, they will be wider and not as tall, which could pose new production problems.

"Org" charts can be used for information other than a company or agency's management structure. Figure 5.55 shows the relationships of different types of integrated circuits.

Schedules

Project schedules are important parts of many technical documents, especially proposals and progress *(interim* or *monthly)* reports. They can also be vital in briefings within the organization and with clients. Schedules show what is to be done on a project (the activities or *tasks*), what has been done, and when the tasks will begin and end. The format allows the readers to see how the tasks are distributed over the duration of a project.

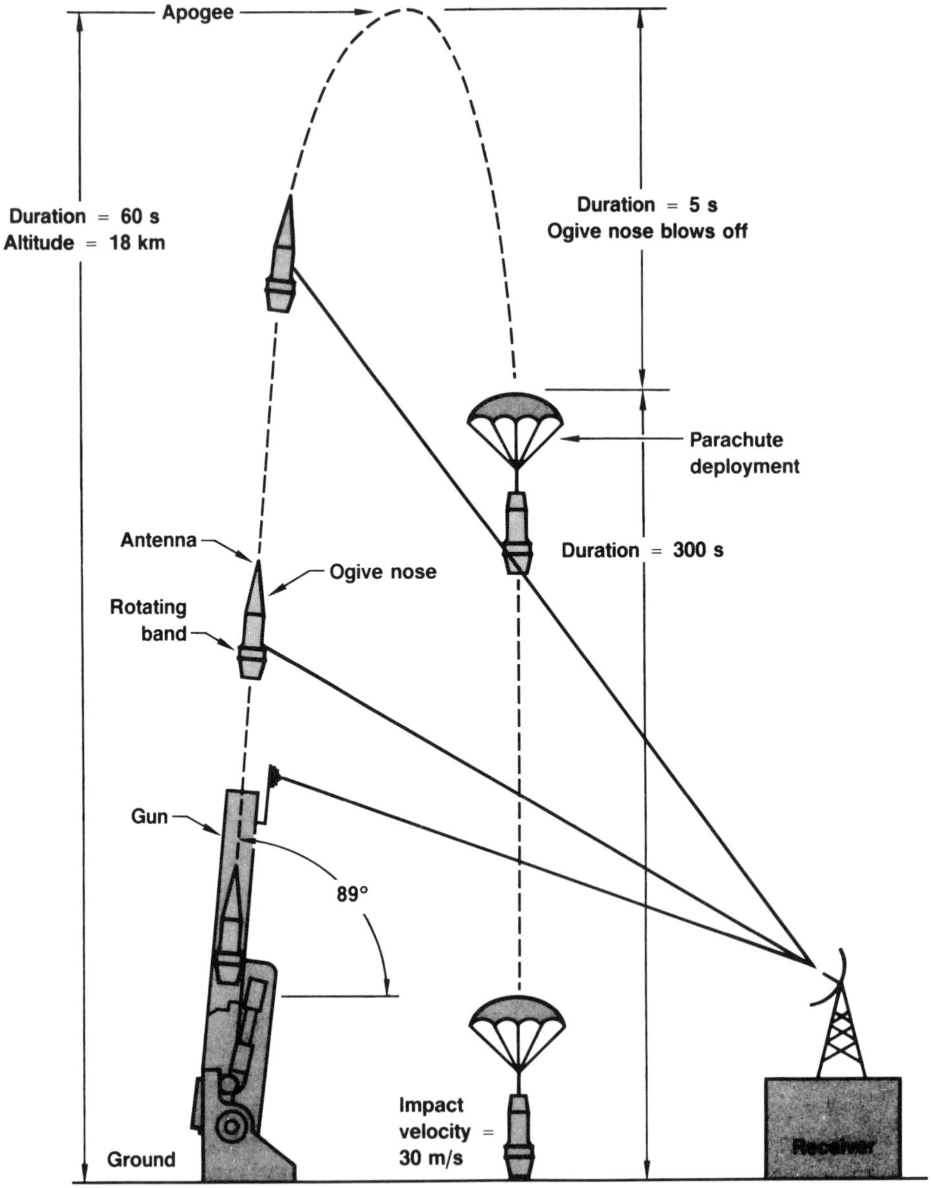

Figure 5.51. A combination block and flow diagram to illustrate current flow in a system. (Source: Lewis, D. Kent, and Ziolkowski, Richard, "Evidence of Localized Wave Transmission," *Energy and Technology Review* [Livermore, Calif.: Lawrence Livermore National Laboratory, November 1988], 25.)

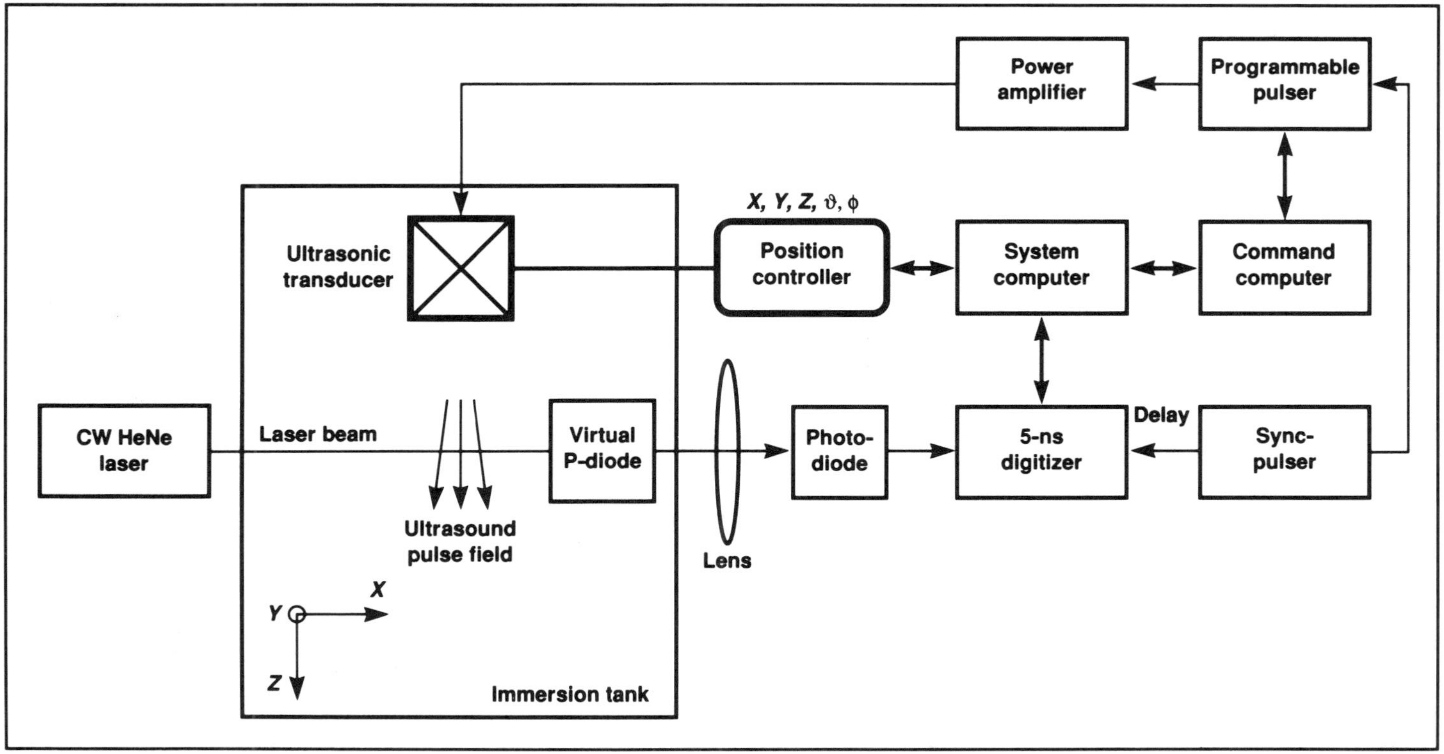

Figure 5.52. A type of flow drawing to illustrate a sequence of events. (Source: "W82 Development Program," in Lawrence Livermore National Laboratory, *Energy and Technology Review* [June/July 1986], p. 9.)

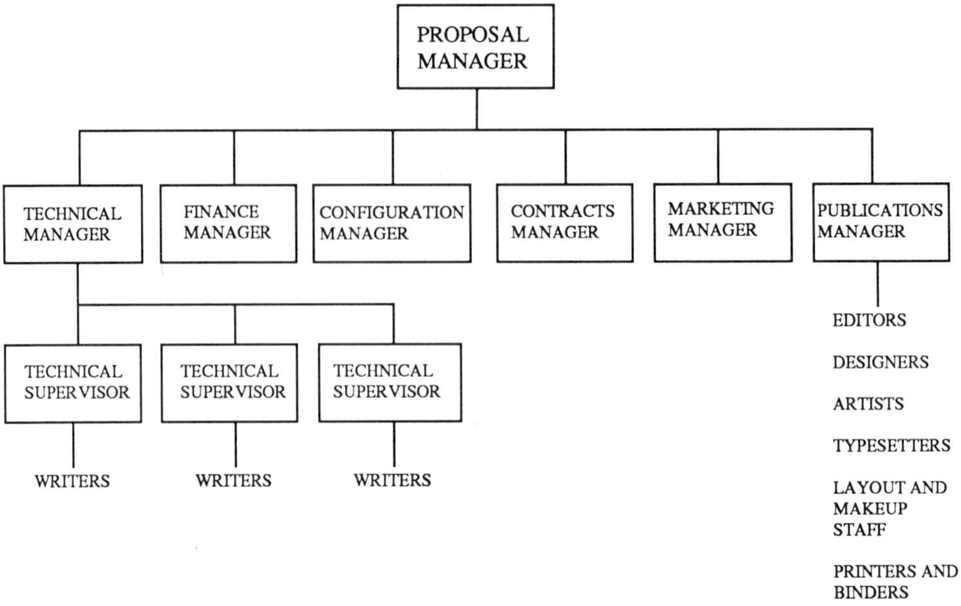

Figure 5.53. An ineffective organization chart.

The most commonly used format for schedules is the Gantt chart, developed by Henry Gantt during World War I. In a Gantt chart, each major activity of the project is listed on a separate line, generally in chronological order. Some Gantt charts are organized not by when tasks occur in the project but by how they are related.

Tasks in a Gantt chart are represented by an empty bar (which might be divided into upper and lower halves) that indicates its beginning and end. As work proceeds, the bar representing that task on the schedule (or the top half of that bar) is progressively filled in. When tasks are postponed or take longer than expected, the bar is extended or a diamond is drawn after the milestone to indicate the new deadline. When more time for a task or milestone is required, the schedule is said to have *slipped* or *moved out.* With Gantt charts, readers can tell easily how much of an activity has been completed and whether a deadline has been met.

A sample Gantt chart for proposal activities is shown in Figure 5.56. This schedule combines the bars of the Gantt chart with the triangles of a milestone (deadline) chart, which focuses on when tasks are to be completed, not when they begin and end. Editors should arrange a schedule's tasks and deadlines in chronological order in Gantt charts or in other formats.

A schedule format such as that in Figure 5.56 allows a writer to show both the duration of activities and the scheduling of milestones (such as "Print review draft") that do not require prolonged effort.

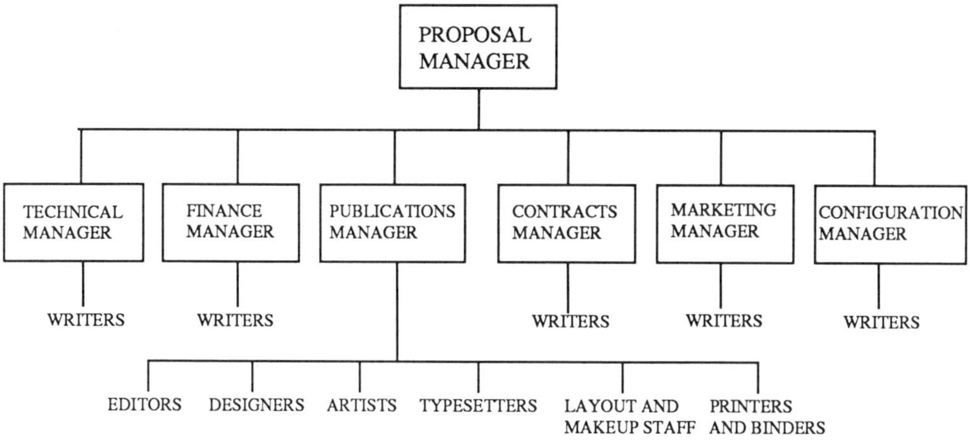

Figure 5.54. The organization chart revised.

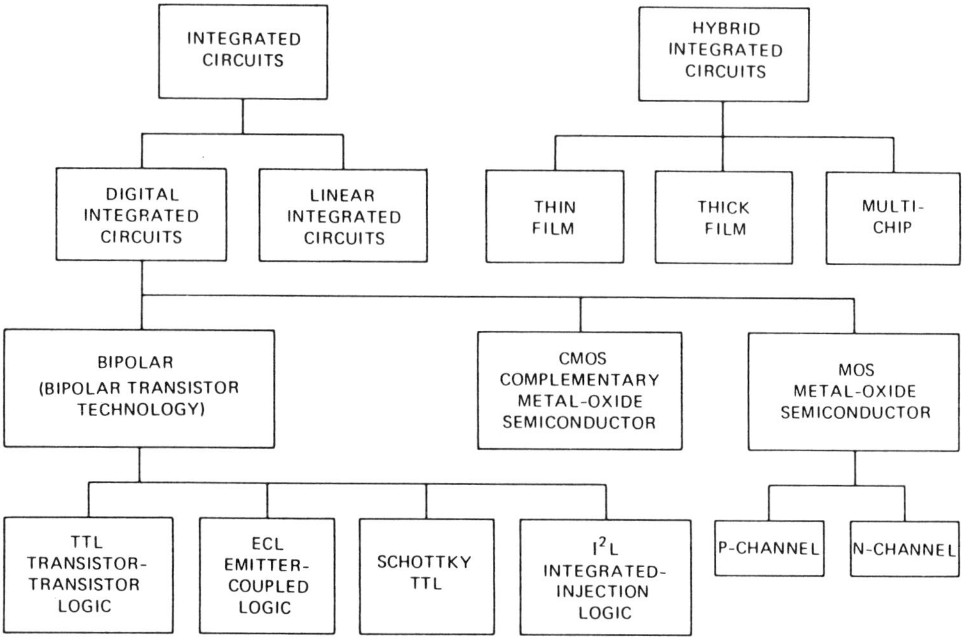

Figure 5.55. A sample organization chart not depicting personnel. (Source: Baer, Charles J., and Ottaway, John R., *Electrical and Electronic Drawing* [New York: McGraw-Hill, 1986], p. 207. Reproduced by permission of the Glencoe Division of Macmillan/McGraw-Hill School Publishing Company.)

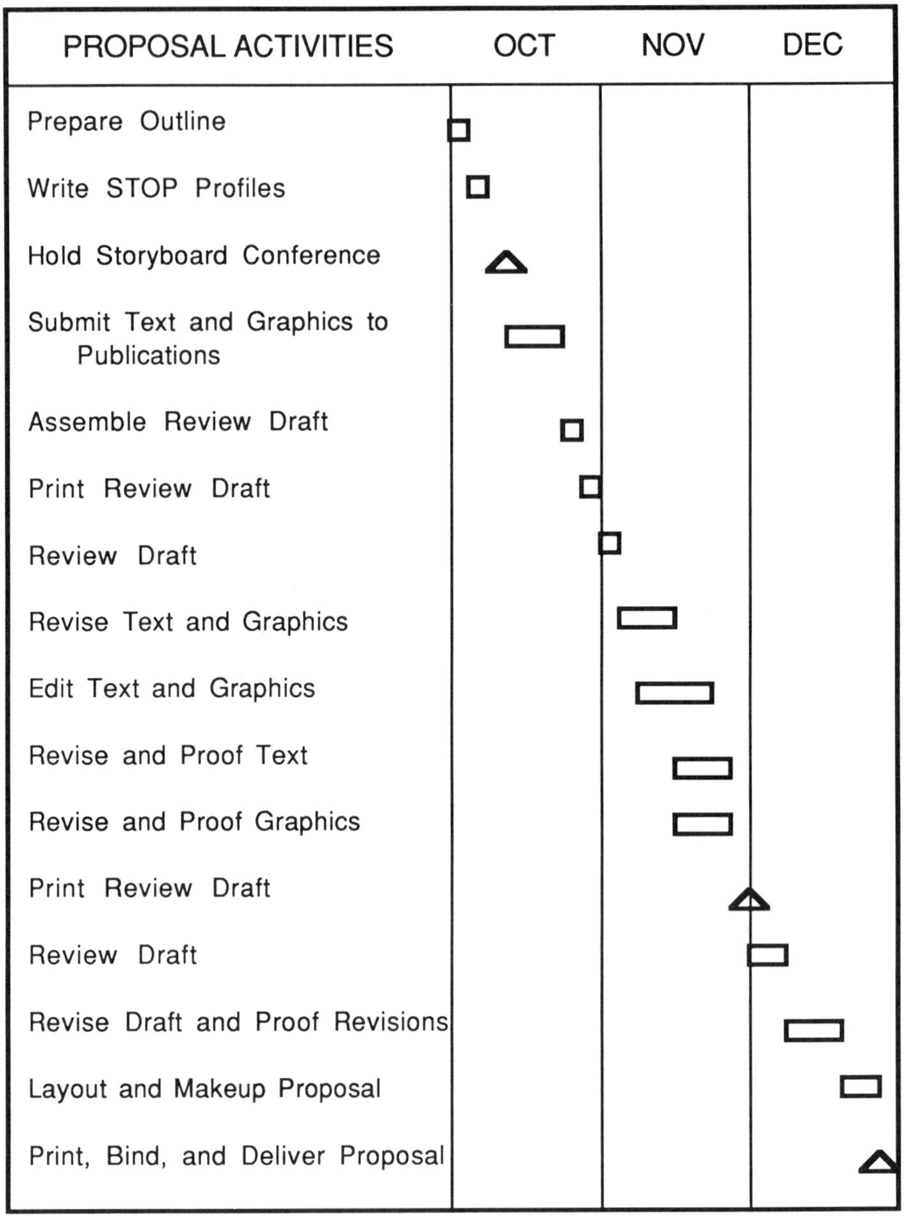

Figure 5.56. A schedule of activities in preparing a technical proposal.

For a major project that involves many actions, a writer can create a master schedule to show the general tasks and the major milestones. Then the components of each task can be laid out in a separate project schedule, often called a *second-level* schedule.

Schedules are also used to present a project's key activities on a time line, as in Figure 5.57.

As with organization and flow charts, editors should try to keep schedules pared down so that activities and their start and completion dates are clear. The main problem in editing schedules is getting technical staff to establish the final version of the schedule. Often this happens late in the work on a document (especially proposals), so with schedules editors can usually expect last-minute changes.

Photographs

Drawings are often better than photographs because artists can omit information that blocks what is to be shown or is unnecessary. However, in technical documents, writers and editors often need to show readers exactly how something or someone looks, so a photograph is called for. In all but the fanciest (and most expensive) documents, editors should use black-and-white photographs. Color photographs do not reproduce well when screened and printed in black ink.

To reproduce photographs with the sharp images and tones in professional documents and magazines, *offset* printing (described below) must be

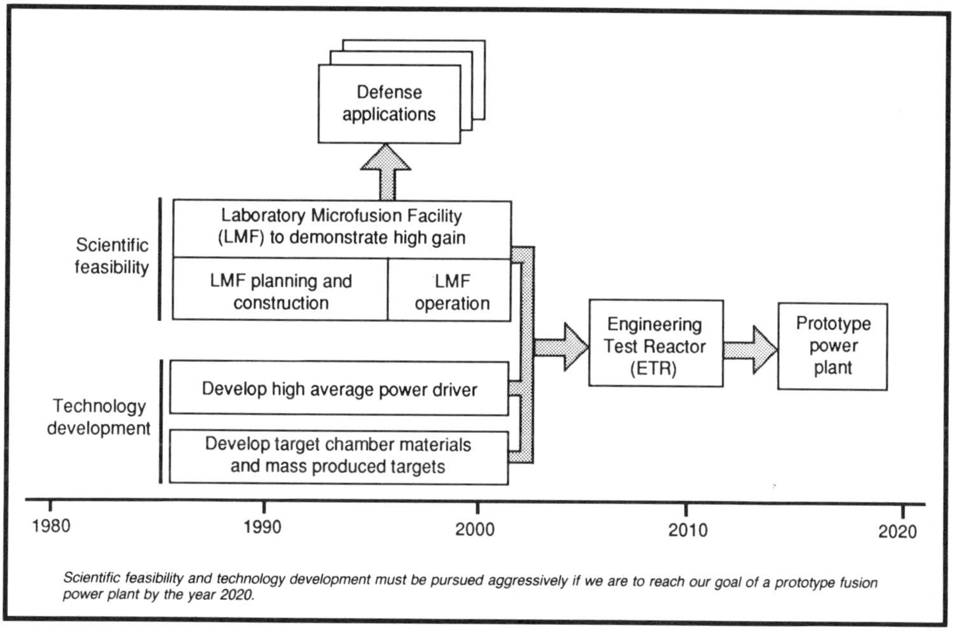

Figure 5.57. A long range, top-level schedule. (Source: "The Future of Inertial Confinement Fusion," in *Inertial Confinement Fusion* [Livermore, Calif.: Lawrence Livermore National Laboratory, 1988], p. 13.)

used. Despite refinements in photocopying, offset printing provides better reproduction of photographs than only the most expensive photocopiers can produce.

Printers or graphic composition staff *screen* the photographs to produce *halftones* for printing. Photographs are *continuous tone;* that is, there are ranges of gray or color that merge to produce lighter or darker shades or colors in the photo. Continuous tones cannot be reproduced on printing presses, because presses cannot apply varying amounts of ink (more in the darker areas, less in the lighter). Presses can only print or not print ink on a particular spot on a sheet of paper. This problem is solved by creating a halftone.

To do this, printers or camera operators photograph the original picture through a piece of plastic or glass that resembles a fine window screen. The image that is then transferred to the camera film is a pattern of dots of differing density concentrations. The negative of the dot pattern is transferred to a printing plate, and when the plate is printed, the areas with a greater concentration of dots attract more ink, printing darker areas on the paper. Fewer dots print correspondingly lighter areas.

Screens of different sizes are used to produce halftone negatives. Screens with 65 to 85 lines per inch are commonly used by photocopy shops and

Figure 5.58a. Effect of an 85-line screen on halftone reproduction.

Figure 5.58b. Effect of a 150-line screen on halftone reproduction. (Source: U.S. Department of the Interior, Bureau of Reclamation.)

newspapers; they produce printed images in which the dot pattern can be seen easily by the naked eye. Most magazines use 120- to 150-line screens, producing sharper images with a less noticeable dot pattern. Screens of up to 300 lines per inch are used in first-quality magazine printing, and art reproductions are sometimes printed with even finer line screens. Photographs or negatives of photographs can be screened, but it is difficult to make a halftone of an already-screened print without getting a distracting geometric pattern called *moire* from the superimposition of the screens. (Moire is also caused by improperly aligning the separations for color printing.)

Figure 5.58 reveals the effect of using different screens. The original print was halftoned with 85- and 150-line screens to produce the two images. Notice the visible dot pattern with the 85-line screen and the improvement in quality with the 150-line screen. To reproduce the original more exactly, a much finer screen would be required.

Figure 5.59 demonstrates a common use of photographs: to enable the reader to see an unfamiliar piece of equipment. Showing the equipment in its customary environment can give readers a better sense of its size, shape,

Figure 5.59. A common use of photographs: to show equipment in its working environment. (Source: Oak Ridge National Laboratory *Review* Vol. 23, No. 3 [1990], pp. 10–11.)

and function. Artists can use airbrushing to add or delete details in photographs.

Editors or graphic artists can affix labels directly onto a photograph or onto a transparent acetate overlay, so that when printed the photograph will contain callouts.

Special care should be taken with negatives and prints. Editors should never write on the front *or* the back of a photograph to be reproduced. Ink can smear or blot onto another print, and the impressions caused by writing on the back of a photograph can show on the front. If possible, editors should label a photocopy of the photograph, being sure to indicate which side is the top if there is any question. Stick-on notes or tape should not be attached to a photograph. If an editor must write on the photograph to make sure it is identified and oriented properly, he or she should write on the back, across the top, in an area to be cropped out. Alternatively, type a label and apply it to the back.

Writers need to indicate if a photograph should be cropped, enlarged, or reduced. *Cropping* means marking a photograph to indicate what should be cut out of the final view. If the editor wanted to use only that part of Figure

5.60 that illustrated the Saturn V vehicle, he or she would place *crop marks* on a photocopy of the original print to illustrate what should be omitted. The composition or camera room staff will then shoot only the areas inside the crop marks. The image area showing the Saturn V would then be 1.5 inches wide by 4 inches tall. To make the photograph 2.5 inches wide to place in a 3-inch column, the editor would use a scaling wheel (described near the beginning of the chapter) to determine that shooting the cropped area at 167 percent is required. When instructed to "Crop and shoot at 167%," the camera room staff would provide the writer with the print in Figure 5.61.

When editors request *enlargements* or *reductions* of photographs or any graphic, they must examine the document's layout carefully before specifying the degree of enlargement or reduction. Of course, remember that enlarging or reducing a graphic affects both its width *and* its height, as Figure 5.61 demonstrates. Increasing one dimension might increase the other so much that the page in the document would have to be relaid. In this case, specify a smaller enlargement so the graphic will fit the vertical column space without major changes to the layout.

Figure 5.60. Uncropped photograph of a Saturn V launch vehicle and the launch complex. (Source: National Aeronautics and Space Administration.)

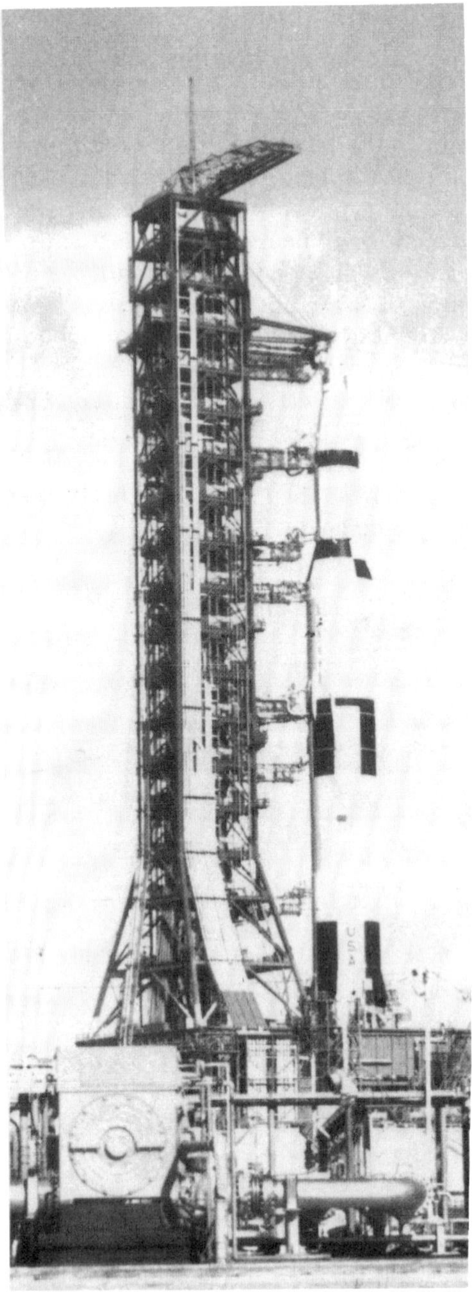

Figure 5.61. Cropped photograph of the Saturn V launch vehicle. (Source: National Aeronautics and Space Administration.)

Editors should use high-contrast, glossy black-and-white photographs in most technical documents (except annual reports and other marketing materials). Color printing is difficult and costly, and it takes far longer than traditional one-color (black on white) printing. Refer to the bibliography for sources that discuss color printing.

CHOOSING THE RIGHT GRAPHIC

The examples in this chapter have presented various uses of the different types of graphics for technical documents. When editors edit a document, they look for material that could and should be presented in graphics and then choose the appropriate graphics to fit the material. No editor says: "I need a line graph at this point in the report." Instead, editors ask: "What kind of graphic should I use for this material?"

To get a better idea of which graphics to use to present certain kinds of information, examine the discussion below, which is organized into people or physical objects (machines, places, plants, animals), and data or representations of data (such as statistics, the organization of groups or processes, etc.).

People or Physical Objects

To show:	Use a/an:
People	Photograph or drawing
A whole object, device, mechanism, or assembly	Photograph or drawing
Where things are located	Photograph, map, or drawing
Part(s) of an object, device, mechanism, or assembly	Photograph, exploded-view, cross section, or cutaway drawing
An object's interior	Photograph, cross section, or cutaway drawing

Data or Representations of Data

To show:	Use a/an:
How a system works	Flow chart, cross section, or cutaway drawing
Steps in a process	Flow chart, photographs, or drawings
How a company, agency, or group is organized	Organization chart
Trends (such as rising costs, revenues, profits, etc.)	Line, bar, or scatter graph
Large amounts of statistical data	Formal table, line or scatter graph
How something is done	Photographs or drawings
General comparisons of data	Line or bar graph
A timetable or schedule for a project	Schedule

Data that adds up to 100 percent	Circle graph
Lists of abbreviations, acronyms, conversion equivalents, symbols, or definitions	Informal table
Sources of revenue or expense	Circle graph, organization chart, formal or informal table

The best way for editors to learn which graphics work well for different kinds of information is to examine the graphics used by technical experts in their documents. When editors examine the graphics experts have chosen to

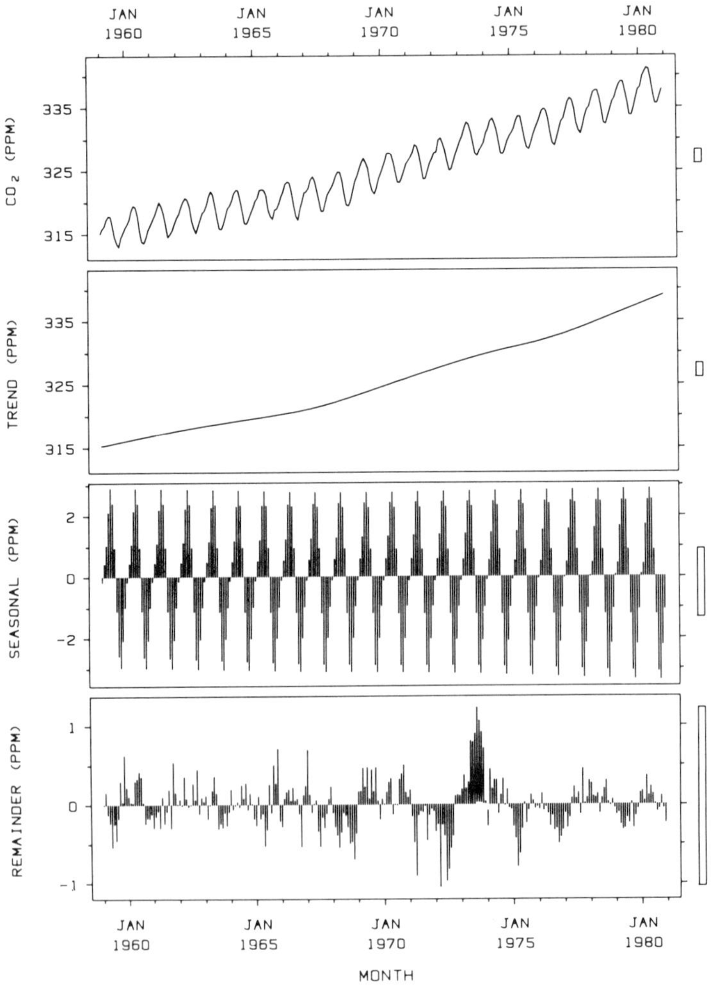

Figure 5.62. "The power of graphical data display." (Source: Cleveland, William S., *The Elements of Graphing Data* [Wadsworth, 1985], p. 11. Used with permission.)

Astronomy and Technical Cross-fertilization

ver the centuries, astronomy and technology have progressed hand in hand. The study of the universe has benefited from improved observational devices and techniques. By the same token, developments in astronomy have led to practical applications in other disciplines.

1500-1600

- Increasingly accurate maps of the sky for navigation

1600-1700

- Christian Huygen's invention of the pendulum clock for navigational time keeping
- Newton's development of the calculus, the laws of motion and the law of universal gravitation as a means to explain the motions of planets and comets

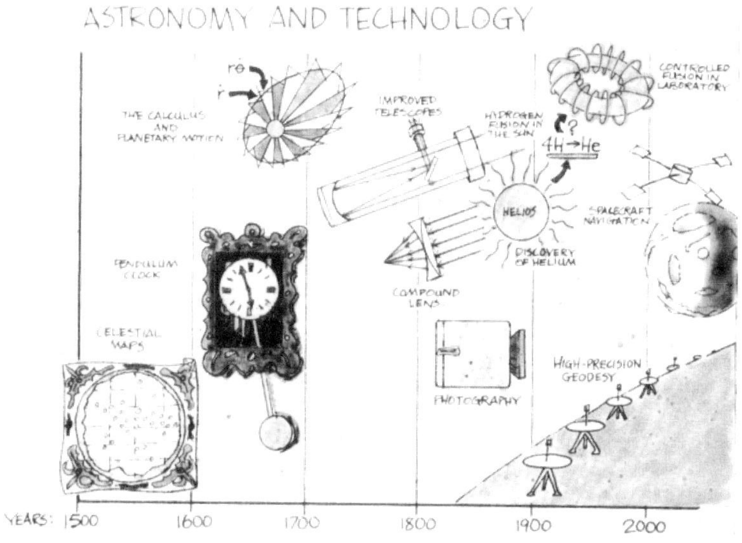

1800-1900

- Increasingly sophisticated optical innovations by astronomers (William Herschel, Fraunhofer, Lord Rosse, Alvan Clark, and many others)
- Development of increasingly sensitive photographic techniques
- Lockyer's discovery of a new chemical element, helium, on the sun before it was known on Earth

1900-NOW

- Hans Bethe's theoretical prediction of hydrogen fusion at the center of the sun, a precursor for all modern fusion efforts
- Lyman Spitzer's development of astrophysical plasma theory, the basis of present devices for releasing energy from controlled fusion
- Very long baseline radio astronomy techniques used in high-precision geodesy to survey the structure of the Earth
- Techniques of celestial mechanics, precursors to the development of accurate spacecraft navigation.

The mutually beneficial interaction between astrophysics and technology continues today.

A Perspective of the Search

Astronomical searches have occupied human thought for millennia. Over the generations, we have succeeded in gaining ever greater insight into the underlying forces at work in the cosmos. In the Space Station era, the family of permanent observatories in space will open the way to new, comprehensive studies of key remaining problems in astrophysics, helping us understand:

- The birth of the universe, its large-scale structure, and the formation of galaxies and clusters of galaxies;
- The fundamental laws of physics governing cosmic processes and events;
- The origin and evolution of stars, planetary systems, life and intelligence.

If we succeed, we will leave a legacy to rank us with the great civilizations of the past.

Figure 5.63. Final layout, astronomy and technology article. (Source: *NASA Activities*, Vol. 21, No. 4 [July–August 1990], p. 14.)

illustrate their technical documents, they should do so critically. You may be surprised to find ineffective as well as effective graphics.

In Figure 5.62, a particularly effective example, four different sets of data about CO_2 concentrations are presented: monthly concentrations at a site in Hawaii; the trend in those concentrations; seasonal differences; and the concentrations minus the trend component and seasonal differences. As William Cleveland points out, the graphs present 2,112 numbers—"No vehicle other than a graph is capable of conveying so much quantitative information so readily."

LAYOUT

The earlier discussion on layout in Chapters 2 and 4 established principles for mixing text and graphics in a document. The final layout of the article on astronomy and technology in Chapter 4, containing the finished graphic for the article, is shown in Figure 5.63. Notice how the final graphic combines perspective sketches of developments such as the pendulum clock with a time line common to schedules, to organize the developments referred to in the article. Also, the "unfinished" nature of the graphic (for example, hand-lettered callouts) might suggest the "unfinished business" of astronomy and technical cross-fertilization, as well as holding down the cost of preparing the graphic. For an in-house publication (this one, *NASA Activities,* is sold to the public by subscription), a "rough" graphic like this, emphasizing early developments in astronomy, is fine. However, the editor should mix finished graphics with rough ones, as is done in *NASA Activities,* to polish the document.

When a document is laid out, the editor should check the layout carefully before printing. Titles of articles or chapters sometimes get transposed, as do captions for graphics (especially names of people under photographs).

CONCLUSION

The discussion of editing graphics in this chapter has examined types of graphics commonly used in technical documents. Other kinds exist as well, and many of the graphics covered here are used in more ways than those discussed. Editors should examine the graphics in documents in their field to see how material is handled and to get ideas on how to use graphics in their own projects. Check the bibliography for more valuable guidance for editing graphics.

As computer graphics programs and systems develop, so too will challenges and opportunities for technical editors as they work with writers and graphic artists. However, the basic principles of effective graphics outlined in this chapter will continue to apply and to inform editors' work with graphics.

CHAPTER SIX

Degrees of Edit

This chapter extends the discussion in Chapters 4 and 5, focusing on how editors apply the principles of editing in situations when different degrees of editing might be required. Sometimes, tight schedules force editors to edit lightly and quickly. At other times, schedules might permit more extensive editing appropriate for the audience or purpose of the document.

In the best of situations, editors have ample resources—enough typists, artists, proofreaders, and other production staff, and enough time and money, to produce an excellent technical document every time. However, in most business and government organizations this is often not the case.

Editorial effectiveness is controlled by the time available for editing, not by the number of available editors. A 100-page report might be edited in ten hours by one editor who edits ten pages an hour, or in one hour by ten editors who edit ten pages an hour. Clearly, one editor is more likely to achieve greater consistency, but editors often have to undermine the notion that the assembly-line approach would be equally effective. Time is the key—and the general rule is that an editor should do as much as he or she can to improve a document *in the time available.*

Different solutions have been suggested for this problem of editing effectively with the resources available for editing. Editors sometimes find it difficult to impress on technical staff the editorial tasks involved in creating a technical document. To clarify and differentiate the types of services editors perform, the levels of edit concept was introduced at the California Institute of Technology Jet Propulsion Laboratory and described by William Van Buren and Mary Fran Buehler in *The Levels of Edit.* Van Buren and Buehler address the range of editorial functions, from merely coordinating publications department work to substantive editing, in which the "meaningful content of the publication" is examined and revised as necessary.

The degree of editing devoted to a document might be dictated by its importance, its budget, and its condition. Also, as Van Buren and Buehler rightly indicate, the degree of editing might be determined by the document's final format. In *The Levels of Edit* (p. 5), they identify three classes of documents that differ in "the level of their physical quality and appearance. A Class A publication is usually typeset, with justified columns, the artwork is integrated with the text, and the publication is printed and bound using high-quality materials and techniques. . . . A Class B publication is usually typewritten, with unjustified columns, the illustrations and text appear on separate pages, and the publication is printed and bound using more economical materials and techniques. . . . A Class C publication is not composed by publications personnel but is processed by them as camera-ready copy" (*The Levels of Edit,* p. 5).

Van Buren and Buehler's concept of levels of edit can help editors determine staffing and budget needs for a project. However, these functions (and the levels of edit based on the combinations as shown in Fig. 6.1) are often new to technical staff outside a publications department. For some, a matrix of nine types and five levels of edit might make the differences in editorial responsibilities unduly complicated. For example, an editor might be asked to perform only a language edit (to polish the phrasing in a document) and a policy edit (to ensure that the document does not violate policies of the

Table 1. Types and levels of edit

Type	Level of edit				
	1	2	3	4	5
Coordination	X	X	X	X	X
Policy	X	X	X	X	X
Integrity	X	X	X	X	
Screening	X	X	X	X	
Copy Clarification	X	X	X		
Format	X	X	X		
Mechanical Style	X	X			
Language	X	X			
Substantive	X				

Figure 6.1. Types and levels of edit. (Source: Van Buren, Robert, and Buehler, Mary Fran, *The Levels of Edit*, 2nd ed. [Jet Propulsion Laboratory, California Institute of Technology, 1980], p. 5, used with permission.)

organization). In this case, it is difficult to determine the applicable level of edit to estimate the job. The language edit is level 2, and the policy edit level 5. To call this a level 2 edit would suggest that it include a mechanical style edit, a format edit, and so forth.

In the levels of edit system, a level 5 edit would not involve a screening edit (for spelling errors, agreement problems, sentence fragments, etc.). To many editors, if a document has not been checked for such problems it has not been edited at all. Differentiating between nine types and five levels of editing might complicate an editor's attempt to define a client or customer's requirements for a document, obscuring what has to be done, by when, and with how much money.

Other, simpler approaches have attempted to describe the degrees of editorial involvement in a document. For example, Martin Marietta Energy Systems, the operator of many U.S. Department of Energy facilities in Oak Ridge, Tennessee, established four levels emphasizing the degree of editorial effort: light, moderate, in-depth, and rewrite/restructure (an extension of the third).

In this system's light editing, editors:

> a. identify and mark the manuscript to conform to proper style and format in regard to headings; reference, figure, and table citations; subscripts and superscripts; display of mathematical and chemical equations; chemical, mathematical, and technical symbols; Greek letters; and italics;
> b. specify and indicate, where appropriate, type or point size and style; spacing; and line lengths for headings, captions, text, references, and all other components of the manuscript;
> c. review figure, table, and reference citations to ensure accuracy of sequence and conformity to listings;
> d. review manuscript to ensure that all components (e.g., abstract, highlights, contents, title page) are present and properly organized;
> e. read the entire manuscript and identify, interpret, and apply rules and guidelines from accepted authorities and from sound, informed judgment (where guidelines are either not available or subject to interpretation) to (i) correct grammar, spelling, punctuation, sentence structure, subject-verb agreement, verb tense, and word usage; (ii) ensure uniformity of units of measure, acronyms, and abbreviations with accepted style; and (iii) ensure correct and consistent usage in all components of the manuscript, including figures, tables, and text;
> f. examine visual materials to assess acceptability for publication or to determine preparatory processes needed to improve their quality;
> g. identify to the author problems related to textual, visual, or tabular presentation of the information, copyrights, and/or conformance of the manuscript with organizational policies or procedures related to the publication of technical information;
> h. for manuscripts requiring copy-marking/light editing, limit rewriting to the minimum necessary to correct the identified errors in grammar, word usage, or style.

In moderate editing, editors also "rewrite as necessary to improve clarity or conciseness," and in in-depth editing, editors:

perform substantive review and editing of the contents of the manuscript to improve clarity of expression, completeness of discussion of topics presented (including the need for additional information or the removal of unnecessary data), and the logical and systematic development and presentation of ideas, and rewrite as necessary to correct awkward sentence structures or ambiguities and to delete unnecessary verbiage."

In the heaviest editing, rewrite/restructure, editors:

rewrite the text and reorganize the components thereof (e.g., sentences, paragraphs, sections, chapters); write abstracts or rewrite text to produce abstracts; [and] restructure tables and revise or reorient illustrations to add or delete information or to shift emphasis as necessary to improve the clarity, succinctness, progression of thought or development of major points within the text, and the overall quality of the presentation of the technical information."*

Given the numerous tasks to be completed even in this system's light edit, it seems strange to call it "light." A simpler approach might be to recognize that editors can have light responsibility for a document (say, merely copymarking for typos), heavy responsibility (such as revising content and organization, and planning and supervising all production activities), or something in between. Some technical documents require very light editing; they are well organized, appropriate in content and presentation for their audience, and fairly correct and polished. Most technical documents, however, need work—often more than the schedule and budget allow. In this situation, the project manager or writer and the editor should discuss the document and decide how much can and should be done to edit it with the resources available. The editor must be able to explain simply what is involved in a light, medium, or heavy edit. This would be the case with the levels of edit concept too, but it is easier for most clients as well as editors to understand three degrees of edit than to differentiate nine types and five levels of edit.

LIGHT EDIT

When the only editing required or possible for a document is a quick scan to catch spelling errors, major grammar and punctuation errors, and major formatting problems such as margins or heading formats, a light edit is done. Its purpose is to correct quickly the noticeable errors and weaknesses.

A light edit does not differ significantly from correctness editing, but it is often used for a review draft of a document that will not undergo a complexity edit before review; it can be a quick cleanup to remove distracting flaws from the draft. Light editing differs also from proofreading in that the text is not yet typeset or near final form, so editing will not disturb document

*Source: Martin Marietta Energy Systems, Request for Proposal J0622-11. Copyright 1986, Martin Marietta Energy Systems; used with permission.

production. With proofreading, we assume that any fine-tuning might affect the schedule for production, so only those problems that would upset clients and company or agency staff are corrected.

With a light edit, editors have more opportunity to make changes that they and the client deem necessary. However, changes are not made for stylistic reasons when there is little time to edit the document. Problems that probably would not be noticed by readers are left unaddressed.

Light edits are used in different situations. For example, an editor might light edit a letter or a short report (especially an internal report). The editor would not be expected to make major changes in the content, organization, or phrasing of the document. Light edits can also be used on the work of important clients who are sensitive to changes in what they have written. The editor light edits the document, correcting errors in the presentation, and returns it to the writer.

MEDIUM EDIT

In a medium edit, text and graphics are revised more thoroughly and methodically. Depending on the length of the document, the editor might read it through first to examine its content and structure and evaluate appropriateness of the level of discussion. In the medium edit, the editor might make minor changes in organization, especially at the paragraph and sentence level, but time would not usually allow for major restructuring. Rather, the editor recommends only minor reorganization of text or graphics when required for clarity or simplicity. The editor then:

- Checks figure and table callouts
- Examines the text for conformance to standards, especially in format
- Edits the text at the sentence level, for clarity and correctness
- Examines graphics for conformance to standards
- Cleans up graphics for draft review (for example, by quickly resketching a messy line graph or drawing)

In a medium edit, time is still limited (although not as limited as with a light edit), so the editor must not linger over word choice or fine points of grammar or sentence structure. For example, restructuring of awkward sentences to avoid unnecessary subordination is possible, but methodical elimination of unnecessary subordination is not attempted. Nor is searching for the exact phrase when the author's intended meaning is sufficiently clear. The medium edit is not to be confused with a half-hearted job. Rather, it is a response to time or money pressures that prevent the thorough edit an editor would ideally perform on every document.

As the medium edit has more impact on a document than a light edit, the medium edit should be used only if the writer will examine the editing changes before the document goes into production, to check the discussion for changes in meaning the editor might have introduced.

HEAVY EDIT

In the heavy edit, the editor strives to make the document as effective as possible. The heavy edit requires more time, effort, and cooperation between writer and editor. The editor completes the macro- and micro-edits discussed in Chapter 4 and tries to accomplish the editing objectives specified in Chapters 4 and 5, considering the following:

- The content and format guidelines for the document (whether internal or external)
- Compliance with the style guide and other standards
- The document's audience
- Its organization
- Its content coverage
- The appropriateness of its graphics
- Technical accuracy of its text and graphics
- Its readability and usability
- Format of text and graphics
- Format of references
- Paragraph organization
- Use of specific detail to support points
- Grammar, syntax, and punctuation
- Sentence clarity, conciseness, and effectiveness

Sufficient time should be available if the document is to receive a heavy edit. The writer should be available for consultation during editing, as described in Chapter 2, and should review all editorial changes. In a heavy edit, the editor has more impact on technical content, and any questions regarding content should be addressed to the writer.

Depending on the complexity of the document and the number of writers involved, a heavy edit may involve extensive rewriting by the editor or writer, with subsequent editing of rewritten passages. Technical experts must check the final content. A heavy edit should be used only when the document will go through a major revision and proofreading cycle before it moves on in the production process.

The heavy edit should be undertaken only when the content of a document is firm. Editors should perform a light or at most a medium edit on review drafts. There is no sense investing time and effort heavy editing a draft that may be rejected by reviewers. This can be a difficult lesson for beginning editors, who often try diligently to make every document as effective as it can be, performing a heavy edit only to find their efforts wasted on a draft.

Occasionally a project manager or the author of an article for a technical journal will ask an editor to examine how well a document draft conforms to a standard such as a request for proposal or the journal's instructions for authors. Editors should welcome such an opportunity to examine the appropriateness of a draft. However, they should not revise the draft to make it fit the standard if major changes in organization or content are required. Rather, they should suggest how the draft should be reorganized, expanded, or cut to fit the standard, waiting to edit the draft until reviewers have made their suggestions and the writers have completed the revisions.

EXAMPLES

To clarify the situations that might call for different degrees of edit, consider the following cases.

Case 1

A technical proposal of four volumes authored by a number of technical writers has first drafts of each volume in preparation. The editor of the technical volume will receive hard copy of 80 rough graphics and 240 pages of text on disk to prepare for review by fifteen evaluators. The schedule calls for delivery of all text to the editor by noon Friday, with some sections delivered Thursday, for a review to begin Monday morning at 10.

It would probably be impossible to edit the text and to process and proof text revisions to meet this schedule, and very difficult (if not impossible) to have graphics processed. It would be inadvisable even if possible. At the beginning of a project like this, when the schedule is being established, a light edit should be planned. Even with a light edit, the editor will have production planning and control responsibilities (see Chapter 2), for example, scheduling the printing and binding of the draft. In light editing the document itself, the editor should:

- Make sure each section is numbered according to the volume outline
- Identify missing sections and negotiate with writers for their submission
- Make sure each graphic is labeled by section and by figure or table number
- Reproduce individual graphics on separate pages after their callouts in the text, and insert a blank sheet with the figure or table number and "To Come" for any missing graphic
- If time permits, pencil in any revisions to section numbers; names of products, projects, or people; and section titles, headings, or summaries

- If time permits, write in spelling corrections to prevent reviewers from focusing on misspelled words rather than the draft's organization and content
- If possible, line edit the introduction for sentence structure, grammar, and punctuation, and have the editing processed

Introductions to important technical documents are usually written by senior technical staff, so it is wise for editors to edit introductions and help staff make a good impression on reviewers, even in a draft. The introduction's condition indicates to reviewers how the entire volume will be edited, thereby establishing the editor's ability.

Case 2

John Stubbs, a contracts officer at Western Corporation, has written a letter explaining why Western will not be able to complete on schedule some of its subcontract work for the Adams Company. Stubbs is concerned that the problem might affect Western's ability to win future work from Adams, and asks an editor to review his letter, which appears in Figure 6.2.

This letter presents the Western Corporation's production problem in a bad light. The detail about the machinists' strike could make the Adams Corporation reluctant to give future work to Western. Also, the information in the letter is not arranged in the most favorable way, given the situation, and the tone of the closing is not conciliatory.

A light edit of this letter might do no more than correct the spelling and punctuation problems, complete the Adams Company's name, and eliminate the reference to the strike. However, in ten minutes, the editor might medium edit the letter for the result in Figure 6.3.

In the medium edit, much has been done with the letter. The editor has placed the good news about the switch panel production before the bad news about the junction boxes. She has glossed over the information about the strike (but has not fabricated an excuse like the clichéd "computer trouble"). Also, she has eliminated what might seem to be an attempt to shift responsibility to the head of production. The institution of a second shift has been disguised to prevent Greene from realizing that the production scheduling required such a drastic measure. (Note how *our* second shift instead of *a* second shift disguises the fact that it was just initiated.) The editor has emphasized Western's effort to meet the original delivery date and softened the closing, making it conciliatory. In a few minutes' work, she has positioned the writer much more positively, and Stubbs has done less damage to Western's chances of future work for the Adams Company.

For a heavy edit of this letter, the editor might discuss the production problem with Stubbs to see if there were some way to cast the possible delay in a less unfavorable light. Also, the editor might inquire for other good news to include.

170 Park Boulevard
Syracuse NY 13210

January 13, 1992

Ms. Mary Greene
The Adams Company
300 Bingham Highway
Ypsilanti MI 48194

Dear Ms. Greene,

I am writing in regard to our work for Adams on the Allied Corporation K125 Controller project.

We won't have the controller junction boxes ready by the end of the month, as specified in our contract (Number AC-K125-WC-JB-1991). Our head of production has told me that we are still trying to fill back orders that plied up when our machinists went out on strike last month. We had several orders scheduled for delivery before yours and we had to complete them first. I hope you understand the problem. We have put in a limited second shift but we are still running behind. Our work on the switch panels for the Allied project is on schedule, however, and we see no delays with it.

I would like to negotiate a new delivery date for the 500 controller junction boxes. I'll call you early next week to discuss the matter and arrange a new delivery date.

Sincerely,

John Stubbs

Figure 6.2. Original letter to be edited.

170 Park Boulevard
Syracuse NY 13210

January 13, 1992

Ms. Mary Greene
The Adams Company
300 Bingham Highway
Ypsilanti MI 48194

Dear Ms. Greene:

I am writing about our work for The Adams Company on the Allied Corporation K125 Controller project (Contract AC-K125-WC-JB-1991).

Our work on the switch panels for the controllers is on schedule, and we see no likelihood of delays. However, we are filling some rescheduled back orders and we may not be able to deliver the junction boxes by the end of the month. We have geared up our second shift, though, and we hope to be able to catch up with the original production schedule.

I'll receive more information from our Production Head on Friday. Would it be possible to arrange a new delivery date for the controller junction boxes? I'll call you early next week to discuss the project.

Sincerely,

John Stubbs

Figure 6.3. A medium edit of the letter.

Case 3

One of the writers in your organization has prepared the following article on auroras. It will be submitted to the local newspaper's science and technology section as part of your organization's educational outreach and public relations efforts.

The Dawn That Glows at Midnight

by Elizabeth Babcock, News Special Writer

Residents of southern lower Michigan may have a rare opportunity this month to see nature's most spectacular light show. Scientists are predicting that solar activity will peak this spring, creating favorable conditions for the aurora borealis (northern lights) to appear in local skies, welcoming the season and delighting viewers with spectacular displays of shimmering arcs and rippling curtains of colored light. Auroras change dramatically throughout the night. In the early evening, they often appear as stationery arcs that hover quietly near the horizon in the northern sky. Toward midnight, they may become suddenly active, billowing into enormous radiant curtains that flicker and pulsate with wave of colored light in a fiery display that moves and changes with astonishing speed.

Jon Wooley, professor of astronomy at Eastern Michigan University, said that aurora displays associated with peak sunspot activity could be visible outside their normal range anytime soon.

The aurora boreallis, which means dawn of the north, appear most frequently in an oval shaped zone centered at the magnetic north pole. The zone stretches across northern Canada, Alaska, Greenland, Iceland, the northern coast of Norway and the Arctic coast of Siberia. People who live in Michigan's Upper Peninsula can see auroras about 50 nights a year. Residents of Michigan's Lower Peninsula are one or two degrees of longitude too far south to see the aurora most of the time. However, the aurora zone expands when the sun becomes unusually active, so the lights become visible at lower latitudes.

Although various theories have long linked Earth's light show to the sun, our current understanding of the sun's role in auroras is relatively recent. "The relationship between auroras and sunspots have been understood since the turn of this century," said Richard Teske, a U-M professor who specializes in solar astronomy. "But the exact mechanism that produces auroras was not understood until space explorers mapped the earth's magnetic field and detected a stream of charged particles coming in from the sun."

This stream of particles, called the solar wind, generates auroras when it enters our atmosphere and interacts with the Earth's magnetic field. Auroras may become unusually intense during periods of maximum sunspot activity. Sunspots that appear as darker regions on the sun's surface become more numerous when the sun periodically increases its energy output, roughly every 11 years. Astronomers refer to the peak of this cycle as "solar maximum," and they agree that its happening right now. Teske explained, "Solar forecasters have made their best honest estimate that solar maximum will occur this month, but we won't know for sure until the end of the year," when statistical analyses are compiled.

The activities that generate auroras originate in the visible layer of the sun, called the photosphere. Convection currents develop in the core of the sun and rise to the photosphere, where they interact with magnetic fields to produce sunspots, cooler regions of the photosphere that show up as dark blotches against the brighter, hotter background.

As solar maximum approaches, and larger, more numerous sunspots develop, the sun's corona becomes brighter and hotter. The corona is the outermost layer of the sun that may be seen during a solar eclipse. When the moon moves across the face of the sun, the corona appears as a halo of light streaming out from behind the moon.

EMU's Woolley explained that the corona "evaporates" to produce the solar wind, a stream of electrified particles pouring out of the sun. When sunspot activity is high, more high energy particles escape from the corona, increasing the velocity of the solar wind.

The photosphere also steps up it's energy output during solar maximum. Wooley said that magnetic fields located near sunspots emit energy which heats up solar gases in those areas and generates solar flares. These explosions on the sun blast electrified particles into space and produce shock waves in the solar wind. If a flare erupts near the center of the sun, the shock wave collides head on with the magnetasphere, the area of space affected by the Earth's magnetic field. The collision compresses the magnetosphere, disturbing magnetic force lines and intensifying magnetic fields. These magnetic disturbances create a geomagnetic storm that can produce highly active auroras over a wide range of latitudes. If a flare erupts near the edge of the sun, the solar wind's shock wave sideswipes the magnetosphere instead of colliding with it directly. When this occurs, the Earth's magnetic field is not greatly disturbed and auroral effects are minor.

The solar wind, moving at supersonic speeds takes two or three days to travel 93 million miles from the sun to the Earth. When this wind reaches our atmosphere, the Earth's magnetic field traps its electrically charged particles and guides them to the polar auroral zones. At an altitude of about 200 miles above the polar regions, charged particles collide with molecules of atmospheric gases which react by giving off light. The predominant colors of the aurora are pale green (emitted by oxygen molecules) and pink or red (given off by nitrogen), although Matt Link, planetarium coordinator at the University of Michigan's Exhibit Museum, said he once photographed a spectacular aurora that displayed hues of purple, yellow and blue.

The northern lights have inspired myths and legends since the beginning of history. According to one legend, said Matt Link, the world was hollow, and inside the earth there was another sun. Before explorers mapped the polar regions, people believed there were holes in the Earth at the North and South poles. They thought the aurora was caused by light shining through these holes from the sun inside the Earth. Linke said that scientists later developed a quasi-scientific theory that explained the aurora as "light reflected from the polar ice caps." "The Algonquian Indian tribes who lived around Lake Superior thought the northern lights were caused when Nanabozoo stirred his fires," said Alithea Helbig, an English professor at EMU who specializes in Native American folklore. Nanabozoo is a heroic "trickster" figure of Algonquan legend.

If you want to see auroras, get away from city lights on a clear night with little or no moonlight. The best time is around midnight, when the aurora typically become more active. To find out if geomagnetic activity is high, listen to Geophysical Alerts at 18 past the hour on short-wave radio frequencies of 2.5, 5, 10, 15 and 20 megahertz. These broadcasts come from the Space Environment Services Center of the National Oceanic and Atmospheric Administration in Colorado. If the summary of the past twenty-four hours reports high or very high solar activity, auroras may be visible within the next day or two, when the blast of solar wind reaches the earth.

The drawing in Figure 6.4 accompanies the manuscript.

For practice with light, medium, and heavy degrees of edit, imagine these three situations and edit the article appropriately.

1. The manuscript reaches you just before it is to be delivered to the paper. You have twenty minutes to clean it up and do whatever you can to increase its likelihood of being accepted for publication. (Light edit.)
2. The author of the manuscript has asked you to "look at it" for him before he has the processed text and art revised and submits the article to the newspaper. You are working on a major project and are close to the final deadline, but you can take thirty to forty-five minutes to edit the article. (Medium edit.)
3. Your supervisor has brought you the manuscript and suggested that you take two or three hours to edit it, then meet with the author to discuss

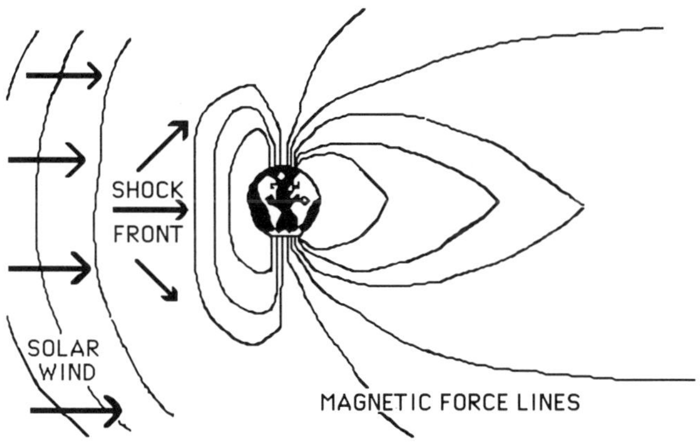

Figure 6.4. The writer's illustration for the aurora article.

your suggestions for improvement. Your supervisor thinks the content is sufficiently accurate, but she suggests that the discussion might be expanded in places and made more appealing to the local newspaper audience. (Heavy edit.)

The version below is the light edit copymarking of the original. The paragraphs have been numbered for reference.

The Dawn That Glows at Midnight

by Elizabeth Babcock, News Special Writer

(1) Residents of southern lower Michigan may have a rare opportunity this month to see nature's most spectacular light show. Scientists are predicting that solar activity will peak this spring, creating favorable conditions for the aurora borealis (northern lights) to appear in local skies, welcoming the season and delighting viewers with spectacular displays of shimmering arcs and rippling curtains of colored light. Auroras change dramatically throughout the night. In the early evening, they often appear as stationery arcs that hover quietly near the horizon in the northern sky. Toward midnight, they may become suddenly active, billowing into enormous radiant curtains that flicker and pulsate with wave of colored light in a fiery display that moves and changes with astonishing speed.

(2) Jon Wooley, professor of astronomy at Eastern Michigan University, said that aurora displays associated with peak sunspot activity could be visible outside their normal range anytime soon.

(3) The aurora boreallis, which means dawn of the north, appear most frequently in an oval shaped zone centered at the magnetic north pole. The zone stretches across northern Canada, Alaska, Greenland, Iceland, the northern coast of Norway and the Arctic coast of Siberia. People who live in Michigan's Upper Peninsula can see auroras about 50 nights a year. Residents of Michigan's Lower Peninsula are one or two degrees of longitude too far south to see the aurora most of the time. However, the aurora zone expands when the sun becomes unusually active, so the lights become visible at lower latitudes.

(4) Although various theories have long linked Earth's light show to the sun, our current understanding of the sun's role in auroras is relatively recent. "The relationship between auroras and sunspots have been understood since the turn of this century," said Richard Teske, a U-M pro-

fessor who specializes in solar astronomy. "But the exact mechanism that produces auroras was not understood until space explorers mapped the earth's magnetic field and detected a stream of charged particles coming in from the sun."

(5) This stream of particles, called the solar wind, generates auroras when it enters our atmosphere and interacts with the Earth's magnetic field. Auroras may become unusually intense during periods of maximum sunspot activity. Sunspots that appear as darker regions on the sun's surface become more numerous when the sun periodically increases its energy output, roughly every 11 years. Astronomers refer to the peak of this cycle as "solar maximum," and they agree that its happening right now. Teske explained, "Solar forecasters have made their best honest estimate that solar maximum will occur this month, but we won't know for sure until the end of the year," when statistical analyses are compiled.

(6) The activities that generate auroras originate in the visible layer of the sun, called the photosphere. Convection currents develop in the core of the sun and rise to the photosphere, where they interact with magnetic fields to produce sunspots, cooler regions of the photosphere that show up as dark blotches against the brighter, hotter background.

(7) As solar maximum approaches and larger, more numerous sunspots develop, the sun's corona becomes brighter and hotter. The corona is the outermost layer of the sun that may be seen during a solar eclipse. When the moon moves across the face of the sun, the corona appears as a halo of light streaming out from behind the moon.

(8) EMU's Woolley explained that the corona "evaporates" to produce the solar wind, a stream of electrified particles pouring out of the sun. When sunspot activity is high, more high energy particles escape from the corona, increasing the velocity of the solar wind.

(9) The photosphere also steps up it's energy output during solar maximum. Wooley said that magnetic fields located near sunspots emit energy which heats up solar gases in those areas and generates solar flares. These explosions on the sun blast electrified particles into space and produce shock waves in the solar wind. If a flare erupts near the center of the sun, the shock wave collides head on with the magnetsphere, the area of space affected by the Earth's magnetic field. The collision compresses the magnetosphere, disturbing magnetic force lines and intensifying magnetic fields. These magnetic disturbances create a geomagnetic storm that can produce highly active auroras over a wide range of latitudes. If a flare erupts near

the edge of the sun, the solar wind's shock wave sideswipes the magnetosphere instead of colliding with it directly. When this occurs, the Earth's magnetic field is not greatly disturbed and auroral effects are minor.

(10) The solar wind, moving at supersonic speeds takes two or three days to travel 93 million miles from the sun to the Earth. When this wind reaches our atmosphere, the Earth's magnetic field traps its electrically charged particles and guides them to the polar auroral zones. At an altitude of about 200 miles above the polar regions, charged particles collide with molecules of atmospheric gases which react by giving off light. The predominant colors of the aurora are pale green (emitted by oxygen molecules) and pink or red (given off by nitrogen), although Matt Link, planetarium coordinator at the University of Michigan's Exhibit Museum, said he once photographed a spectacular aurora that displayed hues of purple, yellow and blue.

(11) The northern lights have inspired myths and legends since the beginning of history. According to one legend, said Matt Link, the world was hollow, and inside the earth there was another sun. Before explorers mapped the polar regions, people believed there were holes in the Earth at the North and South poles. They thought the aurora was caused by light shining through these holes from the sun inside the Earth. Linke said that scientists later developed a quasi-scientific theory that explained the aurora as "light reflected from the polar ice caps." "The Algonquian Indian tribes who lived around Lake Superior thought the northern lights were caused when Nanabozoo stirred his fires," said Althea Helbig, an English professor at EMU who specializes in Native American folklore. Nanabozoo is a heroic "trickster" figure of Algonquan legend.

(12) If you want to see auroras, get away from city lights on a clear night with little or no moonlight. The best time is around midnight, when the aurora typically become more active. To find out if geomagnetic activity is high, listen to Geophysical Alerts at 18 past the hour on short-wave radio frequencies of 2.5, 5, 10, 15 and 20 megahertz. These broadcasts come from the Space Environment Services Center of the National Oceanic and Atmospheric Administration in Colorado. If the summary of the past twenty-four hours reports high or very high solar activity, auroras may be visible within the next day or two, when the blast of solar wind reaches the earth.

In the light edit of the article, the editor has corrected misspellings of people's names (by checking them against a telephone directory) and obvious errors in spelling, punctuation, and grammar. The drawing has been checked but not revised. There is much more that could be done to improve the manuscript, but the changes made here would probably use up all of the editor's allotted twenty minutes.

The version below is the medium edit copymarking. The changes from light editing have already been processed and are underscored for reference.

The Dawn That Glows at Midnight

by Elizabeth Babcock, News Special Writer

(1) Residents of southern lower Michigan may have a rare opportunity this month to see nature's most spectacular light show. Scientists are predicting that solar activity will peak this spring, creating favorable conditions for the aurora borealis (northern lights) to appear in local skies, welcoming the season and delighting viewers with spectacular displays of shimmering arcs and rippling curtains of colored light. Auroras change dramatically throughout the night. In the early evening, they often appear as stationary arcs that hover quietly near the horizon in the northern sky. Toward midnight, they may become suddenly active, billowing into enormous radiant curtains that flicker and pulsate with waves of colored light in a fiery display that moves and changes with astonishing speed.

(2) Jon Wooley, professor of astronomy at Eastern Michigan University, said that auroral displays associated with peak sunspot activity could be visible outside their normal range any time soon.

(3) The aurora borealis, which means dawn of the north, appears most frequently in an oval shaped zone centered at the magnetic north pole. The zone stretches across northern Canada, Alaska, Greenland, Iceland, the northern coast of Norway, and the Arctic coast of Siberia. People who live in Michigan's Upper Peninsula can see auroras about 50 nights a year. Residents of Michigan's lower peninsula are one or two degrees of latitude too far south to see the aurora most of the time. However, the auroral zone expands when the sun becomes unusually active, so the lights become visible at lower latitudes.

(4) Although various theories have long linked Earth's light show to the sun, our current understanding of the sun's role in auroras is relatively recent. "The relationship between auroras and sunspots has been understood since the turn of this century," said Richard Teske, a U-M professor who specializes in solar astronomy. "But the exact mechanism that

produces auroras was not understood until space explorers mapped the Earth's magnetic field and detected a stream of charged particles coming in from the sun."

(5) This stream of particles, called the solar wind, generates auroras when it enters our atmosphere and interacts with the Earth's magnetic field. Auroras may become unusually intense during periods of maximum sunspot activity. Sunspots that appear as darker regions on the sun's surface become more numerous when the sun periodically increases its energy output, roughly every 11 years. Astronomers refer to the peak of this cycle as "solar maximum," and they agree that it's happening right now. Teske explained, "Solar forecasters have made their best honest estimate that solar maximum will occur this month, but we won't know for sure until the end of the year," when statistical analyses are compiled.

(6) The activities that generate auroras originate in the visible layer of the sun, called the photosphere. Convection currents develop in the core of the sun and rise to the photosphere, where they interact with magnetic fields to produce sunspots, cooler regions of the photosphere that show up as dark blotches against the brighter, hotter background.

(7) As solar maximum approaches and larger, more numerous sunspots develop, the sun's corona becomes brighter and hotter. The corona is the outermost layer of the sun that may be seen during a solar eclipse. When the moon moves across the face of the sun, the corona appears as a halo of light streaming out from behind the moon.

(8) EMU's Wooley explained that the corona "evaporates" to produce the solar wind, a stream of electrified particles pouring out of the sun. When sunspot activity is high, more high energy particles escape from the corona, increasing the velocity of the solar wind.

(9) The photosphere also steps up its energy output during solar maximum. Wooley said that magnetic fields located near sunspots emit energy which heats up solar gases in those areas and generates solar flares. These explosions on the sun blast electrified particles into space and produce shock waves in the solar wind. If a flare erupts near the center of the sun, the shock wave collides head on with the magnetosphere, the area of space affected by the Earth's magnetic field. The collision compresses the magnetosphere, disturbing magnetic force lines and intensifying magnetic fields. These magnetic disturbances create a geomagnetic storm that can produce highly active auroras over a wide range of latitudes. If a flare erupts near the edge of the sun, the solar wind's shock wave sideswipes the magneto-

sphere instead of colliding with it directly. When this occurs, the Earth's magnetic field is not greatly disturbed, and auroral effects are minor.

(10) The solar wind, moving at supersonic speeds, takes two or three days to travel 93 million miles from the sun to the Earth. When this wind reaches our atmosphere, the Earth's magnetic field traps its electrically charged particles and guides them to the polar auroral zones. At an altitude of about 200 miles above the polar regions, charged particles collide with molecules of atmospheric gases which react by giving off light. The predominant colors of the aurora are pale green (emitted by oxygen molecules) and pink or red (given off by nitrogen), although Matt Linke, planetarium coordinator at the University of Michigan's Exhibit Museum, said he once photographed a spectacular aurora that displayed hues of purple, yellow and blue.

(11) The northern lights have inspired myths and legends since the beginning of history. According to one legend, said Linke, the world was hollow, and inside the Earth there was another sun. Before explorers mapped the polar regions, people believed there were holes in the Earth at the North and South poles. They thought the aurora was caused by light shining through these holes from the sun inside the Earth. Linke said that scientists later developed a quasi-scientific theory that explained the aurora as "light reflected from the polar ice caps." "The Algonquian Indian tribes who lived around Lake Superior thought the northern lights were caused when Nanabozoo stirred his fires," said Alethea Helbig, an English professor at EMU who specializes in Native American folklore. Nanabozoo is a heroic "trickster" figure of Algonquian legend.

(12) If you want to see auroras, get away from city lights on a clear night with little or no moonlight. The best time is around midnight, when the aurora typically becomes more active. To find out if geomagnetic activity is high, listen to Geophysical Alerts at 18 past the hour on short-wave radio frequencies of 2.5, 5, 10, 15 and 20 megahertz. These broadcasts come from the Space Environment Services Center of the National Oceanic and Atmospheric Administration in Colorado. If the summary of the past twenty-four hours reports high or very high solar activity, auroras may be visible within the next day or two, when the blast of solar wind reaches the earth.

In the medium edit, the editor has made changes that go beyond correction of obvious errors. A number of paragraphs have been divided to create shorter, easier to read blocks. An analogy has been added to explain more clearly why charged particles of the magnetosphere cause atmospheric gases to glow. Nonessential clauses have been set off with commas and introduced by "which" rather than "that." As with the light edit, the drawing that accompanies the manuscript would merely be passed along with the text as a concept for the artist to work from.

The version below is the heavy edit copymarking. The changes from the medium edit have already been processed and are now underscored for reference.

Add subtitle: The northern lights may make a rare appearance in local skies this month

The Dawn That Glows at Midnight

by Elizabeth Babcock, News Special Writer

(1) Residents of southern lower Michigan may have a rare opportunity this month to see nature's most spectacular light show. Scientists are predicting that solar activity will peak this spring, creating favorable conditions for the aurora borealis (northern lights) to appear in local skies, welcoming the season and delighting viewers with spectacular displays of shimmering arcs and rippling curtains of colored light. ¶ Auroras change dramatically throughout the night. In the early evening, they often appear as stationary arcs that hover quietly near the horizon in the northern sky. Toward midnight, they may become suddenly active, billowing into enormous radiant curtains that flicker and pulsate with waves of colored light in a fiery display that moves and changes with astonishing speed.

move ¶ to follow ¶ 15 as ¶ 16

Add here old ¶s 16 and 17, with ¶ 17 broken as marked

(2) Jon Wooley, professor of astronomy at Eastern Michigan University, said that auroral displays associated with peak sunspot activity could be visible outside their normal range any time soon.

(3) The aurora borealis, which means "dawn of the north," appears most frequently in an oval shaped zone centered at the magnetic north pole. The zone stretches across northern Canada,

Alaska, Greenland, Iceland, the northern coast of Norway, and the Arctic coast of Siberia.

(4) People who live in Michigan's Upper Peninsula can see auroras about 50 nights a year. Residents of Michigan's lower peninsula are one or two degrees of latitude too far south to see the aurora most of the time. However, the auroral zone expands when the sun becomes unusually active, so the lights become visible at lower latitudes.

(5) Although various theories have long linked Earth's light show to the sun, our current understanding of the sun's role in auroras is relatively recent. "The relationship between auroras and sunspots has been understood since the turn of this century," said Richard Teske, a U-M professor who specializes in solar astronomy. "But the exact mechanism that produces auroras was not understood until space explorers mapped the Earth's magnetic field and detected a stream of charged particles coming in from the sun."

(6) This stream of particles, called the solar wind, generates auroras when it enters our atmosphere and interacts with the Earth's magnetic field. Auroras may become unusually intense during periods of maximum sunspot activity. Sunspots, which appear as darker regions on the sun's surface, become more numerous when the sun periodically increases its energy output, roughly every 11 years. Astronomers refer to the peak of this cycle as "solar maximum," and they agree that it's happening right now.

(7) Teske explained, "Solar forecasters have made their best honest estimate that solar maximum will occur this month, but we won't know for sure until the end of the year," when statistical analyses are compiled.

(8) The activities that generate auroras originate in the visible layer of the sun, called the photosphere. Convection currents develop in the core

of the sun and rise to the photosphere, where they interact with magnetic fields to produce sunspots, cooler regions of the photosphere that show up as dark blotches against the brighter, hotter background.

(9) As solar maximum approaches and larger, more numerous sunspots develop, the sun's corona becomes brighter and hotter. (The corona is the outermost layer of the sun that may be seen during a solar eclipse. When the moon moves across the face of the sun, the corona appears as a halo of light streaming out from behind the moon.)

(10) EMU's Wooley explained that the corona "evaporates" to produce the solar wind, a stream of electrified particles pouring out of the sun. When sunspot activity is high, more high-energy particles escape from the corona, increasing the velocity of the solar wind.

(11) The photosphere also steps up its energy output during solar maximum. Wooley said that magnetic fields located near sunspots emit energy, which heats up solar gases in those areas and generates solar flares. These explosions on the sun blast electrified particles into space and produce shock waves in the solar wind.

(12) If a flare erupts near the center of the sun, the shock wave collides head-on with the magnetosphere, the area of space affected by the Earth's magnetic field. The collision compresses the magnetosphere, disturbing magnetic force lines and intensifying magnetic fields. These magnetic disturbances create a geomagnetic storm, which can produce highly active auroras over a wide range of latitudes.

(13) If a flare erupts near the edge of the sun, the solar wind's shock wave "sideswipes" the magnetosphere instead of colliding with it directly. When this occurs, the Earth's magnetic field

is not greatly disturbed, and auroral effects are minor.

(14) The solar wind, moving at supersonic speeds, takes two or three days to travel 93 million miles from the sun to the Earth. When this wind reaches our atmosphere, the Earth's magnetic field traps its electrically charged particles and guides them to the polar auroral zones.

<u>(15)</u> At an altitude of about 200 miles above the polar regions, charged particles collide with molecules of atmospheric gases which react by giving off light<u>, much in the same way that an electrical current causes gas in a neon sign to glow.</u> The predominant colors of the aurora are pale green (emitted by oxygen molecules) and pink or red (given off by nitrogen), although Matt Linke, planetarium coordinator at the University of Michigan's Exhibit Museum, said he once photographed a spectacular aurora that displayed hues of purple, yellow and blue.

(16) The northern lights have inspired myths and legends since the beginning of history. According to one legend, said Linke, the world was hollow, and inside the Earth there was another sun. Before explorers mapped the polar regions, people believed there were holes in the Earth at the North and South poles. They thought the aurora was caused by light shining through these holes from the sun inside the Earth.

<u>(17)</u> Linke said that scientists later developed a quasi-scientific theory that explained the aurora as "light reflected from the polar ice caps."

"The Algonquian Indian tribes who lived around Lake Superior thought the northern lights were caused when Nanabozoo stirred his fires," said Alethea Helbig, an English professor at EMU who specializes in Native American folklore. Nanabozoo is a heroic "trickster" figure of Algonquian legend.

Add ¶ on Southern Hem. auroras (attached)

Set ¶ 18 with additions in a boxed sidebar with new title: If you want to see auroras

(18) ~~If you want to see auroras,~~ get away from city lights on a clear night with little or no moonlight. The best time is around midnight, when the aurora typically becomes more active. To find out if geomagnetic activity is high, listen to Geophysical Alerts at 18 past the hour on short-wave radio frequencies of 2.5, 5, 10, 15 and 20 megahertz. These broadcasts come from the Space Environment Services Center of the National Oceanic and Atmospheric Administration (NOAA) in Boulder, Colorado. If the summary of the past 24 hours reports "high" or "very high" solar activity, auroras may be visible within the next day or two, when the blast of solar wind reaches the earth.

The "Geodent" is also available as a recorded phone message at 303-497-3235. Updates are given every three hours in Universal Time, which is five hours ahead of Eastern Standard Time.

Add 2 ¶s attached ("Here's how" & "If the") as separate ¶s

¶ If the forecast for the next 24 hours predicts a geomagnetic storm, it's time to go out and watch the sky.

On a separate sheet attached to the manuscript the editor should type the paragraphs to be added.

(Paragraph on Southern Hemisphere auroras, to be inserted after Paragraph 17):

The Southern Hemisphere also has an aurora, called the aurora australis. When the northern and southern auroras are photographed at the same time from aircraft at similar positions in both hemispheres, the photos look like mirror images of one another. Scientists believe that charged particles from the solar wind zoom rapidly back and forth between the north and south poles along magnetic field lines, producing twin auroras simultaneously.

(Paragraphs to be added to sidebar):

Here's how to interpret the Geoalerts: the broadcasts or messages report two indexes that scientists use to measure geomagnetic activity. Auroras may be seen when geomagnetic activity is high.

If the Boulder A index is 40 or higher (on a scale of 0-400), and the Boulder K index is greater than 5 (on a scale of 0-9), that means auroras are likely to be visible much farther south than normally.

Also, in the heavy edit the editor should consider the graphic submitted by the author, writing suggestions to the artist about size, shading, and typeface and sizes. The editor should determine whether other graphics might also be used in the article. (In the *Ann Arbor News,* a large color photograph of an aurora ran all the way across the page, above the article.)

In the heavy edit, the editor has made macro-edit changes in organization and content. Paragraphs have been moved from the end of the article to near the beginning, to increase the human interest of the piece for lay readers. Information has been added to the sidebar discussion to make it easier for readers to learn about conditions that might produce auroras. Formatting changes include the addition of the subtitle, designed to capture the local audience's interest, and the creation of a boxed sidebar to add variety to the presentation.

Micro-edit changes have been made in the heavy edit, although most of the needed changes were done in the medium edit. The editor has checked proper names and corrected the spelling of the Native American name. Quotation marks have been added to the sidebar discussion to indicate these are NOAA terms. Agency acronyms are added to familiarize readers with them, even though they are not used in the article. Remember, if a heavy edit were undertaken for the document, all of the light and medium edit changes would be made too.

The version below is the final edited version of the article, which appeared in the *Ann Arbor News.*

The Dawn That Glows at Midnight

The northern lights may make a rare appearance in local skies this month

by Elizabeth Babcock, News Special Writer

Residents of southern lower Michigan may have a rare opportunity this month to see nature's most spectacular light show. Scientists are predicting that solar activity will peak this spring, creating favorable conditions for the aurora borealis (northern lights) to appear in local skies, welcoming the season and delighting viewers with spectacular displays of shimmering arcs and rippling curtains of colored light.

The northern lights have inspired myths and legends since the beginning of history. According to one legend, said Matt Linke, planetarium coordinator at the University of Michigan's Exhibit Museum, the world was hollow, and inside the Earth there was another sun. Before explorers mapped the polar regions, people believed there were holes in the Earth at the North and South poles. They thought the aurora was caused by light shining through these holes from the sun inside the Earth.

"The Algonquian Indian tribes who lived around Lake Superior thought the northern lights were caused when Nanabozhoo stirred his fires," said Alethea Helbig, an English professor at EMU who specializes in Native American folklore. Nanabozhoo is a heroic "trickster" figure of Algonquian legend.

Linke said that scientists later developed a quasi-scientific theory that explained the aurora as "light reflected from the polar ice caps."

Although various theories have long linked Earth's light show to the sun, our current understanding of the sun's role in auroras is relatively recent. "The relationship between auroras and sunspots has been understood since the turn of this century," said Richard Teske, a U-M professor who specializes in solar astronomy. "But the exact mechanism that produces auroras was not understood until space explorers mapped the Earth's magnetic field and detected a stream of charged particles coming in from the sun."

This stream of particles, called the solar wind, generates auroras when it enters our atmosphere and interacts with the Earth's magnetic field. Auroras may become unusually intense during periods of maximum sunspot activity. Sunspots, which appear as darker regions on the sun's surface, become more numerous when the sun periodically increases its energy output, roughly every 11 years. Astronomers refer to the peak of this cycle as "solar maximum," and they agree that it's happening right now.

Teske explained, "Solar forecasters have made their best honest estimate that solar maximum will occur this month, but we won't know for sure until the end of the year," when statistical analyses are compiled.

Jon Wooley, professor of astronomy at Eastern Michigan University, said that auroral displays associated with peak sunspot activity could be visible outside their normal range any time soon.

The aurora borealis, which means "dawn of the north," appears most frequently in an oval shaped zone centered at the magnetic north pole. The zone stretches across northern Canada, Alaska, Greenland, Iceland, the northern coast of Norway, and the Arctic coast of Siberia.

People who live in Michigan's Upper Peninsula can see auroras about 50 nights a year. Residents of Michigan's lower peninsula are one or two degrees of latitude too far south to see the aurora most of the time. However, the auroral zone expands when the sun becomes unusually active, so the lights become visible at lower latitudes.

The activities that generate auroras originate in the visible layer of the sun, called the photosphere. Convection currents develop in the core of the sun and rise to the photosphere, where they interact with magnetic fields to produce sunspots. These are cooler regions of the photosphere that show up as dark blotches against the brighter, hotter background.

As solar maximum approaches and larger, more numerous sunspots develop, the sun's corona becomes brighter and hotter. (The corona is the outermost layer of the sun that may be seen during a solar eclipse. When the moon moves across the face of the sun, the corona appears as a halo of light streaming out from behind the moon.)

EMU's Wooley explained that the corona "evaporates" to produce the solar wind, a stream of electrified particles pouring out of the sun. When sunspot activity is high, more high-energy particles escape from the corona, increasing the velocity of the solar wind.

The photosphere also steps up its energy output during solar maximum. Wooley said that magnetic fields located near sunspots emit energy, which heats up solar gases in those areas and generates solar flares. These explosions on the sun blast electrified particles into space and produce shock waves in the solar wind.

If a flare erupts near the center of the sun, the shock wave collides head-on with the magnetosphere, the area of space affected by the Earth's magnetic field. The collision compresses the magnetosphere, disturbing magnetic force lines and intensifying magnetic fields. These magnetic disturbances create a geomagnetic storm, which can produce highly active auroras over a wide range of latitudes.

If a flare erupts near the edge of the sun, the solar wind's shock wave "sideswipes" the magnetosphere instead of colliding with it directly. When this occurs, the Earth's magnetic field is not greatly disturbed, and auroral effects are minor.

The solar wind, moving at supersonic speeds, takes two or three days to travel 93 million miles from the sun to the Earth. When this wind reaches our atmosphere, the Earth's magnetic field traps its electrically charged particles and guides them to the polar auroral zones.

At an altitude of about 200 miles above the polar regions, charged particles collide with molecules of atmospheric gases which react by giving off light, much in the same way that an electrical current causes gas in a neon sign to glow. The predominant colors of the aurora are pale green (emitted by oxygen molecules) and pink or red (given off by nitrogen), although Matt Linke said he once photographed a spectacular aurora that displayed hues of purple, yellow and blue.

Auroras change dramatically throughout the night. In the early evening, they often appear as stationary arcs that hover quietly near the horizon in the northern sky. Toward midnight, they may become suddenly active, billowing into enormous radiant curtains that flicker and pulsate with waves of colored light in a fiery display that moves and changes with astonishing speed.

The Southern Hemisphere also has an aurora, called the aurora australis. When the northern and southern auroras are photographed at the same time from aircraft at similar positions in both hemispheres, the photos look like mirror images of one another. Scientists believe that charged particles from the solar wind zoom rapidly back and forth between the north and south poles along magnetic field lines, producing twin auroras simultaneously.

If you want to see auroras

Get away from city lights on a clear night with little or no moonlight. The best time is around midnight, when the aurora typically becomes more active.

Listen to "Geophysical Alerts" at 18 past the hour on short-wave radio frequencies of 2.5, 5, 10, 15 and 20 megahertz. These broadcasts come from the Space Environment Services Center (SESC) of the National Oceanic and Atmospheric Administration (NOAA) in Boulder, Colorado. The "Geoalert" is also available as a recorded phone message at 303-497-3235. Updates are given every three hours in Universal Time, which is five hours ahead of Eastern Standard Time.

Here's how to interpret the Geoalerts: the broadcasts or messages report two indexes that scientists use to measure geomagnetic activity. Auroras may be seen when geomagnetic activity is high.

If the Boulder A index is 40 or higher (on a scale of 0–400), and the Boulder K index is greater than 5 (on a scale of 0–9), that means auroras are likely to be visible much farther south than normally.

If the summary of the past 24 hours reports "high" or "very high" solar activity, auroras may be visible within the next day or two, when the blast of solar wind reaches the earth.

If the forecast for the next 24 hours predicts a geomagnetic storm, it's time to go out and watch the sky.

Editing to different degrees does not mean that an editor is doing only part of his or her job or not doing the job thoroughly. Rather, it means that the appropriate level of editorial effort is being used on a document at a particular stage in its production. Editors should try to help clients unfamiliar with the concept of the degrees of edit understand this. Also, editors must realize that all clients will expect a document that has received any degree of editing to be free of spelling errors, typos, and obvious format problems. The best way to ensure that is to include a correctness edit in any editing of a document, whatever degree of edit the document is to receive.

Source: *The Ann Arbor News,* Vol. 155, No. 74 (March 15, 1990), pages D1–D2. Copyright, Elizabeth Babcock, 1990; used with permission.

CHAPTER SEVEN

Style Guides

Good editors rely on style guides the way physicians rely on thermometers and x-rays and mechanics rely on tools. Style guides contain editing directions for text, graphics, formatting, and other aspects of technical documents.

Style guides (also called *standards*) do not constrain editors' ability to adapt a document for a particular audience or purpose. Instead, they promote consistency across the documents a company or agency produces, whether project reports, proposals, or manuals, and across articles in a technical journal. Using style guides simplifies editing considerably. For beginning editors, style guides can serve as important authorities to support editorial decisions, reducing the need for "judgment calls" that are sometimes difficult to make and defend.

This discussion of style guides is organized by the following categories:

- Company style manuals for documents produced by a particular company
- Field-specific manuals, such as the Council of Biology Editors' *Style Manual, Mathematics into Type,* the *Publication Manual of the American Psychological Association,* and the American Chemical Society *Handbook for Authors*
- Government style guides, such as the U.S. Government Printing Office *Style Manual* and the once widely used Department of Defense *Military Standard 847B*
- General style guides, such as *The Chicago Manual of Style, Words into Type,* and the discussion of effective writing in textbooks
- Instructions to authors of articles for technical journals, such as those in *Technical Communication, IEEE Transactions on Professional Communication, Nature,* the *Journal of the American Medical Association,* and many others

This text could in no way begin to summarize the valuable guidelines in these guides. Rather, it focuses on the *type* of information included in them, providing selections to prepare readers to explore them on their own.

COMPANY STYLE GUIDES

Most large companies that create documents have one or more specified style guides that help writers, editors, typists, artists, proofreaders, and other publications department staff achieve consistency, especially in long documents that are written or edited and proofread by several people. They save time for writers, editors, artists, and typists by providing guidelines at the outset for formatting, capitalization, abbreviations, punctuation, line thicknesses in graphics, and other aspects.

Style guides often establish the quality levels in documents a company produces, as do the Jet Propulsion Laboratory's *The Levels of Edit* (see Figure 6.1 and the table in Figure 7.1). Such definitions can be expanded to identify the tasks and costs involved, as in Figure 7.2.

Whether or not a style guide discusses the types of documents the company produces, specific guidelines are usually presented about text, graphics, and production activities. Figure 5.2 illustrated sample graphics standards. Figure 7.3 presents a sample from an important section of any style guide, usage of abbreviations and acronyms.

These samples suggest the types of guidelines to be found in company style guides. To gain a full sense of what such standards contain, however, editors should examine the complete guides. Given the cost of their production and the competitive nature of business, most companies' legal and public relations staff are reluctant to allow their style guides to be examined by the public. However, publications department staff may be willing to show other technical communicators their style guides, often at conferences or meetings of organizations such as the Society for Technical Communication. Editors should try to obtain copies of different guides to help solve tough editing problems, to show writers different examples of how text or graphics

Level	Audience	Objective
1—Best quality	Large external audiences Very large internal audiences Senior executive customers or partners	Secure decisions and action critical to achieving major strategic company milestones Maintain strong image with important publics
2—Good quality	External audiences Large internal audiences Executive customers or partners Senior Martin Marietta executives	Secure decisions and action necessary to achieving intermediate or secondary milestones
3—Routine or draft quality	Internal audiences	Secure routine internal decisions and action Rehearse Level 1 or 2 visual aids prior to final graphic production Hold draft review of Level 1 or 2 documents prior to final publication

Figure 7.1. Document quality levels. (Source: Martin Marietta Corporation, *Communication Guide and Standards,* p. 1–2. Martin Marietta Corporation, © 1988, used with permission.)

Level	Editorial or Graphic Function	Standard	Cost Level
1—Best quality	Edit	Do organization and content edit, syntax edit, simple edit, format check, copy check, and proofread by Presentations department (see Appendix C)	Highest
	Type	Use typeset or equivalent serif-style typeface for body copy; serif or sans serif for tables	
	Art	Have illustrations prepared by Presentations graphic professionals to Level 1 (see Tables 5.1-1 and 5.1-2)	
	Cover	Use special or standard cover	
	Color	Use all color, screens, or color highlights as appropriate	
	Photos	Use best quality photos	
2—Good quality	Edit	Do simple edit, format check, copy check, and proofread; syntax edit, organization and content edit by Presentations is recommended (see Appendix C)	Moderate
	Type	Use letter-quality serif-style typeface for body copy, serif or sans serif for tables	
	Art	Prepare simplified engineering or computer drawings to Level 2 (see Tables 5.1-1 and 5.1-2)	
	Cover	Use standard cover	
	Color	Use no screens or color	
	Photos	Use good quality or better photos	
3—Routine or draft quality	Edit	Do format check, copy check, and proofread; simple edit is recommended (see Appendix C)	Lowest
	Type	Dot matrix quality type acceptable	
	Art	Prepare engineering drawings or rough sketches to Level 3 (see Table 5.1-1)	
	Cover	Use standard or no cover	
	Color	Use no screens or color	
	Photos	Use any reproducible photographic image	

Figure 7.2 Quality level tasks and costs. (Source: Martin Marietta Corporation, *Communication Guide and Standards,* p. 3–2. Martin Marietta Corporation, © 1988, used with permission.)

might be handled, or to develop or revise a style guide for their own organization.

FIELD-SPECIFIC STYLE GUIDES

Field-specific style guides are usually prepared by a leading professional organization. For example, the American Mathematical Society publishes two for writers and editors of mathematical material: *A Manual for Authors of*

foot-pound	ft-lb
foot-pound-second (system)	fps
free on board	f.o.b.
freezing point	fp
frequency	
high	hf
low	lf
medium	mf
superhigh	shf
ultrahigh	uhf
very high	vhf
very low	vlf
video	vdf
frequency modulation	FM
full width at half maximum	FWHM
functions (trigonometric)	
cosecant	csc
cosine	cos
cotangent	cot
secant	sec
sine	sin
tangent	tan
coversed sine	covers
versed sine	vers
functions (hyperbolic)	
cosecant	csch
cosine	cosh
cotangent	coth
secant	sech
sine	sinh
tangent	tanh
gallon	gal
gallons per hour	gal/h

Figure 7.3. Sample usage guidelines. (Source: Martin Marietta Energy Systems, *Document Preparation Guide,* p. 6–69. Martin Marietta Energy Systems, © 1989, used with permission.)

Mathematical Papers and Ellen Swanson's *Mathematics into Type*. A sample discussion from Swanson's guide (*the* editorial standard for mathematical material) appears in Figure 7.4. Swanson also provides an appendix with a proof-marked passage and its corrected form (Figures 7.5 and 7.6).

2.4.1 Fractions.

2.4.1a *Stacked fractions* (fractions of the type $\frac{a}{bx}$). Some publishers do not allow these in text because they require spreading of lines in the printed copy, i.e. they cannot be set without adding extra space between lines.

The stacking of fractions may be avoided in at least two different ways. The first is by use of a solidus and the second by introducing a negative exponent. These two methods are shown below; all three of the expressions have exactly the same meaning.

$$\frac{a}{b} \qquad a/b \qquad ab^{-1}$$

Substituting a solidus or changing to negative exponents are operations that should be performed only by a copy editor with a knowledge of the operations involved. A rule of thumb that should be used by a copy editor is: "if in doubt, either don't do it or ask someone who knows." Below are examples showing correct and incorrect ways of changing stacked fractions.

REPLACE	BY	OR	NOT BY
$\frac{a+1}{b}$	$(a+1)/b$	$(a+1)b^{-1}$	$a+1/b$
$\frac{a}{(x+1)^3}$	$a/(x+1)^3$	$a(x+1)^{-3}$	
$\sin \frac{a}{x}$	$\sin(a/x)$		$\sin a/x$
$\frac{\sin a}{x}$	$(\sin a)/x$	$x^{-1} \sin a$	$\sin a/x$
$\frac{a}{x^3}$	a/x^3	ax^{-3}	
$\frac{a}{(b-c)^2}$	$a/(b-c)^2$	$a(b-c)^{-2}$	
$\frac{\partial}{\partial\theta} F(u, k, \theta)$	$\partial F(u, k, \theta)/\partial\theta$	$(\partial/\partial\theta)F(u, k, \theta)$	$\partial/\partial\theta F(u, k, \theta)$
$\frac{\partial}{\partial x}\rho$	$\partial\rho/\partial x$		

Figure 7.4. Sample Guidelines for editing fractions. (Source: Swanson, Ellen, *Mathematics into Type* Revised edition, pp. 14–15. American Mathematical Society, © 1979, used with permission.)

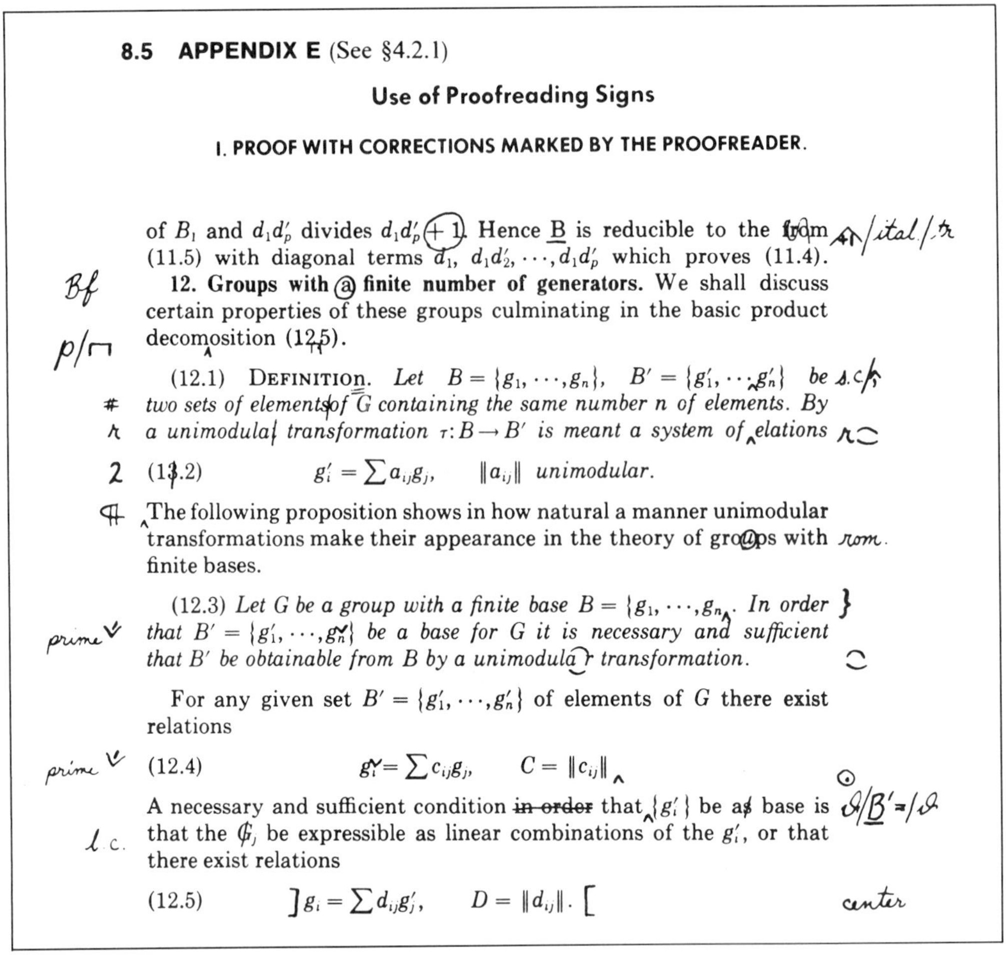

8.5 APPENDIX E (See §4.2.1)

Use of Proofreading Signs

I. PROOF WITH CORRECTIONS MARKED BY THE PROOFREADER.

of B_1 and $d_1d'_p$ divides $d_1d'_p + 1$. Hence B is reducible to the from (11.5) with diagonal terms $d_1, d_1d'_2, \cdots, d_1d'_p$ which proves (11.4).

12. Groups with a finite number of generators. We shall discuss certain properties of these groups culminating in the basic product decomosition (12.5).

(12.1) DEFINITION. *Let* $B = \{g_1, \cdots, g_n\}$, $B' = \{g'_1, \cdots, g'_n\}$ *be two sets of elementsof G containing the same number n of elements. By a unimodulal transformation* $\tau: B \rightarrow B'$ *is meant a system of elations*

(13.2) $\qquad g'_i = \sum a_{ij} g_j, \qquad \|a_{ij}\|$ *unimodular.*

The following proposition shows in how natural a manner unimodular transformations make their appearance in the theory of groups with finite bases.

(12.3) *Let G be a group with a finite base* $B = \{g_1, \cdots, g_n$. *In order that* $B' = \{g'_1, \cdots, g_n\}$ *be a base for G it is necessary and sufficient that B' be obtainable from B by a unimodula r transformation.*

For any given set $B' = \{g'_1, \cdots, g'_n\}$ of elements of G there exist relations

(12.4) $\qquad g_i = \sum c_{ij} g_j, \qquad C = \|c_{ij}\|$

A necessary and sufficient condition ~~in order~~ that $\{g'_i\}$ be as base is that the G_j be expressible as linear combinations of the g'_i, or that there exist relations

(12.5) $\qquad]\, g_i = \sum d_{ij} g'_j, \qquad D = \|d_{ij}\|.\, [$

Figure 7.5. A proofmarked passage containing mathematics. (Source: Swanson, Ellen, *Mathematics into Type* Revised edition, p. 70. American Mathematical Society, © 1979, used with permission.)

Field-specific style guides are, in a sense, applications of general standards to the specific documents and subject matter in that particular field. See the bibliography for a list of field-specific guides for mathematics, physics, chemistry, medicine, and others.

GOVERNMENT STYLE GUIDES

Many United States government agencies have style guides for authors of reports generated by those agencies. The most important government style guide, however, is the United States Government Printing Office *Style Manual.* It presents standards for all documents to be printed by the Government Printing Office, serving as "a standardization device designed to achieve uniform word and type treatment, and aiming for economy of word use" (Preface, vii).

II. THE PRECEDING PASSAGE PRINTED WITH ALL CORRECTIONS MADE.

of B_1 and d_1d_p' divides d_1d_{p+1}'. Hence B is reducible to the form (11.5) with diagonal terms $d_1, d_1d_2', \cdots, d_1d_p'$ which proves (11.4).

12. **Groups with a finite number of generators.** We shall discuss certain properties of these groups culminating in the basic product decomposition (12.5).

(12.1) Definition. *Let* $B=\{g_1, \cdots, g_n\}, B'=\{g_1', \cdots, g_n'\}$ *be two sets of elements of* G *containing the same number* n *of elements. By a unimodular transformation* $\tau: B \rightarrow B'$ *is meant a system of relations*

$$(12.2) \qquad g_i' = \sum a_{ij}g_j, \qquad \|a_{ij}\| \text{ unimodular.}$$

The following proposition shows in how natural a manner unimodular transformations make their appearance in the theory of groups with finite bases.

(12.3) *Let* G *be a group with a finite base* $B=\{g_1, \cdots, g_n\}$. *In order that* $B'=\{g_1', \cdots, g_n'\}$ *be a base for* G *it is necessary and sufficient that* B' *be obtainable from* B *by a unimodular transformation.*

For any given set $B'=\{g_1', \cdots, g_n'\}$ of elements of G there exist relations.

$$(12.4) \qquad g_i' = \sum c_{ij}g_j, \qquad C = \|c_{ij}\|.$$

A necessary and sufficient condition that $B'=\{g_i'\}$ be a base is that the g_j be expressible as linear combinations of the g_i', or t exist relations

$$(12.5) \qquad g_i = \sum d_{ij}g_i', \qquad D = \|d_{ij}\|.$$

Figure 7.6. The corrected passage. (Source: Swanson, Ellen, *Mathematics into Type* Revised edition, p. 71. American Mathematical Society, © 1979, used with permission.)

Section 7, "Guide to Compounding," presents standards for hyphenation of compound terms. (A sample from those guidelines is presented in Figure 3.3.) Although individual items in the GPO *Manual* might be treated differently in some company style guides, section 7 could serve as a base for developing a company style manual's guidelines on compounding.

Section 13, "Tabular Work," is a valuable guide to nomenclature and standards for tables. It should be examined carefully by anyone setting out to work with formal tables for technical documents.

Section 24, "Foreign Languages," is one of the best guides available on the typography of other languages, and the guidelines in that section nicely supplement style guides that are weak on foreign languages.

Other U.S. government style guides are available from particular agencies and installations. The Department of Defense has published standards for use by agencies and contractors such as MIL-STD-847B, a specimen from which appears in Figure 7.7.

While a large government agency may require use of a general standard, branches of that agency may develop guides for writers and editors at a particular installation. Agencies occasionally establish guidelines for a particular purpose, such as proposals for a government contract. Figure 7.8 presents selections from a request for proposal for a Department of Defense contract.

Guidelines can be found in other U.S. government documents as well.

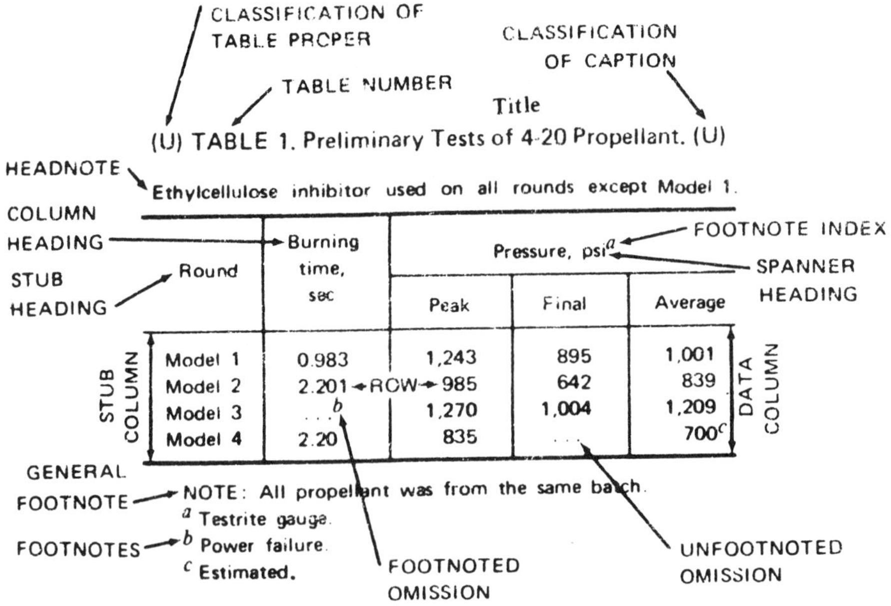

Figure 7.7. Table nomenclature. (Source: U.S. Department of Defense, Military Standard 847B, p. 16.)

17. PREPARATION OF PROPOSALS

a. General Instructions

(1) To aid in the evaluation of competitive proposals, it is desired that all proposals follow the same general format. Therefore, your proposal shall be submitted in three (3) volumes, quantity and page limitations as follows:

Volume 1 - Technical Proposal 8 copies 100 pages

Volume 2 - Management/Personnel 8 copies 50 pages

Volume 3 - Cost/Price Proposal 8 copies 50 pages

. . . .

(2) Pages containing text shall be double-spaced, typewritten, on 8 1/2 x 11 or 8 x 10 1/2 inch paper, in print providing not more than twelve (12) characters to the linear inch, with adequate margins on each side (at least one inch). A page printed on both sides shall count as two pages. A maximum of five foldout pages, no larger than 11 or 10 1/2 x 17 inches, will be permitted. Proposals shall not be supplemented by a package of reference documents.

. . . .

b. Volume 1 - Technical Proposal

As your technical proposal will primarily describe the capability of your organization to participate in this program, it should be specific and complete in every detail. The proposal should be prepared simply and economically, providing straightforward, concise delineation of capabilities to perform satisfactorily the contract being sought. The proposal shall not merely offer to conduct an investigation or perform work in accordance with the Statement of Work, but shall outline the actual investigation or method proposed as specifically as possible. Repeating the Statement of Work without sufficient elaboration will not be acceptable. The Statement of Work reflects the problems and objectives of the program under consideration and, on occasion, some of the possible approaches to the problem as recognized by this agency. Unless otherwise specified, the prospective contractor is not limited to the suggested approaches for equal or even preferred consideration; however, any deviations must be fully substantiated in the proposal. Offerors shall cross-index their proposal elements to the evaluation

Figure 7.8. Sample proposal preparation instructions. (Source: U.S. Department of Defense, *Request for Proposal F08635-84-R-0132,* pp. 2–3.)

criteria listed in Section M. It is the offeror's responsibility to ensure the completeness of his technical proposal; therefore, the technical panel for the Government has been instructed to evaluate on the basis of the information provided in the technical proposal. The Government shall not assume that an offeror possesses any capability unless specified in the proposal.

. . . .

c. Volume 2 - Management/Personnel

This volume shall contain at the minimum:

(1) (a) This section shall identify specific personnel to be assigned for direct work on the program and as direct technical supervisors, plus:

1. Education, background, accomplishments and other pertinent information concerning personnel specified.

2. Estimated personhours each individual will be used on this program (No costs of these personhours are to be included in the technical volume).

(b) Also identify specific additional engineering personnel required for full employment, subcontract, or consultation with source from which they will be available for work on this contract as specified in the proposal. Alternate personnel sources should be listed if assurance of availability cannot be stated. Include full resume of all additional personnel listed.

. . . .

(d). Volume 3 - Cost/Price Proposal

(1) FAR 15.804-6 requires submission of cost or pricing data on Standard Form 1411 in response to any RFP for materials or services whenever cost analysis is involved. It is mandatory that the form be completed in its entirety and without modification except that, with the approval of the Contracting Officer, the Contractor may elect to utilize a different cost summary format. Regardless of the cost/price format utilized, it is necessary that the offeror submit a signed Standard Form 1411 and that he answer the questions on the form and faithfully comply with all instructions and footnotes contained in FAR Part 15, Table 15-3.

Figure 7.8. *(continued)*

Government style guides are designed to facilitate the work of government employees, thereby reducing the cost for publishing or evaluating documents. In many cases, the guides must be followed strictly. For example, proposals prepared to the wrong format or violating page limits may be disqualified from the contract competition.

GENERAL STYLE GUIDES

General guidelines on preparing documents for publication are often used by technical editors as well as by editors in other areas of publication in business, government, and academia. Some guides, such as Turabian's *A Manual for Writers of Term Papers, Theses, and Dissertations,* are obviously more appropriate for college papers than professional documents. Others, such as the *Modern Language Association Handbook,* are more appropriate for nontechnical documents. But two general guides, *The Chicago Manual of Style* and *Words into Type,* are very useful for technical editors.

The Chicago Manual of Style is one of the most widely used guides for writers, editors, and other publication personnel. It developed from one page of style guidelines originated before 1900 for proofreaders, printers, and other employees of the University of Chicago Press and has grown to more than 700 pages in the thirteenth edition (published in 1982). Editing students should examine *The Chicago Manual of Style (Chicago)* firsthand to become familiar with its organization, content, and paragraph numbering scheme, which facilitates reference from the thorough index. Also helpful are the detailed chapter tables of contents.

Although *Chicago* is a valuable tool for technical editors, it underemphasizes the editor's collaborative role in creating a document. Examining the section "The Editorial Function" (pp. 50–52) reveals that *Chicago* considers the editor's role as more of a typemarking function; it suggests that "works that are to be scanned from a clean, well-prepared manuscript and works that are submitted on floppy disks or tape should require minimal attention from an editor" (p. 50). As this textbook has suggested, the apparent cleanliness of a manuscript or its physical format may suggest little about how much editorial effort might be required. In the "Editorial Function" section, the authors examine "mechanical" and "substantive" editing. The former addresses spelling, capitalization, grammar, punctuation, etc.; the latter, "rewriting, reorganizing, or suggesting other ways to present the material" (p. 51). The authors rightly observe that manuscripts vary greatly in how much substantive editing they need, but they state that "no rules can be devised for the editor to follow" (p. 51). This text has attempted to provide guidelines (if not rules) that editors should follow in medium and heavy ("substantive") edits of technical manuscripts (see Chapters 4, 5).

The authors of *Chicago* define copyediting as copymarking, and although they offer good suggestions and examples, the discussions of editing text and graphics are very brief (three pages and one page, respectively). The discussion of working with an author (over four pages) treats its subject more fully. In general, *Chicago* treats all of the subjects a technical editor would expect, but given its concept of the editor's role, many topics may not be sufficiently developed for technical work (as the *Chicago* authors readily admit). In their chapter "Mathematics in Type," an excellent starting point for editing mathematics, the authors recommend Ellen Swanson's *Mathematics into Type* for a fuller treatment of the subject.

The thirteenth edition of *Chicago* has an excellent new chapter on "Composition, Printing, and Binding" that can supplement the discussion of publication production in Chapter 2 of this text. While much of the *Chicago* chapter applies directly to the technical editing work discussed in this text, remember that *Chicago* focuses on book publication. Also, it has somewhat dated information (pre-1982) on phototypesetting and other technological developments in desktop publishing. For coverage of this subject, *Chicago Guide to Preparing Electronic Manuscripts* (University of Chicago Press, 1987), must be bought as a separate booklet.

Another work that serves as a standard for editors is *Words into Type (Words). Words* has an expanded table of contents with chapter tables of contents similar to those in *Chicago.*

The long part 3 on "Copy-Editing Style" (pp. 97–242) examines abbreviations, numbers, capitalization, punctuation, compounding, and other topics. The guidelines and examples are good, but some sections, such as the one on punctuation, require greater familiarity with grammatical terminology to find certain usage rules, and some are subdivided in the table of contents to a degree that seems excessive. For example, one of the comma sections is "Correlative Phrases"—one of over thirty-five uses of commas listed.

Part 4, "Typographical Style," contains a valuable brief discussion of "Details of Page Makeup"; the discussion of "Mathematical and Scientific Writing" (five pages) is not as useful as *Chicago*'s. This section does have, however, fifty pages on the composition of material in foreign languages, while the corresponding discussion in *Chicago* is scattered across several chapters.

Words has a sixty-five page part on grammar that is an extension of some of the material in this text (see Chapter 11). As here, the grammar discussion in *Words* is prescriptive, but occasionally it needs clarification. For example, the authors suggest "expressions using *kind of, sort of,* and so on, should be in the singular, unless the plural idea is overriding" (p. 358). They provide as an example "this kind of cats are native to Egypt." Clearly, this should be "this kind of cat is native to Egypt" or "these cats [or "similar cats"] are native to Egypt." Other aspects of the third edition of *Words* are uncertainly organized. For example, the discussion of verbs begins with voice and then covers mood before it examines tense, person, and number.

Part 7, "Typography and Illustration," contains valuable information on casting off (copyfitting), type, illustrations, and paper, but the discussion of composition is more dated than in *Chicago,* and printing and binding are covered very briefly.

The new edition of *Words into Type* incorporates substantial changes in organization and content of the third edition. For example, the lengthy part 3 is broken into two sections, the first covering copyediting matters such as acronyms, symbols, numbers, and capitalization, and the second punctuation. Part 4 of the fourth edition will combine grammar and usage, with their long lists of items, along with other parts in a glossary at the end. The discussions of typesetting and desktop publishing, among others, are being expanded.

In the category of general style guides fall such books as Karen Judd's *Copy Editing*, a useful textbook, and Judith Butcher's *Copy-Editing: The Cambridge Handbook.* Both are worthy editorial guides that contain sections on editing technical material, as well as copymarking and publication production in general. Other valuable references include newspaper guides such as *The Associated Press Stylebook.*

Editors should realize that style guides sometimes differ significantly. *The Editorial Eye* newsletter (Editorial Experts, Arlington, Virginia) printed a table that illustrates some of those differences (see Fig. 7.9). Below the table,

CATEGORY	GPO STYLE	CHICAGO STYLE	AP STYLE
• Abbreviations			
–Eastern standard time	e.s.t.	EST	EST
–ships	the U.S.S. *Iowa*	the U.S.S. *Iowa*	the USS Iowa
• Capitalization	The Star-Spangled Banner	The Star-spangled Banner	The Star-Spangled Banner
	Lake Erie, Lakes Erie and Ontario	Lake Erie, Lakes Erie and Ontario	Lake Erie, lakes Erie and Ontario
	Washington State, the State of Washington	Washington State, the state of Washington	Washington state, the state of Washington
• Compounding	a well-worn book, the book is well worn	a well-worn book, the book is well worn	a well-worn book, the book is well-worn, the book looks well worn
• Punctuation			
–serial comma	a, b, and c	a, b, and c	a, b and c
–apostrophe	the 1920's	the 1920s	the 1920s
• Numbers	They had many animals: 10 dogs, 6 cats, and 97 hamsters.	They had many animals: ten dogs, six cats, and ninety-seven hamsters.	They had many animals: 10 dogs, six cats and 97 hamsters.
• Spelling	all right subpena	all right, alright* subpoena	all right subpoena
• Word division	ad-verb-i-al pe-ren-ni-al	ad-ver-bi-al pe-ren-ni-al	ad-ver-bi-al per-en-ni-al

*Each spelling has a separate listing in both Merriam-Webster dictionaries that the style guide recommends for spelling.

Figure 7.9. Style guide exercise. (Source: *Editorial Eye* October 1980, p. 2. Editorial Experts, © 1980, used with permission.)

the following passage appeared. Try using the *Chicago, GPO,* and *Associated Press* style guides to edit the passage.

Bridge from Town to Suburbs to be Built

Samuel Smith, the well known Republican senator from the state of New Illiana, announced today the allocation of $152 million in federal funds to build the long planned Smith bridge at the confluence of the Illiana and Westering rivers. The bridge will link the city of Inverness with the communities of Robinwood, Georgeville and Five Corners. and will route a badly needed work force to Inverness's booming shale oil industry.

Answers to the exercise appear in the key at the end of the chapter.

INSTRUCTIONS TO AUTHORS

Technical editors often concentrate in a particular field, for example, microbiology or software documentation. If you do, become familiar with the publication standards of the leading technical journals in that field. Quite often, a general guide like *Chicago* is not sufficient. Rather, technical writers and editors are guided by publications like the Council of Biology Editors' *Style Manual* and the American Chemical Society *Handbook for Authors.* To prepare manuscripts for publication in technical journals, however, editors should follow the particular journal's instructions to authors. Instructions to authors are readily available to teachers and students, so students can work with them in assignments such as the editing exercise in Chapter 4.

Editing students should examine the instructions for authors from a number of leading journals in their field, noting significant similarities and differences among the instructions, to be familiar with journals' standards, and to able to advise writers who ask for help preparing manuscripts for submission to a journal in the field. As it is often a technical editor's responsibility to edit manuscripts for publication in a journal before they are submitted to that journal, it is important for editors to be familiar with the instructions to authors in journals in their field.

In the instructions in most technical journals, attention is paid to format and style. Editors might keep copies of the instructions from the major journals in the field to be able to show writers that journal editors really *do* care whether manuscripts submitted follow guidelines. Also, editors should keep copies of technical journal articles that stress the importance of effective communication in manuscripts, such as Norman Cheville's "Publishing the Scholarly Manuscript" in *Veterinary Pathology* (Vol. 23, 1986, pp. 99–102) and

H. Peter Lehmann, Wanda Townsend, and Philip Pizzolato's "Guidelines for the Presentation of Research in the Written Form" in *American Journal of Clinical Pathology*, (Vol. 89, No. 1, January 1988, pp. 130–136).

CONCLUSION

Style guides are editorial tools, just like calculators, computers, and dictionaries. Using style guides properly can help editors achieve consistency in technical documents and satisfy publishers' requirements. Above all, style guides are resources editors can use to find out how to answer an editing question. Editors should be familiar with a range of style guides and apply the appropriate guides in their work, never slavishly following any single standard but always considering the standard's advice and adapting it deliberately to the document they are editing.

Editorial Style Exercise: Answer Key

GPO Style Manual

Bridge from Town to Suburbs to Be Built

Samuel Smith, the well-known Republican Senator from the State of New Illiana, announced today the allocation of $152 million in Federal funds to build the long-planned Smith Bridge at the confluence of the Illiana and Westering Rivers. The bridge will link the city of Inverness with the communities of Robinwood, Georgeville, and Five Corners, and will route a badly needed work force to Inverness' booming shale oil industry.

The Chicago Manual of Style

Bridge from Town to Suburbs to Be Built

Samuel Smith, the well-known Republican senator from the state of New Illiana, announced today the allocation of $152 million in federal funds to build the long-planned Smith Bridge at the confluence of the Illiana and Westering rivers. The bridge will link the city of Inverness with the communities of Robinwood, Georgeville, and Five Corners and will route a badly needed work force to Inverness's booming shale oil industry.

Associated Press Style Sheet

Bridge from Town to Suburbs to be Built

Samuel Smith, the well-known Republican senator from the state of New Illiana, announced today the allocation of $152 million in federal funds to build the long-planned Smith Bridge at the confluence of the Illiana and Westering rivers. The bridge will link the city of Inverness with the communities of Robinwood, Georgeville and Five Corners, and will route a badly needed work force to Inverness' booming shale oil industry.

CHAPTER EIGHT

Editing Types of Documents

Different types of technical documents pose different editorial challenges. Editing a manual or a progress report, for example, is different from editing a proposal or an important letter. This chapter presents suggestions in addition to those in Chapters 4 and 5 for editing particular types of technical documents, including:

- Manuals
- Proposals
- Progress (interim) reports
- Journal articles
- Newsletters
- Brochures, fact sheets, and capability statements
- Correspondence
- Annual reports
- Briefing materials
- Forms

MANUALS

Technical manuals, like other technical documents, must be well organized, clearly stated, and readable for their audience to accomplish their purpose: to provide information. Technical manuals and other documents are similar in many ways but often different in their formats and production schedules. General considerations for formats, graphics, and production of most types of manuals are examined first, and then instruction, reference, and standards and policies manuals are discussed.

Formats for Manuals

One important difference between manuals and other technical documents is their variety of physical formats. Many reference, standards, and policy manuals are produced in $8\frac{1}{2} \times 11$-inch book-size formats. However, instructional manuals for assembly, operation, installation, troubleshooting, and repair are distributed in a variety of formats designed for the situations in which they will be used.

For example, software documentation may be on full-sized ($8\frac{1}{2} \times 11$-inch) or smaller paper, but it is often bound with looseleaf or mechanical binding. This allows the manual to lie open flat so the user can operate the computer with both hands. New pages can be inserted easily. Troubleshooting or repair instructions can be printed on laminated, pocket-sized cards for use by aircraft engine or automobile mechanics. On-line documentation, in contrast, is far less flexible in format and consequently more challenging to produce.

Graphics for Manuals

Good technical manuals rely heavily on graphics, especially when designed for nonexpert users. Drawings (and to a lesser degree photographs) are used in assembly, installation, operation, and repair manuals to simplify the instructions by showing users what they see when they look at the object or equipment. As discussed in Chapter 5, drawings are often preferable to photographs because the artists can omit details that are not essential to the step being discussed. Drawings are used effectively in the packing instructions in Figure 8.1.

The graphics for manuals may vary from parts lists in informal table format to technical illustrations of parts to flow charts, which can be especially helpful in instruction manuals. Helping the user see what is being depicted in the text can simplify the manual and increase its value.

Because manual users depend on graphics for information, writers and artists should not include extraneous graphics that do not advance the discussion, such as decorative drawings of small animals. Such graphics may add color to a manual (increasing its cost and complicating its timely production) and life to it, but no one reads manuals for pleasure. They can distract users from the text and the informational graphics and weaken the manual. Such "window dressing" doesn't really make a manual or other technical document user-friendly, because it doesn't help users get the information they need from the manual. Apparent, visible organization, clear and thorough content, and appropriate level of discussion do.

Production of Manuals

One of the greatest challenges in producing effective manuals is finishing on schedule. This problem is especially troublesome in software documentation

PACKAGING TIPS

1. **Use a corrugated carton** – The best way to ship most articles is to pack them in a corrugated carton. Choose one in good rigid condition that is large enough to allow room for adequate cushioning material.

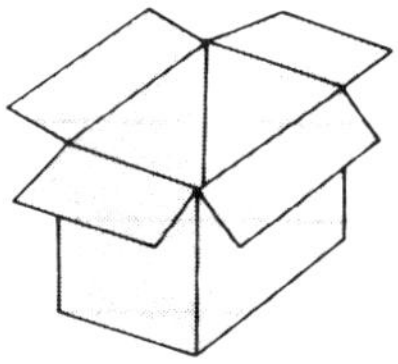

2. **Protect it inside** – Wrap each item to be shipped separately and place it in the center of the carton. Use enough cushioning material so that the contents cannot move easily in transit.

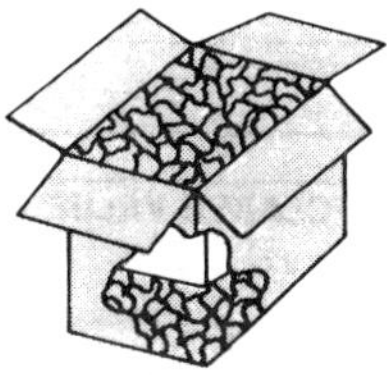

3. **Close it securely** – Use a quality package sealing tape to close your package. Do not use masking or cellophane tape.

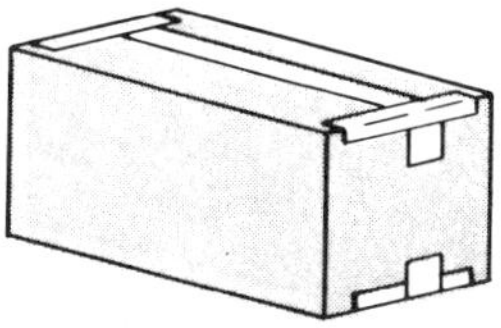

4. **Use proper labeling** – Use one address label that:
- Includes your complete return address including zip code.
- Includes the receivers complete street address including zip code.

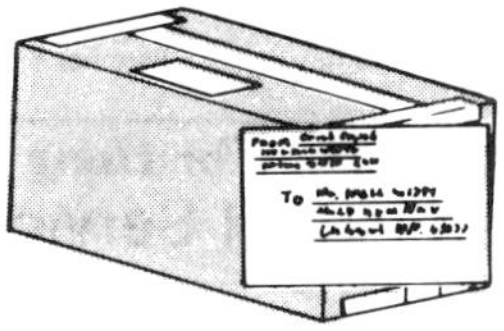

Figure 8.1. Merging of text and graphics in instructions. (Source: United Parcel Service, © 1986, used with permission.)

intended for distribution with an associated product. In such software development efforts, documentation is usually prepared by technical staff (programmers, analysts, or other technical experts) or communication specialists who either join the development project in its infancy or write the documentation when the software is almost ready for release.

Much documentation written by technical staff is good—especially that for an expert audience of analysts, repair persons, and so forth. However, user documentation written by technical staff is often weak because it is not readable by a lay audience. Many software developers assign technical communicators to the project, either early in the development effort or after its completion. Either scenario has assets and liabilities. When the writers are assigned to the project from the beginning, they usually develop an expert's understanding of the software's operation and capabilities. Consequently, the documentation is technically accurate and thorough, but often it is not addressed appropriately to nonexpert users who will install the software and use it in applications. This approach results in documentation that must be edited to fit the audience. However, the method can result in documentation that is complete soon after the software's design, coding, and testing are final.

In the second scenario, when the documentation is prepared after the software is complete, the writers usually have far less familiarity with the software. Consequently, it takes longer to write thorough and accurate documentation, and often there is very little time between when the software design is frozen and when the package is released. One advantage, however, is that the authors of the manual are not as expert and must teach themselves how to use the software as they document it. The result is usually more accessible, requiring less editing. The main problem is that, as with most businesses, short-term marketing goals rather than long-range strategies tend to dominate the software industry. Once the software is ready to market, out it goes to retailers as quickly as possible, often with inadequate documentation. In this second setting, user documentation can be weak because not enough time is allowed the writers to produce good documentation. In the first, the user documentation might be weak if it is written by technical experts and not revised by an editor who could mediate for the users the technical content. In the first approach, software firms avoid paying a technical communicator's salary (say $30,000 a year) to get the documentation written expeditiously and instead have documentation written by a $50,000 a year software engineer or systems analyst. The second approach, however, cuts the technical communicator's position (or obviates hiring one), ignores the need for a technical editor, and produces weak documentation but saves the company the $30,000 plus benefits. Consequently, many user manuals produced by software firms are unnecessarily technical, especially in jargon and unnecessary description of capabilities. They are produced by writers who have become expert in the software, and they are not edited well. The cost-containment argument against editing usually goes "Why pay to have the document edited? It was written by a technical communication specialist. So it must be fine as it is."

Other manuals suffer from this "cost-containment" syndrome, too. The best defense for technical editors is a good relationship with the technical staff. Other strategies include some familiarity with the technology (to gain credibility with technical staff) and examples of edited versus unedited user documentation.

Instruction Manuals

Professionals in all fields frequently find themselves having to explain how to do things—to install, operate, or repair equipment; to conduct tests and experiments; to get from one place to another. Sometimes they have the advantage of being able to demonstrate the task, in person or with videotape. Often, however, they must rely on written instructions, in printed or on-line form.

Whether using print or other media, instructors must analyze their audience carefully, especially their knowledge of the subject.

To edit instruction manuals effectively, editors should remember that many writers were taught to show how much they know about a subject when they write, rather than to inform readers who know less about the subject than they. Sometimes editors must tactfully remind writers that their purpose is not to demonstrate that they understand the process but to teach someone else how to do it.

Writers and editors must define an instruction manual's audience carefully. What do they already know about the process or any equipment involved in it? Are they reading out of interest or because they will perform the process as part of their work? If the audience is a mixed one, it is usually better to include steps that an expert might not need than to omit steps and confuse lay readers. Editors should test the instructions by trying to carry them out or by having a document validation group follow them without help.

Instructions should begin with a general statement identifying the subject and a list of the equipment and materials needed to complete the process.

Editors should consider which types of graphics would best help the readers understand the instructions. For printed instructions, drawings are usually best, but photographs of necessary equipment may be useful.

In any set of instructions, editors should make sure all of the necessary steps have been included in the proper order. All steps should be stated clearly and concisely, with imperative verbs as appropriate and with simple diction. If technical terms, abbreviations, or acronyms must be used in instructions for nonexpert readers, make sure they are defined clearly. Consider using analogy if the instructions are written for a lay audience.

If the steps in the instructions are numbered, each step should have a different number. If the instructions are divided into stages or sections, it is usually better to number sequentially across the sections, as below, to facilitate reference to particular steps:

Right	Wrong
First Stage	First Stage
1 Do A.	1 Do A.
2 Do B.	2 Do B.
3 Do C.	3 Do C.
Second Stage	Second Stage
4 Do D.	1 Do A.
5 Do E.	2 Do B.

Editors should insert cautions and warnings *before* the hazardous step is described. Above all, remember that the most common weakness in instructions is failing to provide all of the necessary information. Although the writers of instructions should be experts on their content, editors of instructions intended for a lay audience are often more effective when they are not experts. Then their familiarity with the content of the instructions is closer to the users'. Technical experts sometimes overlook steps or assume that explanation of a step would be unnecessary, so editors need to recognize what the audience knows to be able to translate the information as effectively as possible to the audience.

As an example of how to edit instructions, see the discussion of "COD Test Procedure" in Chapter 4.

Reference Manuals

Reference manuals may be separate documents, such as parts lists, catalogs, and bibliographies, or sections within a document, such as indexes, lists of illustrations and equations, and tables of contents. Reference manuals may exist on-line or in hard-copy form or on microfilm or microfiche. There is a growing tendency to use computerized reference systems, as with libraries' on-line catalog systems. This range of forms shares a common purpose: to guide users to the information they need.

Reference manuals must be thorough and accurate. If parts are omitted from a parts list or if part numbers are incorrect, the manual can cause mistakes and frustration. If an index lists only some key terms and names mentioned in a work, or only some of the places they are mentioned, the index is as flawed as if incorrect page numbers were listed.

Indexes are time-consuming to edit, even if computers have been used to prepare them. The latest developments in indexing software and descriptions of word processing software's increasing indexing capability can be found in computer magazines and in the proceedings of conferences such as the Society for Technical Communication annual conference. Editors facing the task of preparing an index for a document should consult recent articles on indexing and try to get involved in the document early in its preparation.

Often, codes have to be entered within the text of the document for the software's index program to work. Perhaps the greatest challenge in creating an index is deciding its depth—how detailed it will be. For example, the index of a physics textbook may include only references to topics (gravitation or the Poisson ratio) but not to places (Los Alamos) or people (Newton or Poisson). Generally, the more thorough index is more useful, but document size or publication costs are often restrictive. Whatever the degree of coverage, however, the index must correctly indicate the location of the information.

Reference manuals involving a great deal of numbering are difficult to prepare correctly. Missing or incorrect digits can prevent a reader from finding the information or obtaining the part, especially when the computer system uses reference or catalog numbers only rather than names or descriptions. Reference manuals involving numbers must be checked against sources that are known to be correct. Several inquiries may be required to identify those sources.

Standards and Policies Manuals

Standards and policies manuals are often little more than collections of memos, legal documents, operating procedures, and other materials designed to provide employees with information about what the company or agency requires of them. Sometimes the manuals are merely indexes to several volumes of standards and policies. As these materials are more often considered business or legal documents, technical editors are rarely involved.

Standards and policies manuals are often written with more concern for accuracy than readability. Anyone who reads policies on sexual harassment will see this clearly. The same problem exists with many insurance policies and most tax form instructions. If editors work with standards and policies manuals, it is very important to keep the readers' backgrounds, purposes, and familiarity with the material in mind. Often the editing requires rewriting of legalistic text; in this case especially, editors should ensure that their work is reviewed by the organization's personnel manager, legal staff, or other appropriate expert.

PROPOSALS

Technical proposals are sometimes the most important and difficult publications companies produce. A contract worth hundreds of millions of dollars or even a company's existence may be riding on a proposal. For researchers in universities and business, teachers, and others applying for grants to support projects, grant applications are also extremely important. In both cases, the bidder proposes to do a specified task for a customer. Most of the following proposal guidelines can be translated directly to writing grant applications.

In business, proposals are written to government agencies or other companies to solicit work. (In some large companies, internal proposals are submitted to fund projects.) On major solicitations (also called "invitations to bid," "requests for quotation," and "procurements"), bidders usually receive a request for proposal (RFP), request for quotation (RFQ), or invitation to bid that indicates what the customer wants. In most companies, the proposal is then managed by a technical staff member, written by technical staff, and edited by a technical editor. The editor's role is important; editing a technical proposal well increases its likelihood for success.

The discussion in Chapter 2 of the technical editor's role in planning a document applies especially to work on solicited proposals because major proposals often must be prepared quickly. For U.S. government acquisitions, the turnaround time (from the day an RFP is released to the day the proposal is due) is sometimes as long as 180 days but often only 30–45 days. The longer a company has to prepare a proposal, the more drafts are usually generated for review. Some major proposals may go through three or four drafts, and managers sometimes expect each draft to be edited and perhaps even contain finished art. So for the editors, artists, typists, and other Publications Department staff, a 90-day proposal can become three 30-day proposals.

In addition to planning, the editor must study the RFP carefully and perhaps help the proposal manager define and phrase key marketing arguments. The editor may help determine page limits for sections, of a page-limited proposal. Also, the editor may write or review a compliance matrix (see below). Because many proposals must be submitted by a deadline, the editor must schedule and manage proposal production carefully.

Most companies have a good idea of the content and format to be required in a forthcoming RFP, based on information from their marketing staff. Therefore, much editorial planning can be done even before the RFP is released.

RFPs or other customer instructions about proposal preparation and evaluation are especially important. Evaluation criteria presented in an RFP should guide the proposal manager, editor, and writers in their decisions about content and emphasis. The selection in Figure 8.2 (from the same RFP as the proposal preparation instructions in Chapter 7) establishes evaluation criteria for that proposal.

As soon as work begins, the editor should analyze all RFP statements that affect publications work and inform all support groups of any changes in the preliminary information they were given. Another important reason to study the RFP is to prevent its possible misinterpretation by technical staff.

Reports that the customer will require for the project should be defined. If the publications department will be involved in preparing them, as is usual, the editor often estimates the cost of preparing those reports for inclusion in the company's bid.

Because it is sometimes difficult for proposal writers and editors to know

1. EVALUATION CRITERIA

a. General Considerations

Proposals will be evaluated on a subjective basis for their conformance with the terms of the solicitation. This evaluation will not entail scoring of the proposal but will be considered in determining which proposal is most advantageous for the government. For the purpose of this evaluation, the Government will consider relevant past performance as a general consideration.

b. General Basis for Contract Award

. . . .

(2) This is a technical competition with cost considered subordinate to other factors; therefore, the technical and management areas will be given paramount consideration in the evaluation process. Offerors are encouraged, however, to perform technical-cost tradeoffs to achieve a balance where the requirements can be achieved at the lowest possible cost. The basis of the proposed cost must be entirely compatible with the technical and management volumes. No advantage will accrue to an offeror who submits an unrealistically low cost proposal. Accordingly, the offeror's proposal may be penalized during the evaluation to the degree that the estimated cost is unrealistically low.

c. Specific Areas of Evaluation

(1) The specific areas of evaluation shown in descending order of importance are as follows:

(a) Technical

(b) Management

(c) Cost/Price

(2) Technical Area: The items and factors to be evaluated in this area are listed below. The items are listed in descending order of importance.

(a) Technical Area:

1. Seeker Design. The offeror's proposed design must be technically sound and responsive to seeker performance requirements. The design must be in sufficient detail to provide the evaluators a basis for making an engineering evaluation.

2. Sensor Design. The IIR sensor must be capable of detecting high value targets under a wide range of target characteristics, clutter levels, environmental conditions and employment scenarios. The sensor design shall be evaluated against detection range, minimum resolution, sensitivity, clutter reduction, field of view, and pointing accuracy.

3. Acquisition Capability. The proposed seeker must be capable of autonomous detection and acquisition of targets from the given target classes with the given launch uncertainties. The seeker shall be evaluated against target detection and validation schemes, false target recognition methods, search uncertainty area considerations, and weapon delivery flexibility. . . .

Figure 8.2. Sample proposal evaluation criteria. (Source: U.S. Department of Defense, *Request for Proposal F08635-84-R-0132,* pp. 48–49.)

what key marketing messages or *win themes* to weave into the proposal, editors should meet with the firm's marketing or business development specialists to gather information about key messages to be emphasized at key places in the proposal.

Win themes are the key selling points at the center of the company's argument that it and not a competitor should be awarded the contract. They are based on what the organization thinks makes it a better choice for the work, whether past performance on similar projects, technology unique to the organization, outstanding technical staff, etc. These *discriminators* form the bases for the marketing messages in the proposal, and the content of the proposal exists to support them. Win themes are the most important statements in a proposal, and the editor as a language expert should contribute to phrasing them so they are stated clearly, forcefully (but not arrogantly), and convincingly.

Editors sometimes provide proposal writers with guidelines for the proposal. Win themes are stated, and writers are asked to support them with hard evidence in their discussions. Acronyms are specified to avoid confusion, and abbreviations are listed to cut editing time. Standards for work by publications department support groups should be determined by the editor and proposal manager at this time and discussed with heads of the support groups (see Chapter 2). Such guidelines and standards can save much time, money, and stress during the proposal preparation, making it easier to produce a consistent, attractive, well-edited proposal.

Editors often prepare the *compliance matrix* for technical proposals to ensure that the proposal provides all the discussion called for by the RFP. A compliance matrix should list each subsection of the RFP that calls for discussion and the section of the proposal that provides it. The compliance matrix is very useful to the proposal manager, writers, and reviewers of drafts. It is often included in the final submission as an aid to the proposal evaluators, making it easier for them to see where the proposal discusses each topic called for by the RFP. Any problem with compliance should be pointed out immediately to the proposal manager.

For page-limited proposals, the editor might have to remind managers of the standards they agreed to. Technical managers sometimes want to make room for kitchen-sink write-ups with smaller type and reduced art. Clearly, the editor must oppose this kind of solution. Successful, well-edited proposals are legible, their content controlled to meet length requirements.

Editors should expect the content of a major proposal to change as the company better understands what the customer wants—and as technical staff determine which claims can be supported by evidence. The changes often become more radical late in the preparation process; sometimes they take place during printing. The company's fee for the work outlined in the RFP's statement of work (the "bid" in the proposal) is usually determined at the next-to-the-last minute. Higher company officials often change the bid at the last minute (to thwart competitors from undercutting it, theoretically). Given the tight schedules of many proposal efforts, each milestone must be met, or the final submission can be jeopardized.

Various solutions to meeting proposal schedules exist. An experienced editor can influence the proposal manager to limit the number of review drafts by citing their publication costs. Also, the editor can encourage the proposal manager to control the length of proposals that are not limited. Early drafts can be printed and bound unedited, with rough art. The company can hire temporary staff or freelancers, who can be very helpful but may not know enough about the firm or the proposed work to be truly effective.

Editing a technical proposal is far easier when it is not one long narrative but an arrangement of shorter discussions designed to address the RFP requirements or company outline. One very popular format, "STOP," divides a proposal (or other document) into two-page units on facing pages, the left with text and the right with graphics, each unit focusing on and developing a thesis stated at the beginning of the unit. The STOP format is discussed in several works listed in the bibliography, including Timothy Whalen's useful *Writing and Managing Winning Technical Proposals.*

Editing individual volumes of a proposal requires recognizing the purpose of each volume. The solicitation and award documents for contracting officers, executive summaries, technical volumes for technical experts, management volumes for management specialists, and cost volumes for finance specialists must be appropriate for their audiences. For example, in an executive summary, the bidder summarizes each main section of the proposal, providing an overview of the claims and evidence presented throughout the proposal and stating the bidder's marketing messages. Executive summaries may be read by technical experts to get an overview of a large proposal, but summaries are read by a much broader audience than technical volumes. Agency or company managers who are not technical experts will read the executive summary, so the content of the technical volume must be presented more simply or in less depth in the summary.

Even on classified projects, summaries are usually kept unclassified and made attractive so that contracting officers or executives might display them as examples of the sort of work their contractors produce. Executive summaries for major proposals are often printed in color, and there is usually a greater reliance on graphics, especially photographs, than in the other volumes. Editors working on an executive summary need to estimate and track carefully the production costs. Try to produce an executive summary that suggests the company does solid work but would not spend the customer's money on overly elaborate productions.

Technical volumes are read primarily by technical experts, who score the proposal on the quality of the supporting data. Such volumes should not be ornate, but thorough and clear, with technical discussions carried out in considerable detail. The editor must make sure that the win themes stand out from the mass of supporting evidence in the technical volume in both the text and graphics.

Management volumes require clear organization by subject so that evaluators can find particular sections easily. Certain sections, such as previous experience and project personnel, may be read by all of the reviewers, but others (such as cost and schedule control, facilities and equipment) may not. Organization charts should be carried down only to the appropriate level of detail, and resumes should be edited to emphasize the staff's related experience.

Because management volumes present company policy and philosophy, writers have a tendency to use "motherhood" (general, unsubstantiated claims about the company's expertise). The editor of a management volume must be more concerned than a technical volume editor that specific evidence supports the points argued. Also, there is a tendency to use boilerplate (canned discussions of related experience, facilities, etc.), which must be revised to support the emphases of the particular proposal.

Any technical proposal prepared in response to an RFP or to program announcements from agencies like the National Endowment for the Humanities should follow carefully the instructions about content and organization. The proposal writer(s) and the editor must make sure that the proposal's content is fully responsive—that it provides all the information requested—or it may be disqualified. In the proposal business, the customer is always right. Often, proposal writers or managers provide the information an RFP calls for but deviate from the specified organization. They shouldn't—reviewers often use the RFP's instructions about content and organization as their standard for evaluating proposals. Having parts of the discussion where the evaluators expect to find them creates a better impression of the company or person submitting the proposal than does forcing evaluators to find sections they are looking for. The editor might have to warn against creating an "unnecessarily elaborate" proposal, which U.S. government RFPs and federal acquisition regulations (FARs) discourage.

Proposals succeed when the evaluators are convinced that the bidder's claims are valid. To convince them, proposal writers and editors must be sure that specific, factual details support the claims in the proposal or grant application. A technical proposal should reveal the qualifications of a company or individual, not preach them, and the editor should test each proposal with this goal in mind.

PROGRESS REPORTS

Progress reports are among the easiest technical documents to write, but they are also among the most important. Their content and organization are determined by what has been or will be done, so the subject matter is usually clearly defined. However, progress reports must be submitted "in a timely fashion"—on time and at the appropriate times—usually dates that are established in the contract.

Progress reports are submitted in the midst of a contracted project to inform the customer (and in-house staff) of work completed and remaining, findings, and schedule and budget compliance.

A progress report may be a one-page letter for small projects, or, for a large project, it may be a bound volume of many pages. In this case, the report will have a table of contents, first-order headed sections and perhaps graphics. On some projects, progress is reported by telephone, but any telephone conversation should be followed by a letter recounting the main information presented, promises made, permissions granted, and so forth.

When a project requires publications department support, publications staff and especially the editor should understand the project and examine its schedule carefully. When a contract for a project is awarded, a schedule for progress reports is negotiated if it was not specified in the statement of work.

Progress reports should be submitted at logical points in the project, usually soon after main stages or tasks are completed. However, if progress reports are submitted too soon, there will be too little progress to describe. If submitted too frequently, there will be too little progress to describe since the last report.

Progress reports should be submitted in a timely fashion because customers want to know how the work they have funded is going. The company or agency contracts person may have to report regularly on the project to his or her supervisors.

Consider a twelve-month development project for an electronic component. If design research is to take five months, prototype development four months, and testing three months, a logical schedule for progress reports might be the first at five and a half months (reporting work on design research) and the second at nine and a half or ten months (reporting work on prototype development). The final report, submitted late in month thirteen, would include test results and much of the content of the first two progress reports. (There would not be a progress report following the testing phase—rather, the final project report covers testing.)

As a second example, consider a six-month project to repair two highway bridges, one a four-lane bridge over a river (bridge A) and the other a two-lane bridge over a road (bridge B). Structural repair on bridge A will take five months; structural repair on bridge B will take two months. Surfacing bridge A will take one month, for bridge B, two weeks. If the contract calls for written progress reports on the project, how many would be appropriate, and when might they best be submitted?

Editors should make sure that progress reports include specific statements about the work completed on the project, the work remaining, and the company's adherence to the schedule and budget for the project. Progress reports should emphasize the project accomplishments up to that point in the project.

Consider the adequacy of the progress report in Figure 8.3, on a project to develop a seal for a fishing reel. The form of this letter is correct, and some of the content is good. It begins with a clear indication of the writer's

Jones Seals
400 Smithson Avenue
Pascagoula MS 39567

April 23, 1991

Mr. Robert Bascom
Horner Tackle, Inc.
804 Laguna Street
San Francisco CA 94102

Dear Mr. Bascom:

I am writing to report our progress on the development of a seal for your HT-400 and HT-500 reels.

Work has been going well. We have accomplished a great deal since we started the project March 1. Our findings, though tentative at this point, should help us develop seals for both reels on schedule and below cost.

If you have any questions about the project, please call me at (601) 234-5789.

Sincerely,

Craig Jones

Craig Jones

Figure 8.3. A weak progress report.

purpose, and the closing offer to discuss the project is appropriate. However, there are no details for Bascom to report to his supervisor (who may be wondering if Bascom hired the right firm to develop the seals), nothing that proves the study is getting somewhere and will result in the desired seals.

An editor working with this progress report could suggest that Jones add specific details about the progress thus far, to demonstrate that substantial work has been completed. A medium edit might produce the report in Figure 8.4.

The details about the materials will assure Bascom that significant progress has been made and the study is proceeding as planned.

One way for editors to check the content of a progress report is to put themselves in the place of the audience, imagining what the customer would want to know to feel comfortable with the money committed to the project. Or, what information might the customer need to make decisions now and in the future? Determine if there are problems with the study that the customer should know about and examine the writers' handling of them. Generally, most customers want to know whether they can expect the project to be completed on time and within budget.

Progress reports benefit the companies and individuals preparing them as well as the customers who receive them. The need to submit a progress report on schedule can force technical staff to assess the work done and its schedule and budget. These activities force managers out of the day-to-day routine and make them focus on the project as a whole, evaluating goals, accomplishments, and methods. Having to write a report can force a project to come together.

Progress reports are a proper place to reveal problems. Equipment problems may have delayed research; possible solutions to a problem may have turned out to be dead ends; weather may have delayed testing. Most project managers are reluctant to bring up such problems, but editors should encourage them to. If customers are made aware of the problem soon after it occurs, they will likely be much more agreeable to the extra time or money needed to solve it. Minor problems should be reported as well, because a small problem might become serious enough to affect the schedule or budget later. The customer might wonder why the company did not report the problem when it first became aware of the problem.

A letter progress report should follow proper format and identify the subject of the project at the beginning of the report. Whether or not a letter format is used, editors should ensure that the report includes specific statements of the work completed, findings, and work remaining. Also, the report should address the schedule and budget and express a willingness to discuss the project.

Sometimes an editor is challenged to make a progress report sound like everything is going fine. When a project calls for several progress reports, some might be thin when not much work has been accomplished. An editor might be asked to gloss over the problem creatively. In such a situation, the

Jones Seals
400 Smithson Avenue
Pascagoula MS 39567

April 23, 1991

Mr. Robert Bascom
Horner Tackle, Inc.
804 Laguna Street
San Francisco CA 94102

Dear Mr. Bascom:

I am writing to report our progress on the development of seals for the HT-400 and HT-500 reels.

We have designed and are testing three composites for the seals: a conventional rubber seal, a nylon-reinforced rubber seal, and a plastic seal.

So far, the plastic seal seems the most promising based on predictions of its durability. The plastic seal would be more expensive than a rubber seal (about $.30 compared to $.24), but it has performed much better in corrosion and wear tests than the rubber seal (anticipated life in hard use, salt air, 70 degree environment: 5 years versus 3). The plastic seal and the reinforced rubber seal would have about the same cost, but the plastic again has a better wear expectancy (5 years to 4).

We should finish the testing by May 10 as scheduled, and our expenses are within budget.

We are pleased with our progress on the project. If you have any questions about it, please call me at (601) 234-5789.

Sincerely,

Craig Jones

Craig Jones

Figure 8.4. The progress report revised.

worst mistake the editor could make is to recast the information into clear, concise terms.

JOURNAL ARTICLES

Most technical journals provide instructions to authors that can be used as an editing standard, as discussed in Chapter 7. Also, most authors of manuscripts for technical journals are familiar with the common structures of journal articles.

Guidelines for journal articles are provided by many professional organizations, such as the American Mathematical Society and its *Manual for Authors of Mathematical Papers.* However, editors cannot assume that their writers have read such guidelines carefully. Public or university libraries can provide back issues of journals for current editorial instructions and article samples. Manuscripts for journals are usually short or of moderate length, making it easier to edit for consistency. Many journals that report research results expect articles to follow the traditional organization of Introduction, Materials and Methods, Results, and Discussion.

However, editing manuscripts for submission to journals poses some challenges. Authors of articles are often at work on their next project while the report on previous research is being prepared for publication, rendering it less urgent. While forced to be interested in grant proposals that could result in funding of future (or current) research, authors sometimes must be pressed by editors to finish revising the text and especially completing the graphics for an article.

Editors who do not have strong technical backgrounds may find it difficult to check the accuracy of the study design and results reported. The journal's reviewers can be counted on to address the technical content, but many manuscripts submitted for publication contain gaps and statements that reviewers have challenged as shown in Figure 8.5. In the table, the number in parentheses represents the number of times the comment was made. "Other" includes problems in presentation such as spelling errors and weaknesses in content such as inaccurate data or problems with statistical approach. To increase the proportion of acceptances, more and more organizations will edit journal article manuscripts and grant applications in-house before submission.

For journal article manuscripts, final review is extremely important. However, editors often find such a review difficult to arrange when several authors are involved, as is often the case with technical articles. Usually one author (or a group from a single organization) is deemed responsible for preparing the article.

The Conclusions section of a research article often must be edited carefully to make sure that the Results and Discussion support any conclusions drawn. When firm conclusions are presented, they should be prominent. When the conclusions are tentative, or when the research proved inconclusive, ed-

Table 1. Summary of Reviewers' Comments of the First 50 Manuscripts Received at the AJCP Office in 1985

1. Title
   Misleading or inappropriate for the contents of the text (7)
2. Abstract
   Lacking data pertinent to or descriptive of the work (4)
3. Introduction
   a. Objectives not clearly stated (2)
   b. Information given in introduction not consistent with contents of the rest of the manuscript (3)
   c. Introduction too long and unnecessarily wordy (2)
   d. Other (3)
4. Materials and methods
   a. Details lacking or in some cases irrelevant (36)
   b. Data not reproducible from information given (2)
   c. Procedures not in keeping with accepted procedures (9)
   d. Other (8)
5. Discussion/results
   a. Statements not clear or consistent with the rest of the text (8)
   b. Too long and/or unnecessarily wordy (5)
   c. References cited are old and outdated (1)
   d. Lacks information pertinent to stated objectives (18)
   e. Statements made by the authors challenged by the reviewers (26)
6. Figures
   a. Inferior quality (3)
   b. Number too great or small (2)
   c. Legends unclear or incorrect (1)
   d. Other (5)
7. Micrographs/gross photographs
   a. Poor quality (7)
   b. Number too great or small (2)
   c. Legends unclear or incorrect (1)
   d. Other (8)
8. Tables
   a. Data presented in unclear manner (6)
   b. Number too great or small (5)
   c. Repetitious data (1)
   d. Other (1)
9. General criticisms
   a. Grammar poor (8)
   b. Writing style awkward (8)
   c. Too long and/or unnecessarily wordy (2)
   d. Study design poor (7)
   e. Specific error repeated throughout manuscript (1)
   f. Terms used are inappropriate or inaccurate (10)
   g. Other (10)

Figure 8.5. Common weaknesses in journal manuscripts. (Source: Lehmann, H. Peter, Townsend, Wanda, and Pizzolato, Philip, "Guidelines for the Presentation of Research in the Written Form," *American Journal of Clinical Pathology,* Vol. 89, No. 1 [January 1988], p. 31. Used with permission.)

itors must control writers' desires to make more of those conclusions than is prudent.

Technical journals are sometimes distinguished from professional journals, which present articles of general interest to professionals and to a general audience. The Oak Ridge National Laboratory *Review,* Lawrence Livermore Laboratory's *Energy and Technology Review,* and *IEEE Spectrum* are examples. Professional journals rely on photographs much more than technical journals that report research results. Editors working with manuscripts for professional journals often must arrange photography sessions or obtain appropriate photographs from archives; technical journal authors arrange their own photography. Improperly framed or exposed shots, lifeless "mug shots" of people, and staged shots in which everyone seems ill at ease are common in professional journals, and many types have become clichéd (two people shaking hands as one presents an award). All too often, an event is past before an editor sees the article on the event, or someone without photography training takes a few snapshots that are all the editor has to work with. Photo concepts for a professional journal article should be discussed with the author as far in advance as possible, so that good photographs can be taken or retrieved from files.

Manuscripts often contain graphics and captions that can be easily mismatched. Captions are usually prepared in a list and matched only by number with graphics that might look very similar, such as electron microscope photographs. Even trained editors can have trouble ascertaining that graphics and captions have been matched properly, and the authors must confirm them in proofs.

In addition to following the technical journal's instructions for authors, editors should be aware of prevailing attitudes within a field toward research and the publication of research results. In most fields, dual submission (submission of a manuscript to more than one journal at a time) is condemned, and journal editors caution against the presentation of the same body of material in different manuscripts. In many fields, official statements regarding the ethical treatment of the research subjects are required. Editors should read regularly and thoroughly the journals to which their writers submit manuscripts to be aware of current policies on such issues.

NEWSLETTERS

Desktop publishing has caused a revolution in newsletters. PageMaker and other desktop and electronic publishing software, laser printers, and scanners have made in-house newsletter design and production possible, with text in a variety of type fonts and sizes and graphic reproduction of photographs as well as line drawings.

Newsletters provide information about an organization, for its members and other readers, on a regular or occasional basis. Companies, government agencies, and private organizations often produce them for a general readership, so technical information in newsletters is usually presented at a level appropriate for lay readers.

Newsletters are commonly produced on a small budget by a smaller staff. Sometimes only one person designs, writes, edits, produces, and distributes each issue. Consequently, editing a newsletter encompasses most if not all of the production activities discussed in Chapter 2. Some general points can be covered here, but for more information, consult one of the books devoted to the topic, such as Mark Beach's *Editing Your Newsletter.*

Because newsletters are aimed at a general audience, discussions of technical material must be edited carefully for a lay or middle audience. Often the articles are written from documents provided by technical staff, so the text must be edited to increase readability, and new, less complex graphics might be needed.

Photographs of a technical nature (for example, showing material deformation under stress) should be replaced by human-interest photographs (for example, showing researchers conducting stress tests). A cutline under each photograph should identify its content, as well as tell readers more about what is represented in the photograph. The cutlines should be specific, and actions should be described in present tense. Portrait photos of people showing only the face or the head, neck, and shoulders are appropriate for announcing promotions and awards, but more engaging action photos are preferable.

Editors producing newsletters do more research, interviewing, photographing, writing, and rewriting than usually for other assignments. It can be difficult for beginning editors to estimate how long a newsletter issue will take to prepare. The supposed advantage of not having to spend time communicating with support groups to produce the newsletter may be imaginary. Few people have a realistic idea of how long an issue of a newsletter takes to create—after all, it's only four (or two) pages. However, it may take many hours to research and write, and although page formats can be saved as macros and reused in subsequent issues, copyfitting must be done anew each time. Editors often must work on other projects considered more pressing by managers.

The greatest challenge—and reward—for the editor of a newsletter, though, is to handle all of the writing, editing, and production tasks well. The opportunity and potential gratification lie in becoming a desktop publishing expert who can work independently to create a regular, widely read product.

FACT SHEETS, BROCHURES, AND CAPABILITY STATEMENTS

Fact sheets, brochures, capability statements, and other marketing materials are similar to proposals in their marketing emphasis, and many of the same guidelines apply.

Fact sheets are prepared by companies to describe their products or services. As the name implies, they are often single sheets, printed front only or front and back, that describe the product or service—its purpose, materials, uses, history, and perhaps cost and ordering instructions. Fact sheets are usually illustrated, to show as well as describe the product, and are distributed at trade shows and through other marketing efforts. Federal and state agencies produce fact sheets to inform their constituents.

Not an advertisement, a fact sheet is designed first to inform and secondarily to convince, whereas an ad is intended primarily to sell the product. A fact sheet could be used by an advertising copywriter to prepare an ad for a particular market.

Fact sheets are often printed in full color, requiring the accompanying extra time and cost to prepare. Because they are very visible documents, they must be attractive, attention-getting but not flashy, and interesting, with sufficient technical detail.

Editors contribute to the design of fact sheets, and they often manage their production. They help keep them from becoming brochures or advertisements and revise the presentation of the technical information to make it accessible. Editors also estimate production time and costs and negotiate the schedule and budget with the printers.

Brochures and capability statements, like advertisements, have sales as their primary objective. Brochures, with which everyone is familiar, may describe a company's products or services, but capability statements describe fully the range of a company's services. Technical editors rarely work with brochures or capability statements, because they tend to lack technical detail. Instead, such documents are most often written and edited by marketing, advertising, or public relations staff, although editors might plan or supervise production work on them.

CORRESPONDENCE

If most correspondence gets edited at all, it is correctness edited or proofread by secretaries, who usually know more about grammar and punctuation than the writers. However, editors sometimes edit important correspondence such as a letter to stockholders. Similarly, important memos may be edited before they are sent, and editors should always edit and proofread their own professional letters and memos carefully—typos or other errors are sure to be pointed out.

Letters

Professional letters can be more important to technical experts' reputations than any other writing they do. Letters suggest much about the writers and their company or agency, and writers must try to create good impressions through their correspondence. Sometimes, however, writers create a negative impression through inflated diction, complex sentence structures, and worn-out expressions intended to make the letter impressive or to "sound good." Editors revising such work often need to simplify and clarify the phrasing. A letter's tone and content may need to be revised by someone who has some distance from the subject or situation. Often an editor can better determine whether statements might not be clear enough, or whether the tone is appropriate.

Letter style

To edit correspondence well, editors should try to make letters or memos easy to understand on a quick reading (assuming that the writer intends the discussion to be clear—if not, the editor should know this so he or she can obfuscate the discussion appropriately). For example, a letter explaining that parts weren't shipped because the order was forgotten should not be clear enough that the reader realizes what happened. Editors should apply the guidelines on style discussed in Chapter 4, in particular:

Encourage the use of *I* and *we* as appropriate. Many writers have been taught not to use the first person in their writing, especially in letters. This simple-minded notion inflicted on students by misguided teachers does more to deprive writing of directness, clarity, and force than nearly any other "rule."

Keep paragraphs and sentences shorter and more straightforward than those in other documents.

Use active voice and action verbs to present points clearly and emphatically.

Tighten sentences to make the letter concise, but not so concise as to seem dry, abrupt, or rude.

Examine each statement for clarity, making sure the content of the letter is clear (or unclear, if intended).

Sparingly substitute simple, straightforward diction for clichés and ornate diction, taking care to recognize that letter writers often consider a contrived and flowery (but silly) phrase like Norman Schwarzkopf's "bovine scatology" the best part of the letter.

Consider the willingness of the writer to have his or her writing edited. Often the best writers are the most willing to allow editors to practice their craft. However, many weak writers are certain they write well and resist having their work edited. Always be diplomatic, especially with letters, where many writers are anxious to express a personal style.

Letter format

Occasionally writers mix or fail to use established formats for letters, neglecting to realize that professionals in business and government expect them. A letter that doesn't follow a specific format might suggest the writer is ignorant of them or deliberately idiosyncratic. The most common is the block format, in which all elements of the letter are flush left, as in Figure 8.6. For letters of less than half a page, modified block format sometimes looks better.

5317 Curry Ford Road M-201
Orlando FL 32812

May 16, 1991

Manager of Journals, ASCE
345 East 47th Street
New York NY 10017-2398

Subject: Permission to reproduce material from the ASCE Authors' Guide to Journals, Books, and Reference Publications

I am writing a technical editing textbook to be published by Oxford University Press. In it I plan to reprint graphics and other materials from a number of sources to help students learn to prepare technical materials for publication.

I would like permission to reprint some materials from the ASCE Authors' Guide as examples of the information technical communicators can find in professional societies' style guides (as well as for the information they contain). I have enclosed copies of the materials I wish to reprint, and I have listed them on the enclosed permission form.

The American Society of Civil Engineers' permission to use the materials would be acknowledged in a note under each piece, as described on the permission form, bringing the ASCE to the attention of the users of the text. I hope you will grant me permission to use this material to help technical communication students learn to edit technical material.

Thank you for considering my request and signing and returning one copy of the permission form in the enclosed envelope.

Sincerely,

Donald C. Samson Jr.

(Dr.) Donald C. Samson, Jr.
Enclosures

Figure 8.6. A sample letter in block format.

In modified block format, the return address, date, closing, and signature appear aligned on the right, as in Figure 8.7.

To edit letters to proper format, follow your company or agency's style guide. If there is no standard, consider the following suggestions (see Fig. 8.6):

Center the letter on the page, with 1-inch (or slightly more) margins all around. Keep the top margin less than 2 inches. If the letter is short and seems to occupy only the top half of the page, don't spread the sections with excessive extra spacing. Instead, use modified block format.

Skip a line between the return address, date, inside address, salutation, body, closing, signature, and other information for an uncrowded look on the page.

Edit addresses to conform to United States Postal Service abbreviations for names of states and countries, but write out *Street, Avenue, Boulevard, Suite,* etc.

Make sure that the appropriate title has been used in the inside address and the salutation (*Mr., Ms., Mrs., Miss,* or *Dr.*). If in doubt about which title to use with a woman, use *Ms.* unless the woman has indicated a preference for *Mrs.* or *Miss* in previous correspondence. The honorary titles of some elected officials, clergy, and other persons are used in the inside address. For example, a letter might be addressed to "The Honorable Senator Jane Robinson," but the salutation would be "Dear Senator Robinson:".

Use a subject line if the writer does not have a person to address by name in the salutation. "Dear Sir" and "Dear Madam" are out of fashion; "To Whom It May Concern" is stilted and vague. The writer might address the letter to a department, but "Dear Facilities Department" is awkward. Often the best solution is to use a subject line (without quotation marks) in place of the salutation, such as "Subject: Request to have equipment installed" or "Subject: Information on the Rogers Project." No period is used at the end of the subject line.

Mark the body of the letter for single spacing and an extra line between paragraphs. The skipped line avoids large blocks of text that can hide ideas or information or intimidate readers. Also, indent the beginning of each paragraph five spaces. In a letter, as in other documents, each paragraph should focus on one idea or body of information, and the indentation improves readability by signaling the beginning of a new idea or body of information.

If the letter contains an enclosure (or enclosures), make sure this is indicated by "Enclosure:" two lines under the signature block, sometimes followed by a description of the enclosure.

Encourage writers to limit letters to one page. If a second *(continuation)* page is needed, put the title and name of the person written to (flush left), the page number (centered), and the date of the letter (flush right) on one line

5317 Curry Ford M-201
Orlando FL 32812

January 23, 1990

Copyright Office
Library of Congress
Washington DC 20559

Subject: Request for Circulars

I am writing a technical editing textbook for Oxford University Press. As part of my research, I would like to examine Circulars 1 and 99.

Please send me copies of these circulars.

Thank you.

Sincerely,

Donald C. Samson, Jr.

Figure 8.7. A sample letter in modified block format.

at the top of the new page. This information is valuable if the continuation page is separated from the first page. Do not staple the pages together.

Encourage writers to use a simple closing like "Sincerely," and to sign their name legibly. Malcolm Forbes once called an illegible signature "the biggest ego trip I know."

Letter etiquette

Because a business letter is a formal document, writers should present themselves properly, being tactful, courteous, and considerate. Letter writers sometimes impose on their audience and let letters run on longer than needed. Writers and editors should assume that letter readers are busy people.

Make sure that the letter indicates its purpose simply and straightforwardly, near the beginning.

Make sure the letter indicates clearly (but politely) any action to follow, on either the writer's or reader's part.

Revise statements to make the letter specific but courteous, polite, and brief.

Examine the tone of each sentence and of the letter as a whole to ensure that the writer is communicating the intended impression.

Many good books on business letter writing are available, including some with examples of various types of letters. One such book, Stephen Elliott's *The Complete Book of Contemporary Business Letters,* has sample letters organized by purpose and audience, such as "Managing Your Business." Editors should examine such books carefully before they recommend them; writers can follow examples too slavishly.

Memos

Memos are short letters sent *inside* the writer's company or agency. As letters suggest much about writers and their company or organization, so writers create impressions through their memos within an organization. Memos are less formal than regular business correspondence—too informal to be sent to anyone outside the organization. Memos perform the following functions:

- To request or provide information
- To request, confirm, or deny assistance
- To report on an incident
- To document a conversation
- To initiate or stop action

Business and professional journals, and newspapers such as the *Wall Street Journal,* often carry articles on the importance of good memos. Technical editors are not often called on to edit memos; sometimes, however, they are asked by a company official or manager to "take a look" at an important memo.

Memo style

Memos should be clear. If they are not, the receiver might take the wrong action—or do nothing. Memos should be brief, not more than half a page long.

Editors should be alert for misspellings, grammatical errors, usage problems, and awkward sentences. Take particular care in editing your own correspondence, both letters and memos. Weaknesses, whether in grammar, spelling, or any other aspect, suggest that you are not as expert as you should be.

Memo format

Memos should be prepared on a full-sized sheet of paper, even if the text is short. A smaller sheet might be lost or not noticed in a stack of full-sized papers. The top margin may be up to 2 inches, but left and right margins should be the standard 1 inch. Many organizations have stationery with "MEMORANDUM" printed at the top. With or without such a heading, the following information usually appears at the beginning of a memo:

TO:
FROM:
DATE:
SUBJECT:

This format is shown in the memo in Figure 8.8, which follows proper format but is not a good memo. Why?

The sender should initial after his or her name, to indicate approval. When the sender's or recipient's department or group name must be indicated for routing purposes, it is placed after the person's name, separated by a comma.

Memo etiquette

Memos should be courteous. Sometimes writers trying to be brief seem abrupt or rude, as in the memo in Figure 8.9. Editors should encourage writers to be polite and considerate in their memos, even if doing so adds a sentence or two. In the memo in Figure 8.9, inserting "Please" before "Be there" would avoid the possibility that the readers will feel bossed around.

The memo in Figure 8.10 uses an appropriate format and tone. This memo is informal, clear, and courteous. Also, it illustrates an important function of memos: to record in writing information that might be (or has been) transmitted in a telephone call or conversation. A simple request such as White's might involve a company policy concerning when an evaluation of an employee must be completed. When policies, profits, or reputations are involved, a memo should be used to document a request for assistance, report on an incident, etc.

MEMORANDUM

TO: All Employees

FROM: John Doe JD

DATE: January 16, 1991

SUBJECT: Tardiness

Yesterday over 20% of our work force was not at their job performing their task including servicing our customers. Today the figure stands at approximately 12%. If I were to ask each one of you why you were late, your answer probably would be traffic conditions due to the snow or rain. That excuse prompts me to explain a few facts to you. Snow is not a strange phenomenon here in this state. It is something that we have had to learn how to cope with over the years. The fact that an employee is late due to traffic conditions resulting from the weather or any other condition is clearly not an acceptable excuse. The remedy to all the nonsense excuses is to leave your home earlier when adverse conditions exist. In fact, many of you just barely arrive on time. As a result, you have a high probability of being late. I consider that to be sheer negligence.

You are required to be at your desk working at 8:00 a.m. Socializing or standing in the coffee line at 8:00 a.m. does not qualify you as being on time.

Hereon, as a result of the tardiness being up so high over the last three months, regardless of the weather conditions, each person including officers shall be here at their desk and working on time each day. If not, disciplinary action shall be imposed on each individual in violation of this rule.

Figure 8.8. Improper memo tone.

ANNUAL REPORTS

Annual reports are marketing materials that share some characteristics with capability statements: They reveal what a company does. However, annual reports do not try to sell the company's goods or services; rather, they try to promote the company as an investment. In addition, they contain company financial statements that are prepared by auditors and required by the Securities and Exchange Commission.

MEMORANDUM

TO: All Staff

FROM: I. Payne IP

DATE: June 24, 1991

SUBJECT: Monthly Meeting

The monthly meeting will be in room 12 Friday.

Be there at 2:00.

Figure 8.9. Improper brevity in a memo.

TO: Jane David, Personnel

FROM: Max White, Engineering MW

DATE: June 5, 1991

SUBJECT: Annual Review of James Gates

I have received the evaluation forms for Jim's review. Could you send a copy of his 1989 evaluation?

Also, I'll be in San Francisco all next week, so completing the evaluation by the 14th will be a problem. Would the 18th be OK to complete the evaluation? Please call me at 1310 by Friday afternoon if that's a problem.

Thank you.

Figure 8.10. Proper memo format, style, and tone.

Annual reports are usually prepared by an in-house or contracted public relations or advertising staff and by accountants. The company's publications department is rarely involved directly, but editors may support an annual report effort by providing photographs from the company archives or graphics and descriptions of projects from company publications. Editors may be consulted about designers and printers for the report if public relations staff are not familiar with them.

When involved in editing annual reports, keep in mind the audience—potential investors in the firm—and remember that they want information about the firm's earnings potential. Also, remember that the Securities and Exchange Commission requires certain information (for example, a statement of how much the company earned or lost, and why) but permits any format that allows the readers easy access to it.

BRIEFING MATERIALS

Oral presentations *(briefings)* are common in business and government. A presentation is an opportunity to demonstrate expertise, to show the audience the speaker's (and organization's) technical competence. An oral presentation requires different organization from that of a printed or on-line document.

Editors often edit text and supervise production of visual aids for briefings—viewgraphs, 35 mm slides, transparencies, flip charts, and so forth. Editors contribute to visual aids as they do to graphics for documents: designing and drawing illustrations, providing photographs, editing text, designing formats and layouts, and producing the text and graphics. Editors are often involved in preparing for briefings and other presentations (such as conference papers or poster sessions) because the quality and implementation of the visual aids are usually more vital to the success of briefings than to a printed document.

When staff plan visual aids for a briefing, editors can suggest graphics that would be best for what they want to illustrate. Some types of graphics don't work well in oral presentations, such as formal tables, which are usually too "busy." The audience doesn't need all the data in a table; they need a clear, simple statement of the point the data supports. Editors should encourage speakers to keep all graphics simple and easy to interpret quickly.

Many technical staff commit a common error in trying to put too much material in any one visual aid. Editors can help by examining the visual aid from the audience's point of view. Also, editors often are able to suggest types of graphics and layouts that would be effective for the messages in the visual aids. Once writers create a draft of the text and sketches of the graphics, editors edit them as they would a print or on-line document: for readability, legibility, clarity, conciseness, and the other goals discussed in Chapters 4 and 5.

As graphics software becomes more sophisticated, technical staff will rely more on graphics they can produce themselves. This can cost much less than employing a professional artist, but often the graphics created on PCs are not appropriate for customer briefings or presentations at professional conferences. It may be more cost-effective overall to contract with graphic artists. Visual aids produced on PCs are, however, appropriate for in-house and contract briefings.

Desktop presentation software can be used to create transparencies or 35 mm slides; other software packages can make charts, graphs, and other visual aid formats. Most presentation software can improve the results of graphics programs, and many provide templates and use text formatting options similar to those in desktop publishing. Graphics can be printed out, sent to a film processor for conversion to 35 mm slides, or shown with a projection system connected to the computer.

In skilled hands, these programs can produce effective visual aids. However, most programs can rotate text, insert clip art, and over-label graphics, making it is easy to produce cluttered, poorly designed, ineffective visual aids. Review the principles of effective graphics in Chapter 5, and consult with graphic artists if you have questions about individual visual aids.

Editors often help prepare for presentations by coaching presenters. Encourage them to determine what their audience already knows about the subject and decide how much background information is needed. Technical staff often overestimate their audiences' understanding of their subject. Editors can help by finding out about the audience, listening to a dry run of the presentation, and asking questions at the end.

Beginning technical staff sometimes must be reminded that with an oral presentation, the audience cannot turn back the page to reread a passage they did not understand, or to refresh their memories. If the oral presentation is to be informal, the editor might encourage the speaker to allow (or even encourage) questions as he or she presents the material. However, inexperienced or nonexpert speakers might be advised to ask that questions be held until the end.

Encourage presenters to begin with a clear statement of the talk's subject and organization. Remind them that audiences are more interested in overviews and statements of major work done, findings, conclusions, and recommendations than in raw data. If the presentation is to be coordinated with printed material such as a handout or an abstract printed in proceedings, editors might discuss how to integrate the presentation and the printed materials might support each other.

Editors sometimes have to remind technical staff that briefings should be talks, not strict readings from prepared text. Although the actual word-for-word reading of a paper was once standard at professional conferences, presentations now rely on transparencies, slides, or videos to convey information and capture the audience's attention.

Text slides that list the main points of a speech can be effective at the beginning of a presentation, but text-only aids ("idiot charts") should not be shown to the audience throughout. Instead, they can be used more accurately as speaker's notes.

To help speakers prepare for the chance that they will have less or more time than expected for their presentation, editors can help plan different versions of it. For a twenty-minute presentation, planning a ten-minute and thirty-minute version can help ensure that the speaker will still be effective if other speakers run long or short.

Formats for Briefing Graphics

Several formats exist for briefing graphics. Whatever the formats chosen, control is important. Speakers should be able to show them when they want to without stopping to plug in machines or fumble with sheets of paper. Also, speakers should be able to remove graphics from view so the audience will return their attention to the speaker. The formats below, which are popular for oral presentations, all have advantages and disadvantages that editors should be able to point out to technical staff preparing for briefings.

Transparencies

If the contents are drawn large enough, transparencies *(viewgraphs)* can be seen well by the audience, even with the lights on. They can be easily and quickly shown, discussed, and then removed from the overhead projector. Notes can be drawn with erasable or permanent felt-tipped pens as they are shown. Transparencies can be printed on inexpensive plastic sheets using most photocopiers. Light colors don't show up well, so use black or dark ink. The speaker will need a projector and a screen (or a light-colored wall) and should test the projector beforehand to make sure it works.

The most common weaknesses in transparencies is lettering that is too small to be read by the audience. Titles of transparencies should be 36-point type, preferably bold and in a common face like Helvetica. Subtitles should be 24 to 30 points, text 24 points, and callouts in at least 18-point type. One easy way to tell whether a transparency will be legible to viewers twenty feet away is to place it on a sheet of white paper on the floor, stand over it, and try to read it. If you can read it easily, it will be legible.

35 mm slides

Slides are compact and easy to carry. Excellent image quality is possible, and computer graphics can be used to suggest three dimensions or movement. Slides can be removed from view by turning off the projector or advancing to a black slide (but not to an empty slot in the carousel, which creates a distracting white light on the screen). Slides are relatively inexpensive, but they take time to produce. Also, slide projects requiring photographers or graphic artists can become expensive. Speakers will need a projector, a screen (or wall), and a room that can be darkened. Slide projector bulbs burn out much more frequently than those in overhead projectors—bring a spare.

The most common weakness in slide presentations is backwards or upside down slides. In the middle of a briefing or the presentation of a conference paper, there's usually no way to recover from this disaster. Make sure the technical staff know how to load the slides in the projector's carousel.

Videos or films

Videos and films have a high interest level—people love moving pictures. They can demonstrate clearly how something is done, but videos and films take lots of time to plan and create and are expensive to produce. Proper projection equipment is required, and for films the presentation room must be darkened. Movie projectors frequently malfunction, and video projection equipment can be difficult to procure or expensive to rent at a professional conference. Also, for a large audience several monitors may be needed.

Flip charts

Flip charts are constructed of large pads of paper bound at the top. They can be prepared quickly at low cost and can be seen well if the graphics are drawn large enough (letters 2 inches tall for each 10 feet between the chart and the back row of the audience). Different colors can be used; dark colors are best. One sheet can be left blank between each graphic, so that when the speaker finishes with it that page can be turned, revealing a blank sheet. Flip charts have one disadvantage: They can't be seen by everyone in a large audience. They must be carried or mailed to the conference, and speakers must have a stand for them.

Handouts

Handouts can provide more information than many other aids. They are easy and inexpensive to produce, but handouts have a serious disadvantage: Speakers cannot easily control when the audience looks at them. If speakers distribute them before the end of the presentation, many in the audience will be reading them instead of paying attention to the speaker. If technical staff plan to use handouts in a briefing, the editor might discuss when and how they plan to distribute them, as well as edit and supervise their production.

Posters

Poster sessions allow viewers to interact with the presenter, so they are becoming increasingly common for conference presentations in the sciences and technology. In a poster session, large poster displays incorporating text (mostly headings, top-level data, and conclusions) and graphics are set up in a conference room, and the audience can move from display to display, as at a trade show. Far greater interaction between presenter and audience is possible than with the traditional reading of a paper that will appear in the conference proceedings or a journal.

Poster sessions have drawbacks, however. What might have been a twenty-minute presentation may involve eight or more hours with the poster display. Transporting and setting up large posters is far more complicated than preparing to use transparencies or slides, and the cost of developing poster-session materials can be far greater.

Blackboard drawings

For graphics on blackboards (the old-fashioned black or green chalkboards or the new white plastic boards for use with felt-tipped pens), speakers do not need to bring anything except chalk or a felt-tipped pen and an eraser. However, speakers cannot take much time during a presentation to write on the board, and they shouldn't turn their back on the audience. If the graphic is drawn in advance, the board must be covered so the audience won't look at it when the speaker doesn't want them to.

Models and demonstrations

Models and demonstrations can show an audience an actual object or procedure, and consequently they can be very valuable. Editors rarely help technical staff prepare models or demonstrations, but some large companies have a model shop that editors occasionally work with. Editors can help staff rehearse the presentation and offer suggestions on it. Help presenters make sure the entire audience will be able to see the model or demonstration. Realize that if a model is passed around, some people will be looking at it when they should be listening to the speaker, and the model might be damaged.

Editing Presentations

A Chinese saying goes, "It is better to see something once than to hear it a hundred times." This principle lies behind the use of visual aids in presentations, and it should be kept in mind when editing presentation materials. Because the audience can only hear the text being presented, it should be edited so that the subject, organization, and purpose are stated clearly early in the presentation, even more apparently than in a written document. Shorter sentence structures, simpler diction, and transitions for emphasis should be used in texts for presentation to help the listeners follow the presentation.

Humor is valuable in a presentation, but editors must be diplomatic in criticizing a speaker's attempt at humor. Chances are the speaker tried the joke on his or her friends or co-workers, who may have responded positively in part out of a sense of obligation. When the humor fits the situation and seems impromptu, it can work. At an International Technical Communication Conference session in Philadelphia, one speaker was significantly underdressed in jeans and a day-old shirt. As he began his presentation, he told the audience: "I'm glad I was able to make it to Philadelphia to be here today. I just wish my suitcase could be here." The laughter that followed answered the question of his appearance, relaxed the audience, and established a candid tone for the presentation.

Editors often arrange to have the transparencies, slides, or other visual aids reproduced in a *briefing book* used as a handout during or after the presentation. Production duties for the visual aids and printed materials are similar to those for other documents, but editors must concentrate even more stringently on preparing simple, straightforward graphics that communicate information clearly rather than impress through flashiness.

Presentations, like airplanes, are most likely to crash on takeoff and landing, so editors should help speakers should work most carefully on the beginning and ending of their presentation. For more information on presentations, refer to Larry Venable's valuable chapter on "Making Effective Oral Presentations" in the American Chemical Society *Handbook for Authors.*

FORMS

Poorly designed forms crop up everywhere—in unnecessarily complex Internal Revenue Service instructions, in employment applications that force users to squeeze large amounts of information into tiny spaces, in order forms that have no space for catalog or part numbers. Bad forms are frustrating to users. They can also be costly, as the IRS discovered when it had to scrap the revised instructions for withholding because they were too complex for taxpayers to interpret.

The form in Figure 8.11 was sent to an applicant for a teaching position at a university. It illustrates a common problem with producing forms: Because few people pay much attention to forms, it is difficult for the designer to get a draft of a form reviewed properly by an editor or a manager. The main problem in the form in Figure 8.11 is a glaring omission, but notice too how the design of some sections is weak.

Poor form design can create problems for those filling it out and for those who must interpret the responses. A poor form can also create the impression that the organization that produced it does sloppy work. Men filling out the form in Figure 8.11 might write in "Male" or skip the sex section, and interpreters will probably assume the applicant is male if the box is not checked, but the university may be creating an unwelcome impression.

The same validation testing used for computer documentation should be used for forms as well. Once a form is created and reviewed by a designer, an editor, and the appropriate technical supervisor, the editor should try to have the form validated by a test group similar to the people who will actually use the form. Without such careful editing, forms can be confusing, bothersome, or counterproductive.

———— University
City, State

CONFIDENTIAL EMPLOYMENT APPLICANT DATA COLLECTION FORM

———— University is committed to providing leadership in the attainment of equal employment rights for all persons. Accordingly, every effort will continue to be exerted to end discrimination against women and members of all minority groups and to develop affirmative action programs to involve them at every level of employment and decision-making. This effort is in compliance with all federal and state laws.

As part of the affirmative action program, applicants are requested to answer the questions listed below. This information, provided by the applicant in personally non-identifiable form, will be used only for statistical compilation and report filing. This information will not be part of the institution's employment decision with regard to the applicant. Since race/sex identification is a critical reporting factor, data forms with no race/sex identification will not be processed.

Position
Applied for: ______________________ Department/Area: ______________________

Sex: _____ Female

Race/Ethnicity:
_____ American Indian or Alaskan Native
_____ Asian or Pacific Islander (Includes member of Indian sub-continent)
_____ Black - Non-Hispanic Origin
_____ Hispanic
_____ White - Non-Hispanic Origin
_____ Yes U.S. Citizen
_____ No
_____ Yes Handicapped
_____ No
_____ Yes Veteran Status
_____ No

How did you learn of this opening?
Advertisement (indicate)
University announcement: If so, where posted
Other, please indicate

Please return this form to:
Mr.
Affirmative Action Officer
———— University
City, State, ZIP

Employment Applicant's Name:
Upon receipt and check off by the institution's affirmative action officer, the applicant's name will be detached and this part of the form destroyed.

Figure 8.11. A poorly designed form.

CONCLUSION

As editors specialize their work in a field, other types of documents too will require their attention. However, beginning editors will find that the editing guidelines presented in Chapters 4 and 5 and in this chapter's discussion of particular types of documents will prepare them for work with most technical documents.

CHAPTER NINE

Proofreading

To many book and magazine publishers, and to others in professional communications, *proofreading* means checking galley or page proofs, hard-copy versions of text and graphics, in their final stage before printing. The National Information Standards Organization defines the term this way in the ANSI Standard *Proof Corrections.* However, most technical communication professionals use it more broadly, to refer also to earlier stages in the creation of a document.

TYPES OF PROOFREADING

As a document moves through the production process, the proofreading changes. For example, when new text is first typed on disk, the disk version is compared to the writer's original draft in a *one-to-one proof.* A one-to-one proof is also used to check a writer's original sketches of graphics against what typists and artists created from them. Much proofreading is done before galley and page proofs are created, corrected, and proofed again in the *final proof.*

A document to be proofed may be a preliminary draft or the final version ready for printing; it may be on disk or tape and viewed on a screen, or it may be a hard copy. Proofreaders examine a document to make sure that it is attractive *and* correct. When a document isn't correct, it can frustrate users, increase costs to cover corrections, and embarrass the organization that produced it.

For example, the Greater Boston Convention and Visitors Bureau produces a guidebook for tourists with information on landmarks and activities in Boston. According to a United Press report, the 1988–89 guide had a number of errors. The map in the guide showed the Ritz-Carlton Hotel three blocks away from its actual site and claimed that the Four Seasons Hotel is on Newbury Street, although it is on Boylston. The tourism bureau sold the guide in its office in the Prudential Center, which the map also showed several blocks from its true location. Errors such as these render a guidebook useless, countering the intention of the tourism bureau to promote the city.

STAGES OF PROOF

Written documents are proofread at three stages: original, revision, and final.

In the original stage, processed text typed from manuscript is proofed against the original copy letter by letter to ensure that no typing errors were made. Word processing and desktop publishing are changing the way this proofreading stage is done. When documents are created, edited, and made up on-line, there is often no hard copy until printing masters are produced. There are no originals for comparison with the processed text, so the writer must check for errors and omissions that arose during typing, editing, and corrections by the editor. In this setting, what the writer has proofed and corrected might be printed without examination by a proofreader. This is a mistake. If allowed the time and pressed for accuracy, most writers can proofread effectively for content. However, writers (and editors who have worked extensively with a document) are often so familiar with the text that they read right past errors without noticing them. To a proofreader unfamiliar with the document (but preferably familiar with the subject matter), the material is fresh, and there is greater likelihood that errors will be caught as the proofreader reads each word carefully, with no preconceptions about what should be (or is) said.

One-to-one proofing is very laborious and time-consuming, but it must be done carefully to ensure that nothing has been omitted by a typist who may have skipped lines or even a page. The most efficient way to do such a proof is to place the original and hard copy of the disk version side by side, original on the left (for right-handed proofreaders), covering each with an off-white sheet of paper to aid concentration. The proofreader compares each sentence in the original to the disk version, moving the cover sheets down the pages as he or she reads. Some proofreaders use intricate methods like reading backwards or dividing a page or screen into quarters, but proofreaders should do so only if such methods work well for them. Graphics are checked line by line and number by number against the original sketches and the organization's standards.

Many people believe that computer programs can proofread documents acceptably, but its an erroneous nation (see?). No matter the capacity the program has for checking spelling, punctuation, and grammar, some expressions and constructions will always create problems. For example, the sentence "Clearly written instructions are more useful" would have a comma placed after "clearly" by most grammar-checking programs, as might "however" in the following sentence: "I don't care how you solve the problem;

however you solve it is fine with me." Similarly, no program will object to the statement that water boils at 100 degrees Fahrenheit.

Few errors in a document are noticed more quickly than spelling errors, so proofreaders should check spelling carefully. They should not rely on spelling checkers, which will only flag words not in their dictionary. Spelling checkers will not recognize correct words used improperly, so proofreaders should check each word in the document. Also, unless a proofreader runs the spelling checker he or she cannot be sure it was run.

The spelling of all proper nouns should be checked carefully. Typists are least familiar with these words, and misspelling of names of products or people can cause embarrassment. Spelling in titles, headings, and captions for graphics should also be checked carefully. Errors in these parts of a document are often overlooked by writers and editors, but as these parts stand out, errors are often noticed quickly by readers.

To proofread effectively, proofreaders should understand what the author or editor wants done and do only that. They should not edit. They should change only what needs to be changed to correct an error, and they should mark changes as neatly as possible.

In the revision stage, processed text and graphics have been altered by the author or an editor, so the revisions are proofed to ensure that changes were made correctly and no other errors were introduced. When graphics and the disk version of the text are revised, only the changes need to be proofed against the original disk version. However, when a text change is made, hyphenation for the rest of the paragraph must be checked since new word breaks may not be correct. This step is called *proofing changes.*

The final proofreading of a document does not involve comparing the text or graphics with earlier versions. Rather, the text and graphics are read straight through word by word and line by line to catch any errors in content and presentation. Technical details in the material are checked against reference information provided or identified by the editor. For example, proofreaders encountering different numerical values for a tolerance should be able to consult a standard and determine the proper value. Final proofing should be more painstaking than earlier passes.

In thorough proofreading, the text and graphics are checked for correctness, accuracy, and conformity to standards. This involves checking both presentation (grammar, spelling, punctuation, type size and spacing in text and graphics, etc.) and content (names, numbers, etc.). Proofreaders should watch for inconsistencies in data and for errors in labeling units of measure. Greek letters, formulas, and mathematical notation must be checked very carefully, as many typists are unfamiliar with such notation and it is easy for them to misread unfamiliar symbols. Usually, a thorough proofing does not take place until the final stage, as revisions would change the document and necessitate another final proof. If there is not enough time to complete all the proofreading steps presented below, as much as possible is done. The proofreader should use a nonreproducing blue pencil to complete steps 4, 9,

13, 10, and 11, and then, if possible, begin with step 1 and complete the remainder in numerical order.

If time is limited, more proofreaders can be called in for the final proof of a document. More sets of eyes can mean a greater number of errors are detected. However, more proofreaders can also mean less consistency because last-minute help is usually unfamiliar with the content and standards of the document. They can, however, relieve the main proofers of some of the straightforward chores (such as proofing headings and captions), allowing them breaks so they will be more alert, and speeding up the job.

Because the final proof is one of the last stages in production and schedules are usually very tight, proofreaders should never make a change at this point unless it corrects something that is clearly wrong and will create problems because it is wrong. Late changes to typeset text or processed graphics can jeopardize a publication schedule in an organization, and authors' alterations ("AAs") are charged to writers publishing in a journal or book. If the changes are agreed to be absolutely necessary, mark them neatly to avoid any further errors by the printer or typesetter ("PEs").

STEPS IN PROOFREADING

Chapter 3 gives a full discussion of proofreading and editing symbols, shown again in Figure 9.1. This list of steps outlines what proofreaders and editors should cover when proofreading a document. If time permits, follow the order presented here.

1. Establish the purpose and audience of the document. You should not make editorial changes in the level of discussion, but knowing the purpose and audience can help orient you quickly.
2. Establish whether you are proofing originals, revisions, or finals, and determine when the work must be completed. With this information, you can determine how best to mark the hard copy and how thorough the proofreading can be.
3. Establish whether the author (or editor) wants a thorough or a limited proof. Is everything to be checked, including content (such as technical information, proper names, etc.)? Or is the presentation alone to be checked (spelling, grammar, punctuation, etc.)? If it is a limited proof, plan to report to the writer or editor what you did with the document *and* what you did not do.
4. Establish the standards to be used. Will a company style guide determine spellings, punctuation, and terminology? Are instructions to authors for a technical journal or publisher to be followed? If none is specified, obtain the author's or editor's approval to use your choice of standard.
5. Establish how the editor or author wants the text marked. For hard copy of text to be corrected by typists, fine-tipped dark red pen stands out well. For graphics to be corrected by artists, photocopy the graphics (unless

1.26. Corrections in proofs read by authors or department readers must be indicated as follows:

Mark	Meaning	Mark	Meaning
⊙	Insert period	*rom.*	Roman type
∧,	Insert comma	*caps.*	Caps—used in margin
:	Insert colon	≡	Caps—used in text
;	Insert semicolon	*c+sc*	Caps & small caps—used in margin
?	Insert question mark	≡	Caps & small caps—used in text
!	Insert exclamation mark	*l.c.*	Lowercase—used in margin
=/	Insert hyphen	/	Used in text to show deletion or substitution
∨'	Insert apostrophe	₰	Delete
∨“ ”∨	Insert quotation marks	₰͡	Delete and close up
1/N	Insert 1-en dash	*w.f.*	Wrong font
1/M	Insert 1-em dash	⁀	Close up
#	Insert space	]	Move right
ld>	Insert ( ) points of space	[	Move left
shill	Insert shilling	⊓	Move up
∨	Superior	⊔	Move down
∧	Inferior	‖	Align vertically
(/)	Parentheses	=	Align horizontally
[/]	Brackets	][	Center horizontally
□	Indent 1 em	⊔⊓	Center vertically
□□	Indent 2 ems	*eq.#*	Equalize space—used in margin
¶	Paragraph	√√√	Equalize space—used in text
no ¶	No paragraph	....	Let it stand—used in text
tr	Transpose[1]—used in margin	*stet.*	Let it stand—used in margin
∽	Transpose[2]—used in text	⊗	Letter(s) not clear
sp	Spell out	*run over*	Carry over to next line
ital	Italic—used in margin	*run back*	Carry back to preceding line
——	Italic—used in text	*out, see copy*	Something omitted—see copy
b.f.	Boldface—used in margin	*S/?*	Question to author to delete[3]
∼∼∼	Boldface—used in text	∧	Caret—General indicator used to mark position of error.
s.c.	Small caps—used in margin		
═	Small caps—used in text		

[1]In lieu of the traditional mark "tr" used to indicate letter or number transpositions, the striking out of the incorrect letters or numbers and the placement of the correct matter in the margin of the proof is the preferred method of indicating transposition corrections. (See rule 2.88.)

[2]Corrections involving more than two characters should be marked by striking out the entire word or number and placing the correct form in the margin. This mark should be reserved to show transposition of words.

[3]The form of any query carried should be such that an answer may be given simply by crossing out the complete query if a negative decision is made or the right-hand (question mark) portion to indicate an affirmative answer. (See example, p. 6.) (See rule 2.84.)

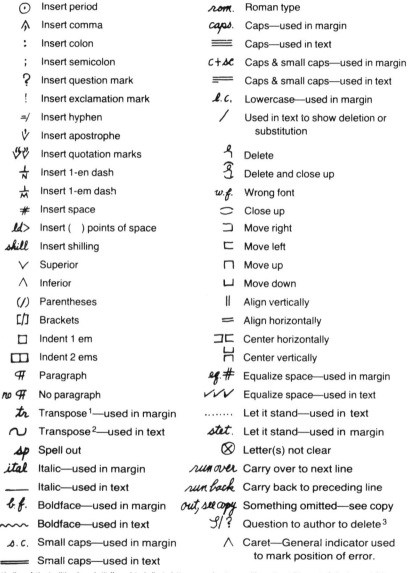

Figure 9.1. Proofreading symbols. (Source: Government Printing Office *Style Manual*, p. 5.)

they are classified) and mark the copies with red pen. If copying is undesirable or not allowed, attach a sheet of transparent vellum if the artist hasn't done so already, and mark the necessary changes *lightly* on it.

On galley and page proofs of text, use red pen, but for corrections to printing masters use only nonreproducible blue pencil. The necessary correction might be made by correcting the error and "cutting in" a new piece of text rather than printing out a new page, so non-repro pencil should be used on the proof. Similarly, nonreproducible blue pencil should be used to mark final versions of graphics. Mark lightly, as some supposedly nonreproducible blue pencil marks can be picked up by a copier or a platemaker, especially if the marks are heavy or dark. Red or any other color ink should not be used on printing masters because the printing schedule may require that the document be printed without having the corrections made.

6. Find out for certain whether the text has been spell-checked. Do not ask the author or editor; ask the typist who processed the text. Make no assumptions.

7. Assemble the necessary guides and materials. Proofreaders need the relevant technical dictionary; a style guide; related company documents to verify spellings of names and technical details; the proper pens, pencils, white-out, and line tape; a pica ruler; plain paper; and perhaps a calculator and gummed note paper for questions.

8. If the document is short and there is time, read it all the way through without marking anything. If it is longer, read the contents page, the summary, and the first and last two paragraphs of each section without marking anything. Try to get a sense of the content.

9. Examine the spelling of the title, all headings, all table titles, and all figure captions and callouts. Writers and editors tend to be so anxious to examine the body of a document that errors in titles and headings slip by them. However, readers of a document focus on the title and headings as signposts, and errors in them stand out. Readers are especially drawn to graphics, so captions should be checked carefully. Also, in many documents the title, headings, and captions are not typed with the text, so even if the text was spell-checked, they might not have been.

In the text, check the spelling of all names of people, 70oducts, places, and technical terms. Carefully check spelling in organization charts against the text. Refer to company documents for correct spellings only if you establish that those spellings are correct. Also, make sure that any tricky spellings or formats are handled properly; one like BIRDiE (a missile) will be missed almost every time.

Perhaps the best way to check spelling is to use a blank sheet of paper as described earlier: Move it down the page line by line, forcing the eyes to focus only on the isolated line above the edge of the paper. Look at each word separately, not as part of a sentence.

10. Check the layout of the document—its line lengths, leading, and word-

and letter-spacing. Check the photographs to make sure they have been stripped in correctly for platemaking. Photographs are sometimes placed upside down or reversed *(flopped)* when strippers do not have clear instructions or are rushed. Check graphics against their captions to make certain that they have not been transposed. At the same time, check traditional paste-ups for elements that might have come loose and been lost. If the document is classified or sensitive, make sure the classification marking on each page matches the highest classification of text or graphics on that page.

11. Check each sentence for structure problems, punctuation errors, and grammatical errors. Make sure each sentence is complete, with phrases and clauses joined properly. Check the punctuation against usage rules in this book or your company or agency style manual. Check for errors in subject–verb agreement, pronoun–antecedent agreement, and case.

12. Wherever technical details such as size, speed, or capability are mentioned, check the numbers against specifications you know are accurate. Watch for problems with decimals. Check all simple calculations, which often don't get checked well by authors because they are simple.

13. Check documents in the final stage for appearance problems (spots, broken type, and especially loose edges in traditional paste-ups). Proofreading standards can help editors to see problems in the appearance of a document. The ANSI standard on *Proof Corrections* provides instructions on letter spacing, "rivers," and other proofreading concerns.

14. Check the front and back matter of final documents for proper page numbering in the table of contents, list of illustrations, and index.

Completed properly, these fourteen steps can help make a document correct and attractive. The steps might be broken down further, and some proofreaders might choose to change the order somewhat, but they are the basics.

Editing students should also keep in mind that proofreaders must be expert in spelling, grammar, and punctuation. They must be allowed to concentrate without distractions and without being rushed. They can't catch every error, and they can't proofread effectively more than a few hours a day. Finally, proofreaders can't be expected to check something they wrote or edited themselves.

EXERCISE

The following discussion contains far more errors than the usual copy ready for final proof, but consider it final typeset text that will be corrected and then printed. Try to identify all of the errors and proofmark the passage correctly and neatly.

Tips Ten on Proofreading

Proofreading is an extremely important step in producing a document, but it is often assigned to new, inexperienced staff with little training in how to proofread effectively.

Careful proofreading can insure that printed or online material is correct and attractive, and that it creates a good impression of the company or agency that produced it. No one wants errors or sloppyness in a document to suggest that the company did not care enough about the user to produce a correct, attractive document.

Shoddy documents suggests that a company does shoddy work in other areas beside documentation.

In order to produce better documents more efficiently, here are ten tips for proofreaders to help them work more effectively.

1. When you get a proofreading assignment, understand what the author or editor wants you to do, then do it. Don't altar your assignment.

2. Know traditional grammer, punctuation and sentence structure. If you don't, you wont be able to spot and correct the errors that creep into documents. If you need to review grammar, use your company style guide or a good handbook.

4. Don't rely on fancy proof reading methods like reading backwards or dividing a page or screen into quarters. Read down the page or screen line by line, holding white sheet of paper under the line you are reading to help you focus on that line.

3. Use the proper reference books and materials when you proofread, the appropriate technical dictionary, company style guide, and writing instruments.

5. Don't rely on a spelling checker. Unless you run it yourself. You can't be sure it was run. And no spelling checker will mark the error on this sentence.

6. Check titles, headings, and illustration captions very carefully. Errors in these parts of a document are often overlooked by writers and editors.

7. Check spellings of all proper nouns carefully. Typists are least familiar with these words, and misspelling of names of products or people can cause great embarassment.

8. Check all technical details against a reference that *you* are sure is correct. Watch for inconsistenceis.

9. Stay alert. Don't proofread more than four hours a day, and take short brakes frequently.

10. Above all, proofread don't edit. Change only what needs to be changed to correct an error, and mark changes as neatly as you can.

Proofreading is a very valuable step in making a document communicate well and create a positive image of its producer. Because proofreading is so important, and because it is difficult, demanding work, proof- readers should be trained well and allowed enough time to do their work effectively.

Answer Key

Tips Ten on Proofreading

Proofreading is an extremely important step in producing a document, but it is often assigned to new, inexperienced staff with little training in how to proofread effectively.

Careful proofreading can insure that printed or online material is correct and attractive, and that it creates a good impression of the company or agency that produced it. No one wants errors or sloppness in a document to suggest that the company did not care enough about the user to produce a correct, attractive document.

Shoddy documents suggests that a company does shoddy work in other areas beside documentation.

In order to produce better documents more efficiently, here are ten tips for proofreaders to help them work more effectively.

1. When you get a proofreading assignment, understand what the author or editor wants you to do, then do it. Don't altar your assignment.

2. Know traditional grammer, punctuation and sentence structure. If you don't, you wont be able to spot and correct the errors that creep into documents. If you need to review grammar, use your company style guide or a good handbook.

4. Don't rely on fancy proof reading methods like reading backwards or dividing a page or screen into quarters. Read down the page or screen line by line, holding white sheet of paper under the line you are reading to help you focus on that line.

5. Use the proper reference books and materials when you proofread the appropriate technical dictionary, company style guide, and writing instruments.

5. Don't rely on a spelling checker. Unless you ran it yourself You can't be sure it was run. And no spelling checker will mark the error on this sentence.

6. Check titles, headings, and illustration captions very carefully. Errors in these parts of a document are often overlooked by writers and editors.

7. Check spellings of all proper nouns carefully. Typists are least familiar with these words, and misspelling of names of products or people can cause great embarassment.

8. Check all technical details against a reference that *you* are sure is correct. Watch for inconsistencies.

9. Stay alert. Don't proofread more than four hours a day, and take short brakes frequently.

10. Above all, proofread don't edit. Change only what needs to be changed to correct an error, and mark changes as neatly as you can.

Proofreading is a very valuable step in making a document communicate well and create a positive image of its producer. Because proofreading is so important and because it is difficult, demanding work, proof readers should be trained well and allowed enough time to do their work effectively.

CHAPTER TEN

Staffing, Scheduling, and Estimating Costs

Establishing the staffing needs, schedule, and budget for publications work on a document is usually done by an experienced editor. Beginning editors rarely face these challenges but should understand how to schedule projects and estimate costs. Even new editors may need to plan and control a document production schedule and stay within a budget allocated by a project manager or writer.

Many companies contend that excellence is the bottom line in their publications, but in most the true bottom line is cost. Company and agency managers want the best possible documents at the lowest possible cost, and they will pressure publications staff to meet that objective. The cost of preparing documents is high in any organization; the only way an editor can estimate costs accurately and control them is to plan the publications effort very carefully.

To be able to understand staffing, scheduling, and budgeting, it is important to keep in mind the different steps in creating a document, presented in Figure 10.1.

To staff a project appropriately, editors must consider the document's schedule and budget. Each of the three influences the others, so they must be planned concurrently. However, staffing can usually be done first, using preliminary information about the document(s) to be created.

STAFFING PUBLICATIONS EFFORTS

Publications department staffing for a typical technical document involves at least an editor, typist, artist, proofreader, layout and makeup specialist, printer, and binder. (This assumes that the writers are from technical groups in the organization, not the publications department.) If any part of the document

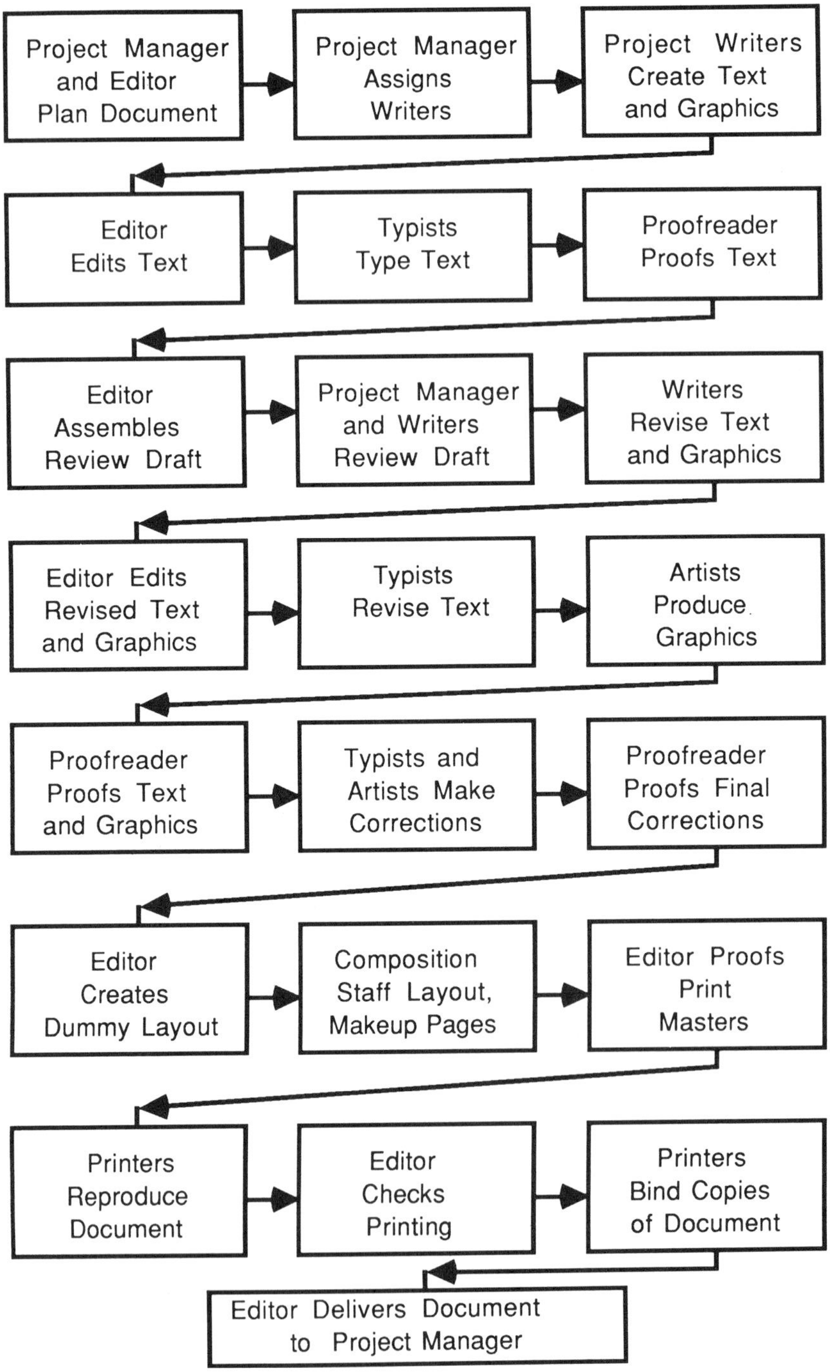

Figure 10.1. Creating a technical document.

is to be typeset, a typesetter will also be required. The effort may also require a photographer.

Often the editor assumes some of these roles. For example, in small companies or in work on a short document, the editor may also be the proofreader. If desktop publishing is used to produce the document, the editor may also serve as proofreader, artist, and layout and page makeup specialist. If the document is to be merely photocopied and stapled or spiral bound, the editor may serve as printer and binder as well. The availability of publications staff and the nature of the document determine what tasks the editor must do.

Once the editor has gathered information about the project and informed the supervisors of the publications support groups, as described in Chapter 2, the supervisors can consider their shoploading, schedule the work, and determine their staffing needs.

Levels of Technical Editors

Within most organizations, there are ranks of technical editors. These gradations go by different names (GS levels, labor grades, etc.) but nearly all divide the editorial staff by their experience and responsibilities. One company's divisions are shown in Figure 10.2. Entry-level editors are characteristically assigned to the lowest editorial grade for two or three years, while they learn the company or agency's procedures, methods, and organization, develop editing skills, strengthen their technical backgrounds, and demonstrate their abilities. They proofread, edit documents (beginning with light edits and moving up to heavy edits), coordinate production activities, and get to know publications department and technical staff.

As editors gain experience, they are assigned to more challenging projects, and if successful they advance in labor grade and salary. The documents and often the technical staff they work with become more important and sometimes more difficult—proposals for major contracts instead of monthly project reports. Experienced editors supervise beginning editors, plan and supervise production on documents, and establish budgets for major publications efforts. Although some exceptionally good editors are able to advance solely as copy editors, most do so by moving into positions with increased supervisory and administrative responsibilities.

Some editors find that in publication management positions, they no longer edit at all. Instead, they manage publications departments or support groups, assign staff to projects, plan the training of new staff, investigate equipment, argue for and stretch publications budgets, and fight the never-ending battle of cost containment. Their responsibilities are broadened and stress level often increased, but the financial reward is much greater.

Because work on a technical document involves collaboration, editors must be flexible and able to work productively with people of different backgrounds and interests, especially late in the production process, where time

PRESENTATIONS WRITER/EDITOR PERFORMANCE PROFILES

	Content Impact	Edit	Research/Writing	Art/Graphics Edit	Scheduling, Planning Administration
Associate Writer/ Editor (Grade 41)	Impact limited. Handles structured/well-written input with no supervisory guidance. Given assistance on more complex jobs and input. (Given input).	For responsiveness to specification. Has good grammatical, spelling, sentence structure ability. (With supervisory guidance, gains familiarity with graphic media, etc.).	Simple jobs such as resumes, facilities. (Personnel profiles, fact sheets, contract and task summaries.)	Edits for grammar, spelling, style manual and proofreads; exhibits ability to grasp effectiveness of art (or lack of) and works with source to modify. (Dependent on graphic approach, but shows critique for effectiveness).	Instructs and coordinates supporting groups on requirements of job; knows types of reproduction and other graphics.
Writer/Editor (Grade 43)	Impacts. Ensures responsiveness and determines need for major rewrite/edit. Produces content that moves reader to take desired action. (Given guidance and some input).	Edits/rewrites without losing or distorting technical meaning.	Volume summaries and other moderately complex jobs such as related experience sections. (Researches technical or other data to write first draft on brochures, etc. Implements rewrites — including necessary research — with minimum supervisory assistance.)	Determines need for and restructures art as needed to support messages. (Leads designer rapidly to good graphic solution — more than just a visually pleasant presentation.)	Same as 41 except that the jobs handled are more complex.
Senior Writer/ Editor (Grade 45)	Heavily impacts. Translates overall strategies into outline, key messages, etc. Convinces volume leader and task leader of need for edit/rewrite. Little input. Develops communication strategy and key points with minimal guidance from Mktg., Eng. or others.)	Edits/rewrites to implement key messages and strategy.	Complex assignments. Researches information on customer, etc., to develop key messages. (Major brochures, films, exhibits. Has strong research ability and technical grasp.)	Generates or has generated art and photos necessary to implement strategies and key messages. (Stimulates artist and others to above average creative solutions.)	Good planner to assess need and controlling factors (RFP, budget, etc.) to estimate manpower, cost, establish milestones and schedules to ensure smooth execution of major jobs. (Jobs on schedule within cost with minimal supervision.)
Presentations Specialist (Grade 46)	Acts as catalyst with management, technical and Presidents staff to ensure that Mktg. and communication strategies are developed and implemented. Convinces mgr/ director levels of need for rewrites. (Solves complex communications problems i.e., identifies and defines problem and develops total plan to solve. Works confidently with top mgmt to "sell" proposed solutions.	Leads efforts involving supervision and coordination of multiple editors. (Same, but writers.)	Communications plans. Researches requirements and information necessary to develop total communication plan strategies. (Strong marketing and business understanding; does communication plans and supervises implementation.)	Leads efforts involving supervision of multiple artists and editors. (Familiar with all types of graphics to develop total campaign strategy.)	Excellent ability to plan and lead group effort. Supervisory skills. Understands aerospace, customer organization and practices.
Senior Presentations Specialist (Grade 47)	Same as 46 except deals with VP and directors on jobs of critical importance to division. (Same as 46.)	Same as 46.	Same as 46.	Same as 46.	Same as 46 except deals in face-to-face negotiations and problem solving with high level customer and in-house management; gets budget to do the job.

Figure 10.2. Editorial responsibilities by labor grade. (Source: Martin Marietta Corporation, © 1984, used with permission.)

and fuses are short. The editor must realize that too many consultants on one document can be detrimental; the mass-production approach to editing and proofreading doesn't work.

In most publications departments, one challenge for managers is to have enough editors to handle the "peaks" but not so many that some are idle during the "valleys" of activity. Cost-conscious managers employ fewer staff than the busiest times require, to meet budget restrictions and avoid later layoffs.

In many organizations, editors are paid for their overtime work or earn time off *(comp time)* for it. This is not always the case. Editors are "professional staff" or "exempt" employees rather than hourly employees in most companies and agencies, so working overtime without compensation is often expected. Editors interviewing for positions should inquire tactfully, perhaps in the second interview, about the amount of overtime they might expect and whether they would be compensated for it. If they are told that overtime work is compensated on some projects but not all, they can assume that work on those projects will first be offered to experienced staff, to reward them for unpaid overtime work they did in the past.

SCHEDULING PUBLICATIONS WORK

It is impossible to say that any document of *x* pages will take *y* days to create, even in an organization that has created hundreds or thousands of documents. Even when an experienced editor can predict quite accurately how long each step might take, shoploading (the amount of other work to be done by staff assigned to this document), equipment failure, employee illness, and other factors necessitate that all schedules and budgets be expressed as *estimates.*

The best way for an editor new to scheduling to get an idea of a realistic timetable is to prepare a checklist similar to the one in Figure 10.3. Note that the delivery date to the client is first on the list. Subsequent deadlines *(milestones)* are listed in reverse chronological order, because schedules are developed backwards from the final deadline.

Once a final due date is set, each support group supervisor can consider any other scheduled projects and determine when the material must be submitted to finish it in time for the work in the next shop to be completed on schedule. For example, when the editor has determined the document size, number of copies, printing process to be used, binding method, and day and time of delivery to the client, the print shop supervisor can then estimate when the printing masters must be ready to print and copies bound.

The editor then provides the composition supervisor with the document's typesetting requirements and indicates when the job must be in the print shop. The amount of time needed for all publications activities is judged back to the beginning, when the writers submit text and graphics to the editor.

DOCUMENT SCHEDULE PLANNER

Activity	Date/Time Complete
Deliver to client	______________
Bind and check	______________
Print	______________
Compose (layout and makeup)	______________
Receive final art corrections	______________
Receive final typesetting corrections	______________
Edit and proof final	______________
Receive text and graphics changes	______________
Print review draft	______________
Process corrections for review draft	______________
Receive graphics from Art Department	______________
Receive text from typing	______________
Edit text and graphics	______________
Submit 100% of text and graphics to editor	______________
Submit 50% of text and graphics to editor	______________
Begin inputs of text and graphics to editor	______________

Figure 10.3. Document schedule planner.

The preliminary estimate of the schedule for the document should be used to create the final schedule, because the support group supervisors need more specific information about the document to refine their estimates. The cost planner in Figure 10.4 gives that detail (see below for how to fill one out).

Support group supervisors can use the information the editor enters in the left half of this form to determine how long their work will take. Many supervisors prefer that their production rates and labor and material costs not be public, even within the organization. For that reason, each support group's Cost Planner might be used only by the editor and the appropriate support group supervisor.

The milestones in the schedule planner are deadlines, but steps can overlap or take place concurrently. For example, edited text and graphics are submitted to typing and the art department as soon as possible; they are not held until after the "Edit text and graphics" deadline. Similarly, printing often begins before all of a long document is made up.

Be honest when discussing deadlines with your supervisor and with production group heads. Don't incorporate "breathing room" into the schedule unless told to by your supervisor. In most publications departments, support staff cooperate as well as they can to help editors meet deadlines. If support staff learn that unwanted overtime became necessary because an editor had set up a false deadline, that editor cannot expect much cooperation in the future.

Note that the Schedule Planner does not include any labor or material costs. The final project schedule (see Figure 5.54 for an example) can be established with information from the schedule and cost planners, and material costs can also be estimated at this time.

ESTIMATING PUBLICATIONS COSTS

Estimating the cost for publication department work on a document requires the editor to work closely with the supervisors of all the publications groups involved, providing them with enough information about the project to enable them to determine costs for work in their groups.

Many support group costs are not readily apparent to an editor, especially for materials. For example, the editor might remember to include the print shop labor cost for making plates for offset printing while overlooking the materials cost for the plates themselves. Be sure to describe the document requirements and schedule fully to the support group supervisors and ask them for estimates to cover *all* costs of work on the document in their shop. File the resulting estimates for future reference. As you gain experience, you will be expected to provide rough estimates of publications cost on your own.

No blanket formula for price can be developed for technical documents. Most organizations differ slightly in how the work is done and in their pricing factors. For example, most publications departments *burden* their rates to

DOCUMENT COST PLANNER

Publications Group and Activities	Labor Hours	Labor Cost	Material Cost	Total Cost
Print Shop				
Bind and check	___	___	___	___
Print pages				
# Regular press ___	___	___	___	___
# Offset press ___	___	___	___	___
# Oversize pgs ___	___	___	___	___
Number of copies ___	___	___	___	___
Bind review draft	___	___	___	___
Print Review Draft	___	___	___	___
# Regular pgs ___	___	___	___	___
# Oversize pgs ___				
Composition				
# Pages laid out ___	___	___	___	___
# Line reductions ___	___	___	___	___
# Photoreductions ___	___	___	___	___
# Pages typeset ___	___	___	___	___
# Captions, heads ___	___	___	___	___
# Tables	___	___	___	___
Simple ___				
Medium ___				
Hard ___				
Art Department				
# Graphics	___	___	___	___
Simple ___				
Medium ___				
Hard ___				
Typing				
# Pages ___	___	___	___	___
# Captions ___	___	___	___	___
# Tables	___	___	___	___
Simple ___				
Medium ___				
Hard ___				
Editing				
# Text pgs ___	___	___	___	___
# Graphics ___	___	___	___	___

Figure 10.4. Document cost planner.

help meet some of their overhead costs: They multiply labor rates for each support group by a factor (such as 1.2), thus charging the client an extra 20 percent in order to recover some of the group's operating expenses, such as maintenance and repair of equipment and the department's share of utilities costs. Burden rates vary widely and are rarely made public, even within a company. When editors receive and provide estimates of labor costs for publications work, they should be aware of and indicate whether the rates are burdened.

The document cost planner in Figure 10.4 is similar to the schedule planner in that it lists pertinent support groups in reverse chronological order. However, it must contain details about the document, break down support group activities, and provide space for materials as well as labor estimates. Often, organizations have estimating sheets for each support group so that the editor can simply enter the necessary information about the document (including the preliminary schedule), and the support group supervisors then fill in their cost estimates.

In the following discussion, the estimated document production rates and material costs are based on several companies and printing firms. The dollar figures assigned are arbitrary round numbers, and the labor rate for each group can for simplicity be used for all work in that group. Of course, the real rates vary by skill and experience within groups; any attempt here to present true rates for labor and materials costs would be misleading. Specific numbers are given so you have an idea of rates and costs and can use the numbers in the estimating exercise at the end of the chapter. In your work, merely substitute your organization's production rates and costs for those presented here. All costs presented in this chapter are burdened.

Estimating Writing Costs

If the document is written by technical staff, as is increasingly common when describing complex technology, costs for writing text and sketching graphics are estimated by the technical manager for the project because they are not part of the publications department's budget. Publications department activity begins with planning work on the document before the writers submit text and graphics to the editor. In this situation, editors are not responsible for assigning writers or determining schedules and costs for writing. This is fortunate, because no reliable production rates can be established for technical writing. Some types of information are intrinsically easier to gather, organize, and present than others.

Whether technical writers are staff in a technical department or a publications department, the project manager estimates the number of hours the writer will be needed, divides it by 2,000, and budgets that percentage of the cost for the writer's salary and benefits.

Estimating Editing and Proofreading Costs

As with production rates for technical writers, it can be difficult to establish production rates for technical editors. The quality of writers' work can vary greatly, from disorganized, confusing, and error-ridden to focused, clear, and correct. Manuscripts might be in messy handwriting in light pencil, or in word processed, double-spaced text with sketches of graphics that are nearly print quality. Assuming a middle ground, with some need for tightening organization and copyediting text and graphics, a reasonable estimate for editing, proofing the texts and graphics against the original, and light editing and proofing of revisions might be 1.5 pages per hour, at a labor cost of $40 per hour (which will vary according to the levels of editors involved). For light editing, estimate 5 pages per hour. For logging and transferring classified materials, add 4 hours for each 100 pages in the document.

Estimating Typing Costs

Editors can reduce typing costs by encouraging project managers to ask writers for clean, typed manuscripts; editing them neatly and clearly; and having typists use scanners with good optical character recognition (OCR) software.

For typing neatly handwritten material, estimate 2.5 pages per hour, and for typewritten material 4.5 pages per hour, at a rate of $15 per hour. Both rates include one correction cycle. Pages with complicated mathematics or large tables will of course take much longer. If there are many in the document, the editor should alert the typing supervisor before the text is submitted for a more accurate estimate of typing time. For simple (small) tables, estimate 3 per hour. For medium tables (half a page), estimate 2 per hour; for hard tables, 1 per hour.

Estimating Art Production Costs

Estimating the cost of creating figures is more difficult than for typing because of their range of complexity. Also, material costs must be estimated, such as 10¢ per sheet for drawing vellum (used on the boards) and 40¢ per sheet of typesetting paper for computer-drawn figures. It takes much longer to create illustrations than new editors and technical staff think, especially for scenarios, complex flow or organization charts, isometrics, and other intricate art. Using computerized equipment can save much time, on hard pieces especially, but minor corrections to a computer-generated piece can take far longer than the same correction would take "on the boards."

Specify the number of simple, medium, and hard pieces in the document, following your art department's standards for classification. For creating figures, estimate 1 hour per simple piece, 2.5 hours per medium piece, and 5 hours per hard piece, at a labor cost of $50 per hour. These rates include one correction cycle. Oversized graphics take about 50 percent longer. Hard pieces, which might take days to produce, should be brought to the attention of the art department supervisor as soon as possible.

For figures reprinted from other documents (*pick-up* figures), estimate 8 per hour at $25 for locating and sizing the figures, and 40¢ each for paper.

Estimating Photography Costs

Photography costs for a document can range from a few dollars for a resume photo taken in the office by an editor to thousands of dollars for a professional photographer covering a rocket test. This discussion assumes the organization's photo lab does the work.

For photographs of staff at the organization's facility, estimate 1 hour for shooting up to 10 people on one 36-exposure roll (or part of a roll), at $30 per hour. For photos of facilities or equipment, which often involve moving lighting or backdrop equipment from the photo lab, obtain from the photographer an estimate of the time required. Plan on the photographer taking 3 or 4 shots for each photo needed.

For lab work (retrieving negatives and developing and printing film), estimate $25 per hour per 36-exposure role, including materials. For standard 8 × 10-inch black and white (b/w) prints from negatives in the organization's photo files, estimate 0.5 hour per negative for retrieval from the files and processing of up to 10 prints per negative. Estimate $5 per roll or part of a roll of film.

Color processing costs more, as will services like airbrushing that can be estimated only when the photographer sees the original and has the changes defined. For photographs taken away from the organization's facility or by outside commercial photographers, costs can be far higher. Consult with other editors who have experience in estimating freelance photography costs.

Estimating Composition Costs

Typesetting, layout, and page makeup costs depend largely on the format required, the placement of graphics, and any page limitation for the document. If the document is being laid out differently from the standard company page format, especially in an unfamiliar format, layout and makeup will take longer. If graphics are *dropped in* (embedded within text) rather than placed on the page following their callout, makeup time can increase significantly (based in part on the number of graphics). If the layout must be tight to meet page limitations, additional time will be required to reduce graphics and rearrange text. For composition group activities, estimate $20 per hour for all costs.

For typesetting text and captions on disk, estimate 10 pages per hour. For headings and graphics captions, estimate 0.05 hour each or 20 per hour. For typesetting tables already on disk, estimate 0.5 hour per simple table, 1.0 hour per medium table, and 2.0 hour per hard table, and, again, 40¢ per page. For typesetting text not yet on disk, estimate 0.5 hour per page and 40¢ per sheet of paper.

For a tight layout and makeup of pages with dropped-in graphics and a familiar format, estimate 5 pages per hour; if the graphics are not dropped in, 8 pages per hour. For a loose layout, 10 pages per hour might be appropriate. Estimate 15¢ per page for each backing sheet (the *page frame*). For applying classification or proprietary markings at the top and bottom of each page, estimate 40 pages per hour. For page-numbering by hand any pages not already numbered by desktop publishing or typesetting equipment, estimate 50 pages per hour.

For camera shots to size figures, estimate 0.1 hour each, and $1 for materials. For sizing, cropping, and identifying photographs, estimate 20 per hour. Composition labor rates vary, but for convenience here estimate $20 per hour.

Estimating Printing and Binding Costs

Printing costs depend on the process and equipment to be used, whether printing is done on the front only or on both front and back, the number of oversized pages to be printed, the number of negatives to be stripped in, the type of binding, and several other factors.

Labor time

For black-and-white (b/w) offset printing of regular-sized pages, estimate 50 copies of 35 originals, or 100 copies of 30 originals, per hour, and 50 copies of 1 original in 0.05 hour. For b/w printing oversized pages offset, estimate 50 copies of 10 originals, or 100 copies of 7 originals, per hour, and 50 copies of 1 original in 0.1 hour. (The back of each oversized page also costs the oversized rate.)

For making offset plates, estimate 6 plates per hour. For stripping negatives, estimate 0.1 hour per negative.

For four-color printing of regular-sized pages, estimate 10 to 100 copies of 1 original in 2 to 3 hours. For full-color printing of 11 × 17-inch pages, estimate 10 to 100 copies of 1 original in 2 to 3 hours. For color separations, estimate 1.5 per hour.

For photocopying, estimate 50 copies of 25 originals, or 100 copies of 15 originals, per hour, and 50 (or fewer) copies of 1 original in 0.05 hour.

For folding, estimate 6,000 pages (one fold) and 5,500 pages (two folds) per hour.

For collating and spiral (plastic) binding, estimate 20 copies of a 100-page document, or 10 copies of a 200-page document, per hour. For collating and looseleaf binding, estimate 25 copies of a 50-page document, or 15 copies of a 100-page document, per hour, and 0.5 hour to affix covers to 10 binders. For collating and saddle-stitching, estimate 75 copies of a 16-page document, or 50 copies of a 32-page document, per hour.

For die-cutting covers, estimate 0.3 hour per setup and 0.005 hour to cut each cover.

For page-checking a document to make sure no pages are missing or duplicated, estimate 1,500 pages per hour (1 hour minimum).

Labor costs

Estimate $15 per hour average for b/w stripping, platemaking, and offset printing; $10 for photocopying; $20 per hour for color printing; $25 per hour for color separating; and $15 per hour for bindery work.

Materials costs

For 20-pound bond 8.5 × 11-inch paper, estimate 3¢ per sheet; for 11 × 17 inches, 5¢. For 60-pound paper, estimate 7¢ per sheet. For coated paper, estimate 12¢ per sheet.

For plates for offset printing, estimate $3 per plate. For three-ring binders, estimate $3 each for a 50- or 100-page document. For spiral (plastic) bindings, estimate 22¢ for each binder for a 50-page document and 36¢ for each binder for a 200-page document. For preprinted corporate covers, estimate 40¢ per set, either 8.5 × 11 inches or 11 × 17 inches. Budget cost of all covers for a multivolume document under one volume.

Remember that pages of a document can be *ganged* on a press for printing. For example, each printed 11 × 17-inch sheet to be saddle-stitched into an 8.5 × 11-inch document can have 4 pages printed on it. An 8-page document formed by two 11 × 17-inch sheets will have pages 2 and 7 on the front of one sheet and pages 8 and 1 on the back; pages 4 and 5 will be on the front, and pages 6 and 3 on the back, of the second sheet. When the sheets are folded, collated, and bound, the eight-page document will be formed by two sheets of paper.

CONCLUSION

Staffing, scheduling, and estimating the costs of publications work on a document are very important activities for editors. Beginning editors should discuss these editorial responsibilities with experienced editors to learn how projects are planned within the organization. Production rates and labor and material costs within publication department support groups may change regularly with increased volume, new equipment, and changing burden rates. Editorial training in most companies and agencies addresses these issues, as well as others. Beginning editors should concentrate on them, because they will become more important to editors as they gain experience and take on broad editorial responsibilities.

ESTIMATING EXERCISE

As an introduction to estimating, use the figures provided in this chapter to estimate the publications department budget for producing the proposal described below. For convenience, round dollar estimates to the nearest number divisible by 5, and assume a 1-hour minimum for each labor activity and a $5 minimum for each cost. Although overtime is a fact of life in technical publications work, assume no overtime is required on this proposal.

The proposal is in three volumes—Executive Summary, Technical and Management, and Cost—to be written by staff outside the publications department. The specifications follow.

Executive Summary

Unclassified full-color volume, 16 pages front and back, saddle-stitched. Twelve pages of neatly handwritten text to be typed; the resulting 8 pages of word processed text are to be typeset. Eight pages of neatly handwritten captions to be typed. Layout includes 14 new computer-drawn figures (5 simple, 5 medium, and 4 hard) and 20 b/w photographs (16 pick up from previous publications, 4 new, by staff), to be sized and laid out to standard company format. Two medium tables to be typed and then typeset (1 page each).

Printed offset in full color on 60-pound coated stock. A commercial printer will size the graphics; lay out and number the pages; do separations, stripping, and platemaking; and print the copies for $3,950 (including related materials but not including collating, binding, and page-checking). Preprinted company covers; 50 copies.

Technical and Management

Secret volume limited to 200 pages printed front and back offset on 20-pound bond. Spiral (plastic) bound. Word processed text from 130 pages of typed text. Title page, section headings (200 on 20 pages), and captions (125 on 16 pages) to be typeset.

Total of 40 photos: 30 pick-up (20 from Executive Summary, 10 from photo lab files), 5 new photos of facilities and equipment requiring 2 hours to shoot, and 5 new b/w photos of staff. Eighty new figures (40 simple, 30 medium, and 10 hard), and 40 pick-up figures. Ninety figures and 30 photos to be sized. Twenty-five new tables (10 simple, 10 medium, and 5 hard) to be typed; 5 pick-up tables. Five oversized pages (11 × 17 inches), double folded.

Tight layout to familiar format. Die-cut preprinted company covers revealing proposal name and volume number on title page. Appropriate classification markings on every page. Twenty copies, page-checked.

Cost

Unclassified but proprietary information volume of 40 pages photocopied front only on 20-pound bond, looseleaf bound. Four figures picked up from Executive Summary; 16 pages of handwritten text to be light edited and word processed, to form 8 pages of text and 2 pages with 3 medium tables.

One oversized page (11 × 17 inches), single folded. Loose layout. Thirty pages of computer printout requiring page-numbering only. Die-cut preprinted company covers to reveal typeset proposal title page with proposal name, volume number, and date between the cover and the front of the binder. Proprietary information markings on every page. Covers affixed to binders. Ten copies.

Answer Key

Executive summary

Editing and Proofreading	
16 pp @ 1.5/hr = 24 hrs @ $40 =	$ 960
Typing	
20 pp of text and captions @ 2.5/hr = 8hrs @ $15 =	120
2 medium tables @ 2/hr = 1 hr @ $15 =	15
Art Production	
Labor	
5 simple figures @ 1/hr = 5 hrs	
5 medium figures @ 2.5 hr/each = 12.5 hrs	
4 hard figures @ 5 hr/each = 20 hrs	
Total: 37.5 hrs @ $50 =	1,875
Materials	
14 sheets of paper for computer-drawn figures @ 40¢ =	5
Photography	
Labor	
Print 16 photographs from negatives in photo lab files @ 0.5/hr = 8 hrs @ $25 =	200
Shoot 4 new photographs @ 36/hr = 1 hr @ $30 =	30
Develop and print 4 new photos @ 1 roll/hr = 1 hr @ $25 =	25
Materials	
1 roll of film @ $5 =	5
(Cost of materials for developing and printing is included in the labor charges.)	
Composition	
(Labor and materials for sizing of figures and photographs, layout, and page numbering included in printer's estimate.)	
Labor	
Typeset 8 pages of word processed text @ 10 pp/hr = 1.2 hrs @ $20 =	25
Typeset 34 word processed captions @ 20/hr = 1.7 hrs @ $20 =	35
Typeset 2 medium tables @ 1/hr = 2 hrs @ $20 =	40
Materials	
14 pp of typesetting paper @ 40¢/page =	5
Printing and Binding	
(Labor and materials for separations, stripping, platemaking, and printing in printer's estimate.)	

Labor	
Stripping, separations, platemaking, and printing =	3,950
Collate and bind @ 75 copies of a 16-page document/hr = 1 hr @ $20 =	20
Page check 800 pages @ 1,500 pp/hr = 1 hr (minimum) @ $20 =	20
Materials	
400 sheets of coated paper @ 12¢ =	50
80 preprinted corporate covers (includes the covers for the other two volumes) @ 40¢ set =	35
Total	$7,415

Technical and management volume

Editing and Proofreading	
Edit and proof 200 pp @ 1.5/hr = 133 hrs @ $40 =	$5,320
Log and transfer 200 pages of classified text and graphics @ 25 pp/hr = 8 hrs @ $40 =	320
Typing	
130 pages of text @ 4.5/hr = 29 hrs @ $15 =	435
10 simple tables @ 3/hr = 3.3 hr @ $15 =	50
10 medium tables @ 2/hr = 5 hrs @ $15 =	75
5 hard tables @ 1/hr = 5 hrs @ $15 =	75
Art Production	
Labor	
40 simple figures @ 1/hr = 40 hrs	
30 medium figures @ 2.5 hr/each = 75 hrs	
10 hard figures @ 5 hr/each = 50 hrs	
Total: 165 hrs @ $50 =	8,250
Reprint 40 pick-up figures @ 8/hr = 5 hrs @ $25 =	125
Materials	
120 sheets of typesetting paper for computer-drawn figures @ 40¢ =	50
Photography	
Labor	
Retrieve and print 10 photos @ 0.5/hr = 5 hrs @ $25 =	125
Shoot 5 new staff photos @ 10/hr = 1 hr (minimum) @ $30 =	30
Shoot 5 new facilities photos in 2 hrs @$30 =	60
Develop and print 10 new photos @ 1 roll/hr = 1 hr @ $25 =	25
Materials	
1 roll of film @ $5 =	5
(Developing and printing materials included in labor charges.)	
Composition	
Labor	
Typeset 165 word processed captions @ 20/hr = 8.25 hrs @ $20 =	165
Typeset 200 word processed headings @ 20/hr = 10 hrs @ $20 =	200
Size, crop, and identify 30 photos @ 20/hr = 1.5 hrs @ $20 =	30
Size 90 figures @ 10/hr = 9 hrs @ $20 =	180

Lay out 200 pages @ 5 pp/hr = 40 hrs @ $20 =	800
Classification mark 200 pages at 40 pp/hr = 5 hr @ $20 =	100
Number 200 pages @ 50/hr = 4 hr @ $20 =	80
Materials	
236 sheets of typesetting paper (for text, title page, headings, and captions) @ 40¢/page =	95
120 figures and photos sized @ $1 =	120
Printing and Binding	
Labor	
Strip 40 photo negatives at 10/hr = 4 hrs @ $15 =	60
Make 200 plates for offset printing @ 6/hr = 3.3 hrs @ $15 =	500
Print 190 8.5 × 11″ pages @ 50 copies of 35 originals /hr = 5.4 hrs @ $15 =	80
Print 10 11 × 17″ pages @ 50 copies of 10 originals /hr = 1 hr @ $15 =	15
Setup diecut of preprinted covers @ 0.3 hr/setup = 1 hr (minimum) @ $15 =	15
Diecut 30 covers (including covers for Cost volume) @ 0.005 hr/each = 1 hr (minimum) @ $15 =	15
Double fold 200 11 × 17″ sheets @ 5,500 pp/hr = 1 hr (minimum) @ $15 =	15
Collate and bind 20 copies @ 10 copies of a 200-page document/hr = 2hrs @ $15 =	30
Page-check 4,000 pages @ 1,500 pp/hr = 2.7 hrs @ $15 =	40
Materials	
3,800 sheets of 8.5 × 11″ bond (for printing front and back) @ 3¢/ sheet, and 200 sheets of 11 × 17″ bond @ 5¢ =	125
200 offset plates @ $3/plate =	600
(Cost of preprinted corporate covers budgeted under Executive Summary.)	
Total	$18,210

Cost volume

Editing and Proofreading	
16 pages of text and tables @ 5 pp/hr = 3.2 hrs @ $40 =	$ 130
Typing	
16 pages of text @ 2.5/hr = 6.4 hrs @ $15 =	95
3 medium tables @ 2/hr = 1.5 hr @ $15 =	25
Art Production	
No costs.	
Composition	
Labor	
Typeset title page @ 10 pp/hr = 1 hr (minimum) @ $20 =	20
Size 4 figures @ 10/hr = 1 hr (minimum) @ $20 =	20
Lay out 40 pages @ 10/hr = 4 hrs @ 20 =	80
“Proprietary Information” mark 40 pages @ 40 pp/hr = 1 hr @ $20 =	20

Number 40 pages @ 50/hr = 1 hr @ $20 =	20
Materials	
4 figures sized @ $1 =	5
Printing and Binding	
Labor	
Photocopy 40 pages @ 50 copies of 1 original in 0.05 hr (3 minutes) = 2 hrs @ $10 =	20
Single fold 40 oversized sheets at 6,000 pp/hr = 1 hr (minimum) @ $15 =	15
Paste covers on 10 binders @ 20 binders/hr = 1 hr (minimum) @ $15 =	15
Collate and looseleaf bind 10 copies @ 25 copies of a 50-page document per hr = 1 hr (minimum) @ $15 =	15
Page-check 400 pages @ 1,500 pp/hr = 1 hr (minimum) @ $15 =	15
(Diecutting of covers budgeted under Technical and Management volume.)	
Materials	
360 sheets of 8.5 × 11″ bond paper @ 3¢ and 40 sheets of 11 × 17″ paper @ 5¢ =	15
10 looseleaf binders @ $3 =	30
(Cost of preprinted corporate covers budgeted under Executive Summary.)	
Total	$ 540

Total estimate

Executive Summary	$ 7,415
Technical/Management volume	18,210
Cost volume	540
Total	$26,165

Remember, the labor rates, labor costs, and material costs in this exercise are based on estimates from several commercial printers and inhouse print shops. In your work, substitute for these numbers those your organization uses. If your organization does not have rates set, use the description of activities in this chapter to help you create a set of rates for publications department activities.

CHAPTER ELEVEN

Grammar

Editors must know correct grammar to be able to edit technical materials effectively. Grammatical correctness is important because errors can distract readers from the content of the discussion. Also, grammatical errors can create a negative impression, leading readers to question the care with which the document was prepared and perhaps even the technical accuracy of the document.

A second reason for reviewing grammar and sentence structure is that editors must understand grammar and be able to talk about grammar with writers and other editors. Editing involves changing a writer's work, and very few writers like having their work changed. Editors must be able to explain to writers why they changed what the writers wrote. Editors cannot defend a change with a statement like "It sounds better *my* way." To most writers, it sounds better *their* way. Editors must be able to explain (and perhaps justify) their changes, especially to writers who are sensitive about anyone changing what they have written.

To most people, grammar is not an exciting subject. Some writing teachers say there's no point in studying grammar because it cannot be shown to improve writing. Some people point out the ungrammatical constructions everywhere around us and ask, "Why bother?"

College and university writing instruction in the last fifteen or twenty years has focused on the process approach to writing. It has concentrated on developing critical thinking skills, on *invention* (how to come up with what to say in the first place), on *composing* (getting the words down), on writing as self-expression. In the process approach, sometimes little attention is paid to correctness.

Concern for correctness is a concern for the final document, not for the process that created it. Consequently, some writing teachers dismiss correctness editing as proofreading, which to them is not part of the writing process. For many instructors who concentrate primarily on writing as process, proofreading, and even the editing that should precede it, are not included.

In business and government, there is more concern for the quality of the final product than the process. Professionals expect documents to be well organized, thorough, and accurate (well "written" in the teacher's sense), but they also expect documents to be *correct.*

Technical editors should not be surprised at writing problems they encounter. Rather, realize that little attention is sometimes paid to sentence-level correctness in writing instruction, especially for students who choose to major in technical fields rather than English, journalism, or communications. Unfortunately, part of an editor's job is to make up for the shortcomings of writing instruction in some colleges and universities.

Teachers have visited on students some foolish notions about effective writing. They have forbidden students to begin sentences with coordinating conjunctions or to end them with prepositions. Editors should realize that many technical writers were forced to learn those rules and punished if they broke them. Writers often mention them (especially when editors break them). You must know the rules before you can demolish such pedantry.

For these reasons, and because a growing number of technical experts speak English as their second language, writers in business and government commit errors like "between you and I" or "send it to John or myself." They make verbs agree with objects of prepositions instead of their subjects. They use vague or ambiguous pronoun references. They misuse words, using, for example, *hopefully* in place of *I hope.* Such expressions are like fingernails on a blackboard to some listeners or readers, especially editors. However, when the problem is pointed out, some speakers or writers merely ask, "Who cares?"

Editors care—or they should. It is no one's duty to fight for the preservation of the purity of English (or any other language) against the barbaric onslaught, but neither can editors adopt a purely descriptive stance, letting a grammatical irregularity pass so long as someone somewhere uses the construction. Editors should occupy a middle ground in which correctness in grammar, spelling, and punctuation establishes the writer or speaker as a person who is educated, observes decorum, and deserves to be listened to. Correctness creates an impression of intelligence, competence, self-control, and professionalism.

Some writers and many editors have a good command of grammar. They have an "ear" for grammar, recognizing that something is wrong when they come across an ungrammatical construction in someone else's writing or their own. Often they can fix the problem even if they cannot explain it. Although this instinct can help editors solve problems in a text, it is not sufficient. Editors must be able to explain why the new reading is preferable, in standard terminology.

TRADITIONAL GRAMMAR

Many approaches to grammar exist: traditional, structural, transformational-generative, among others. Most writers are familiar with traditional grammar, and the discussion in this chapter is based on it. To discuss words in terms of parts of speech rather than class I, II, and III words means more to writers who in freshman English or in middle or secondary school used a traditional grammar handbook.

In traditional grammar, words are grouped into parts of speech that have established basic uses. *Nouns* and *pronouns* indicate who or what; *verbs* express action or state of being; *adjectives* modify nouns; *adverbs* modify verbs, adjectives, and other adverbs; *conjunctions* join words or phrases; *prepositions* introduce phrases; and *interjections* form short, emphatic statements. Rules govern the conjugation of verbs, declension of nouns, and comparison of adjectives and adverbs. Words are grouped into phrases and/or clauses within sentences.

Most technical writers may not remember much beyond these terms. However, with them editors can take a writer over charted ground in explanations of grammatical changes in the writer's work, and this increases the likelihood that the writer will cooperate.

Editors who conduct writing workshops for technical writers and help individual writers with writing problems often must explain the grammar of example sentences. Also, much good advice teachers and editors give writers requires familiarity with some of the terminology of traditional grammar. For example, writers are often advised to limit the use of passive voice, which is good advice if they know what passive voice is. Also, teachers advise students to avoid comma splices by using a coordinating conjunction instead of a conjunctive or transitional adverb when joining main clauses with only a comma. Again, that's fine if students can define comma splices, conjunctive adverbs, and coordinating conjunctions.

Remember that very few writers in business and government enjoyed studying grammar, and even fewer, want to "go back to school" in the form of listening to an editor lecture on grammar. So while you should understand traditional grammar and use it when necessary to explain changes in a text or points in a writing conference, avoid dwelling on grammar for its own sake.

A solid grounding in grammar can instill confidence in your judgment and abilities. Some years ago, an editor was having the printing masters of a proposal volume reviewed by a company director who had not seen it previously. The director drew attention to a number of phrases he considered split infinitives, and he wanted several pages corrected and remade before printing. The editor knew that the changes could not be made without missing the deadline for printing. So he collared a more experienced editor, quickly explained the situation, and introduced him to the director. The second editor listened to the director's comments, agreed that some of the expressions could have been smoother, and complimented the director on his awareness of grammar. He discussed how some of the expressions would be difficult to state smoothly if the infinitives were not split, suggested that most readers would not recognize the expressions as errors, and explained that there is some disagreement over how egregious split infinitives really are. The director was satisfied that he had been painstaking in his review and let the project go to press on time without revision.

Most technical staff will not press their desire for correct grammar to this degree, but many are quite concerned that their writing be correct, especially in journal articles, where reputations are made.

A concern for grammar can result in much more than grammatical correctness, however. Understanding phrases and clauses can help editors and writers create more effective prose that focuses better on key points and presents its information in a variety of structures that hold readers' attention. Much good advice on writing style is based on familiarity with phrases and clauses. For example, key statements should be placed in main clauses, not in subordinate clauses or phrases, to receive greater emphasis. Also, rules of punctuation make little sense until one can recognize phrases and clauses in sentences and see how they work together.

PARTS OF SPEECH

In English, there are eight parts of speech:

- Nouns
- Pronouns
- Verbs
- Adjectives
- Adverbs
- Conjunctions
- Prepositions
- Interjections

The dictionary definition of a word lists the part(s) of speech as it is commonly used. A confusing property of English is that many words can function as more than one part of speech. *Light* can be a noun ("He turned off the light"), an adjective ("He carried a light package"), a verb ("He tried to light the fire"), and an adverb ("The sky was light blue"). Editors can often determine the part of speech only by the context.

Nouns

A noun is the name of a person, place, or thing. In its most common use, as the subject of a sentence, a noun is the who or what that acts.

> Louis deBroglie, semiconductors,
> Chicago, hypertext

Nouns have four main uses in English sentences:

- Subject of a statement

  Physics was her favorite subject in college.
  Mr. Adams arrived early for the meeting.

- Completer of a statement

  His goal is a 20 percent *increase* in production.
  Susan wanted to be a financial *analyst.*

- Object of an action

  This press brake can bend 1-inch cold drawn *steel* easily.
  Ms. Johnson tried to call *her* this morning.
  Sally gave *John* the *report.*

(*Report* is the direct object of the action; *John* is the indirect object.)

- Object of a preposition

  On the *table* were three models of the *prototype.*
  The codes can be displayed on the *screen.*

Nouns have other less common uses. Sometimes they define another noun more clearly, as the noun *typesetting* does in "We plan to order new typesetting equipment." Also, nouns sometimes indicate time, as in "I'll have the report finished Wednesday."

Pronouns

Pronouns are substitutes for nouns, used to avoid an awkward repetition.

> The President vetoed the appropriations bill because *he* thought *it* would be inflationary.

The two main types of pronouns are *personal* and *relative.* Personal pronouns refer to people and things. They have singular and plural forms, for the first, second, and third persons. When I speak of myself, I use the singular form *I* or the plural *we,* if I include another person; I use *you* when I address one or more persons; and I use *he, she,* or *it* (in the singular) or *they* (in the plural) for statements about people or things other than *I* or *you.*

	Singular	Plural
First person	I	We
Second person	You	You
Third person	He, she, or it	They

I saw the Joneses at the party. (First person, singular)
You should have been there. (Second person, singular or plural—the context must make clear which)
The rare-earth elements are those from cerium to lutetium. *They* are also referred to as the lanthanide series. (Third person, plural)

Personal pronouns also have forms to indicate possession. These forms, such as *my, mine, your, yours, his, her, hers, its, our, ours,* and *their,* function as adjectives and are discussed below. The objective form of a personal pronoun (below) indicates that the person is the object of an action (a verb) or a preposition.

	Singular	Plural
First person	Me	Us
Second person	You	You
Third person	Him, her, it	Them

Please send me a copy of the contract they sent you; I need it for my meeting with them.

Personal pronouns have compound forms that add *self* or *-selves,* emphasizing that the verb acts on the subject of the sentence: "John hurt himself when he tried to pick up the package." The compound form can add emphasis and clarity. In "To celebrate his promotion, John bought himself a CD player," *himself* establishes that John is the indirect object of the verb *bought.* If *him* were used instead of *himself,* a reader would assume *his* and *him* refer to a person other than John: "To celebrate Richard's *[his]* promotion, John bought Richard *[him]* a CD player."

In addition to personal pronouns, the relative pronouns *who, whom, which, that,* and *what,* and the related compound forms *whoever, whomever, whichever,* and *whatever* join dependent clauses to main clauses, as discussed below.

The interrogative pronouns *who, which,* and *what* introduce direct questions (sentences ending with a question mark). They also introduce indirect questions, in which a question is repeated indirectly in a declarative sentence. For example, in "She wondered who broke the beaker," *who* introduces the indirectly stated question: "Who broke the beaker?"

Indefinite pronouns such as *anyone, somebody, everyone, nothing, each,* and *anything* refer to no particular person or thing, whereas the demonstrative pronouns *this* and *that* (plurals *these* and *those*) refer to specific persons or things.

Verbs

Verbs are used to express an action or a state of being. Verbs are the heart of an English sentence. They can be more forceful and emphatic than any other part of speech.

writes
will be held
to call
running
having been stopped

Finite verbs can make a complete statement about a subject by themselves.

Alex edits very carefully.
We called her.

Both of these sentences are grammatically complete assertions about subjects; they contain finite *(defined)* verb forms. *Nonfinite* verbs cannot make a complete statement about a subject.

> Alex editing very carefully.
> We to call her.

These sentences don't make sense because they are incomplete. In each example, there is not enough information in the verb form to make a complete statement. Even if the meaning can be understood, as in the nonstandard "We be trying to finish the tests on schedule," the incorrect verb form creates a bad impression in a professional environment.

A complete sentence needs a finite verb, which may be stated ("John will drive") or understood ("Who will drive?" "John"). Nonfinite verb forms are often combined with finite forms to make complete statements. In the examples below, nonfinite forms are italicized.

> We plan *to finish* the airbag study on schedule.
> *Having finished* the study on schedule, we brought in the windshields *to be tested.*
> *Finishing* the airbag study took less time than we thought it would.

The examples above demonstrate the three types of nonfinite verb forms:

- *Infinitives*—to test, to be tested, to have tested (The *to* is considered part of the infinitive.)
- *Participles*—testing, being tested, having finished
- *Gerunds*—testing, finishing

Nonfinite verb forms are used in verb phrases, discussed later in this chapter.

Finite verbs have five main properties: *person, number, tense, voice,* and *mood.* Different verb endings or spellings express these properties. To review how to generate the finite and nonfinite verb forms, consult a grammar text.

Person and number

As personal pronouns have individual forms for the first, second, and third persons in both singular and plural, so do finite verbs. Conjugations (that is, the forms for the different persons, numbers, tenses, voices, and moods) can be found in good grammar texts.

Tense

The tense of a verb tells when the action occurred or will occur, or when something is, was, or will be. Verbs have six tenses:

Present, for an action or a state of being taking place now. Present tense can describe habitual action or a customary state of being.

> John types his letters at work.
> Anne drives 80 miles a week.
> Bill is late for work occasionally.
> The results of the study are surprising.

Past, for an action or a state of being that began and ended before the present time.

> John typed his letters at work yesterday.
> Anne drove 80 miles last week.
> Bill was late for work three times last month.
> The results of the study were surprising.

Future, for an action or a state of being to occur after the present time.

> John will type his letters at work.
> Anne will drive 80 miles next week.
> Bill will be late for work tomorrow.
> The results of the study will be surprising.

Present perfect, for an action or a state of being that began in the past and continues in the present.

> John has typed his letters at work for years.
> Anne has driven 80 miles this week.
> Bill has been late for work twice this month.
> The results of the study have been surprising.

Past perfect, for an action or a state of being that began and ended in the past.

> Before John's group hired a typist, he had typed his own business letters.
> Ann had driven 80 miles a week before she was transferred.
> Bill had been late for work five times before his supervisor talked to him about his tardiness.
> The results of the study had been surprising to people who were unaware of the situation.

Sometimes the simple past is used instead of the past perfect, as in "Before John's group hired a typist, he typed his own business letters." Also, sometimes the state-of-being verb and its completer are combined to simplify the sentence, as in "The results of the study surprised people who were unaware of the situation."

Future perfect, for an action or a state of being that will begin in the future and will end before some point in the future.

> John will have typed his letters by the time he leaves work.
> Anne will have driven 80 miles by 9:00.
> If Bill continues to be late as often as he has so far this year, he will have been late nearly twenty times by the end of the year.
> The results of the study will have been surprising to many people even before they are fully understood.

Note: Sometimes the simple future is used instead of the future perfect.

> John will type his letters before he leaves work.
> Anne will drive 80 miles before 9:00.

> If Bill continues to be late as often as he has so far this year, he will be late nearly twenty times by the end of the year.
> The results of the study will surprise many people even before they are fully understood.

Writers choose the verb tense according to what they have to say. Editors need to make certain that tenses are appropriate for the intended meaning, especially when the sentence contains actions or statements of being that occur at different times. If two tenses seem correct, choose the simpler form.

Voice

In English, verbs can be in *active* or *passive* voice. In active voice, the subject of the verb performs the action or exists in the state of being described by the verb.

> We store all project documents in Room 306.
> The freight train struck our delivery van at the railroad crossing.
> His wife is a neurosurgeon.

We and *the train* perform the action of their verbs; *His wife* exists in the state of being described by *is*. The verbs *store, struck,* and *is* are therefore in the active voice.

In passive voice, the noun or pronoun receives the action of the verb.

> Project documents are stored in Room 306.
> Our delivery van was struck at the railroad crossing.

Documents and *van* receive the action and are thus in the passive voice. (There is no passive voice for *being* verbs; *be* is used to form the passive voice of action verbs.)

The distinction between active and passive forms is often made in discussions of how to write well. To make documents more direct and effective, writers and editors should rely on active voice verb forms, which are shorter, more familiar, and more forceful. Also, active voice emphasizes the subject of the statement (the doer of the action, or the embodiment of the being verb).

Passive voice has its place in technical documents, however, and many teachers (and editors) mislead writers by encouraging them to avoid it. When a writer wants to emphasize what was done and not who or what did it, passive voice is appropriate. In the sentence "Project documents are stored in Room 306," who stores them isn't stated, suggesting that who does the storing is not important. The statement focuses on the subject, *documents,* which is actually the object of the action. In "Our delivery van was struck at the railroad crossing," what struck the van does not need to be stated. Readers will assume it was a train.

Mood

The *indicative* mood is what native speakers use almost always to make statements and ask questions. *Imperative* mood gives commands or instructions.

> Solder the lead to the second post.
> Never fill the tank above the red line.

The imperative form is the verb's infinitive without the *to*. The imperative mood is used extensively in sets of instructions.

The *subjunctive* mood is used to emphasize necessity, requirement, and insistence, or to indicate that a statement is contrary to fact.

> Our supervisor insists that we be at our desks by 8:00.
> If I were John, I'd decline the promotion.

The subjunctive forms *be* and *were* are used instead of the indicative *are* and *was* to emphasize the insistence in the first sentence and the untruth of the second. The most commonly used subjunctive forms are the infinitive (without the *to*), and a past tense form instead of the usual present tense. Subjunctive mood is not used frequently in technical documents other than contracts, where it may be common. Editors should be aware of situations that call for the somewhat rare subjunctive mood, lest they wrongly substitute an indicative form.

Adjectives

Adjectives provide further information about nouns. They may be one word or a combination of words (a phrase or a clause, discussed below).

> The white wire
> The resistors in the circuit
> The resistors that Smithby's manufactured

The adjective *white* tells which wire. The phrase "in the circuit" and the clause "that Smithby's manufactured" tell which resistors and are therefore also adjectives. Adjectives may come before or after the nouns they modify. In a sentence with a form of the verb *be,* an adjective may follow the verb and complete the statement about the subject.

> All the switches in the top row are white.

Adjectives modify nouns by telling whose, which, what kind, and how many. They can provide other kinds of information about nouns as well.

The personal pronouns discussed earlier also have adjective forms to indicate possession:

	Singular	Plural
First person	My, mine	Our, ours
Second person	Your, yours	Your, yours
Third person	His, her, hers, its	Their, theirs

> My computer has a 30-megabyte hard disk; so does theirs.
> Our office is down the hall from hers.

A very frequent error in documents is the confusion of *its* and *it's*. *Its* is a possessive adjective, as in "The Earth has a hole in its ozone layer." *It's* is a contraction, usually for "it is," as in "It's time everyone stopped using chlorofluorocarbons."

Adverbs

Adverbs provide further information about verbs, adjectives, and other adverbs. They may be one word or a combination of words (a phrase or a clause). Adverbs tell when, where, how, why, and if.

The motor performed well in the tests.
(The adverb *well* modifies the verb *performed,* telling how the motor performed.)
Paint the warning signs bright yellow.
(The adverb *bright* modifies the adjective *yellow.*)
Add the acid to the beaker very slowly.
(*Very* modifies the adverb *slowly.*)
The meeting will be held on Tuesday.
(The prepositional phrase *on Tuesday* modifies the verb *will be held,* telling when.)
The speed of light is slower than Michaelson thought.
(The dependent clause *than Michaelson thought* modifies the adjective *slower.*)

Adverbs that modify verbs usually come after the verb. Adverbs modifying adjectives and adverbs are usually placed before if they are a single word and after if they are phrases or clauses.

The adverb *there* sometimes confuses writers and readers because it can be used to change the normal word order in a sentence without telling "where." In "There are ten people waiting in the office," the normal sentence order ("Ten people are waiting in the office") has been reversed by using *there.* In this sentence, *there* is not an adverb that tells where the people are waiting; the prepositional phrase "in the office" does that. *There* is an *expletive,* a word that inverts normal word order, putting the subject of the sentence after the verb.

It can function in the same way. More common as the neuter pronoun, *it* can be an expletive, for example, "It is too late to change the design." Here *it* does not refer to a place or thing. Rather, using the expletive *it* enables the writer to avoid making an ungrammatical statement like "To change the design, not enough time exists." Of course, the idea could also be stated "There is not enough time to change the design."

Prepositions

Prepositions are used in phrases to describe a relationship to other words in the sentence. A preposition and its object (the noun or pronoun that follows it and completes the idea) form a prepositional phrase.

Place the report on the desk.
(*On* is the preposition in the prepositional phrase "on the desk," which modifies the verb *place* by telling where.)

The meeting will be held at 2:00.
(*At* and its object *2:00* form a prepositional phrase modifying the verb "will be held" by telling when.)

The position requires experience in programming.
(*In* and its object *programming* form a prepositional phrase modifying the noun *experience,* telling what kind of experience.)

They regretted her retirement because they were all fond of her.
(*Of* and its object *her* form a prepositional phrase modifying the adjective *fond,* telling how they were fond.)

Words and phrases commonly used as prepositions include:

above	despite	onto
according to	down	out of
after	during	outside
against	for	over
around	from	since
at	in	through
because of	in addition to	to
before	inside	under
below	into	until
by	near	within

Many words used as prepositions can also be other parts of speech, especially adverbs and conjunctions.

Look up.
(*Up* is an adverb, telling where to look. Note that it has no object and is not a preposition.)

Do it after you get home.
(*After* is a conjunction that joins the two clauses.)

Conjunctions

Conjunctions join words. They may be one word, such as *and, if,* and *therefore,* or a phrase, such as "either . . . or" and "as . . . as." Examples of how conjunctions join words, phrases, and clauses follow:

- *Nouns:* Either Janet or Tom will write the report.
- *Verbs:* Ms. Jones planned and conducted the research.
- *Adjectives:* The parts will be shipped in blue and white boxes.
- *Adverbs:* Mr. Smith will leave for Detroit today or tomorrow.
- *Prepositions:* The disks slide easily into and out of the drive.
- *Phrases:* The parts will be shipped by truck or by plane.
- *Clauses:* Production is ahead of schedule; as a result, we will be able to complete some modules early.

Different conjunctions can state a particular relationship (a thesaurus can provide others):

- *For adding an idea or example:* and, moreover, either . . . or, neither . . . nor, likewise, accordingly
- *For stating a contrast:* but, except, however, or, nor, yet
- *For indicating a result:* thus, therefore, as a result, consequently, accordingly, because, for, so
- *For establishing a condition:* if, whether, unless
- *For indicating time:* while, before, after, as, until
- *For indicating manner:* how, as, likewise

Conjunctions are often divided into three types:

- *Coordinating conjunctions:* and, but, or, nor, for, so, yet
- *Subordinating conjunctions,* such as: after, because, although, when, while, if, before
- *Conjunctive adverbs,* such as: however, therefore, thus, moreover, accordingly, as a result

The uses of these different types of conjunctions are discussed below, in the section on phrases and clauses.

Interjections

Interjections are words or very short phrases that make a quick, strong statement.

> No! I did not order a desk.

(The interjection *No* emphasizes the denial of the order.)

> What? You want to transfer me to New York?

(The interjection *What* emphasizes surprise or disbelief.)

> Note: These maintenance steps must be performed after each flight.

(*Note* emphasizes the statement that follows it.)

Interjections are usually set off in a separate sentence, followed by a question mark, an exclamation mark, a colon, or a period.

EXERCISE 1: PARTS OF SPEECH

In the following passage, label the part of speech of each word according to its primary dictionary definition, not according to its use in the sentence. For scoring convenience, use N, PRO, V, ADJ, ADV, PREP, C, and I for noun, pronoun, verb, adjective, adverb, preposition, conjunction, and interjection, respectively. For infinitives, consider *to* part of the verb.

> Good writing is clear, concise, and correct. To write well when we try to present information, we should first be sure that we understand what we want to say. Then we should decide how much our audience knows about the subject, so we can determine how much explanation they will need. All of the information that the audience needs should be presented, in language the readers can understand. The sentences should be simple and clear, for easier understanding. They should be as concise as we can make them and correct in grammar, sentence structure, spelling, and punctuation.

Answer Key

ADJ N V ADJ ADJ C ADJ V ADV C PRO V

Good writing is clear, concise, and correct. To write well when we try

V N PRO V ADJ V ADJ PRO PRO V

to present information, we should first be sure that we understand

PRO PRO V V ADV PRO V V ADV ADJ PRO N

what we want to say. Then we should decide how much our audience

V PREP ADJ N C PRO V V ADV ADJ N

knows about the subject, so we can determine how much explanation

PRO V V ADJ PREP ADJ N PRO ADJ N V V

they will need. All of the information that the audience needs should

V V PREP N ADJ N V V ADJ N

be presented, in language the readers can understand. The sentences

V V ADJ C ADJ PREP ADJ N PRO V V

should be simple and clear, for easier understanding. They should be

C ADJ C PRO V V PRO C ADJ PREP N N

as concise as we can make them and correct in grammar, sentence

N N C N

structure, spelling, and punctuation.

In this exercise, some words are used as a different part of speech than that listed for the first dictionary definition of the word. For example, *first* is an adjective in its most common use, but here it is used as an adverb. *Sentence* is a noun, but here it is an adjective that tells what kind of structure. Words are commonly "shifted" in this way to serve as other parts of speech. The next exercise focuses more on how the words are used in their sentences.

EXERCISE 2: PARTS OF SPEECH

Label the part of speech of each word in the following passage according to its use in the sentence (not according to its dictionary definition). Use the same abbreviations as in exercise 1.

> To write well, you should follow the rules of grammar and punctuation. Much writing that is correct in grammar and punctuation is still weak because the writer did not know enough about the material, did not organize the material effectively, or did not use appropriate wording to present the information.
>
> If a writer does not know the material or think carefully enough about it, the ideas or information may be unclear, disorganized, or illogical. Readers might lose their way or stop reading because they question the writer's abilities. If a writer does not use appropriate phrasing, some readers might not understand the discussion, and other readers might think it is oversimplified.
>
> One good way to avoid these writing problems and improve your writing is to increase your reading. Reading can give you more to say, and it can present models of organization for your material. If you consider how other writers develop their content, control their organization, and choose the right words, you can learn how to write better.

Answer Key

V ADV PRO V V ADJ N PREP N C N
To write well, you should follow the rules of grammar and punctuation.
ADJ N PRO V ADJ PREP N C N V ADV ADJ
Much writing that is correct in grammar and punctuation is still weak
C ADJ N V ADV V ADV PREP ADJ N V ADV
because the writer did not know enough about the material, did not

V ADJ N ADV C V ADV V ADJ N
organize the material effectively, or did not use appropriate wording to
V ADJ N
present the information.

C ADJ N V ADV V ADJ N C V ADV ADV
If a writer does not know the material or think carefully enough
PREP PRO ADJ N C N V V ADJ ADJ C
about it, the ideas or information may be unclear, disorganized, or
ADJ N V V PRO N C V N C PRO
illogical. Readers might lose their way or stop reading because they
V ADJ N N C ADJ N V ADV V ADJ
question the writer's abilities. If a writer does not use appropriate
N ADJ N V ADV V ADJ N
phrasing, some readers might not understand the discussion,
C ADJ N V V PRO V ADJ
and other readers might think it is oversimplified.

ADJ ADJ N V ADJ ADJ N C V PRO
One good way to avoid these writing problems and improve your
N V V ADJ N N V V PRO N V C
writing is to increase your reading. Reading can give you more to say, and
PRO V V N PREP N PREP ADJ N C PRO V
it can present models of organization for your material. If you consider
ADV ADJ N V ADJ N V ADJ N C
how other writers develop their content, control their organization, and
V ADJ ADJ N PRO V V ADV V
choose the right words, you can learn how to write
ADV
better.

In this exercise, some words are used as parts of speech other than their customary one. *Disorganized* and *oversimplified* are verb forms, past participles used here as adjectives. In the first use of *writing* in the third paragraph, the noun is used as an adjective to tell what kind of problems.

Also, a number of words in this exercise are not used in their most common meaning. The first dictionary definition of *well* is for the noun; the adverb use here is the second listed. The first and second uses of *enough* are as an adjective and pronoun; the third use is as an adverb, as it is used here, telling how much for the verb *know*. The first use of *more* is as an adjective, but here the second use as a noun occurs in the direct object of "can give."

SENTENCE STRUCTURE

Sentences are made up of groups of words known as *phrases* and *clauses,* which have one main grammatical difference: clauses have both a subject and a finite verb, and phrases do not. As the finite verb form is the heart of an English sentence, analyzing sentence structure should begin with attention to its finite verb form(s). Consider the following groups of words:

1. in the conference room
2. talking on the telephone
3. to call her Monday
4. finished the report early
5. will take a cab from the airport

Which groups contain finite verb forms? If you can't tell, review the discussion of verbs above before you read on. Of the five groups of words above, numbers 4 and 5 contain finite verb forms that with a subject could make clear, grammatically complete statements.

> Sharon finished the report early.
> Ms. Luong will take a cab from the airport.

The difference between a phrase and a clause is determined by the finite verb form. A *clause* has both a *subject* and a *finite verb,* as in "Jim studied botany in college" and "The meeting Wednesday will be held in Room 4." A phrase can have a subject (Jim, the meeting Wednesday) or a finite verb with or without modifiers (studied botany in college, will be held in Room 4), but it does not have both.

We distinguish phrases from clauses, but we also recognize two kinds of clauses: *main* and *subordinate* (dependent). A main clause can stand alone as a complete sentence.

> Jim studied botany in college.
> The Monday meeting will be held in Room 4.

A dependent clause cannot function as a complete sentence.

> when Jim studied botany in college
> because the Monday meeting will be held in Room 4

Neither of these clauses can stand alone as a complete statement. Each subordinate clause depends on more information—that in a main clause.

> When Jim studied botany in college, he went on field trips all over the Northeast.
> We can't reserve Room 4 for the entire week because the Monday meeting will be held in Room 4.

In each of these complete sentences, the dependent clause provides more information for the statement in the main clause, telling when Jim went on field trips and why we can't have Room 4 for the entire week.

Remember:

- A clause has both a subject and a finite verb.
- A phrase has either a subject or a finite verb.
- A subordinate clause cannot stand alone as a complete sentence.
- A main clause can be a sentence.

PHRASES

A phrase is a group of words with either a subject or a finite verb. There are six kinds of phrases:

- Prepositional phrases
- Noun phrases
- Verb phrases
- Participial phrases
- Infinitive phrases
- Gerund phrases

Prepositional Phrases

A prepositional phrase begins with a preposition (see above) and usually ends with the object of that preposition. The object, a noun or a pronoun, may be modified by one or more adjectives.

> in the laboratory
> after lunch
> because of her extraordinary courage

Prepositional phrases most often modify nouns and verbs.

> The building with the statue in front of it is the administration building.

The prepositional phrase "with the statue" is an adjective modifying the noun *building,* telling which building. The prepositional phrase "in front of it" is an adjective modifying the noun *statue,* telling which statue.

> The meeting will be held after lunch.

The prepositional phrase "after lunch" is an adverb modifying the verb *will be held,* telling when.

> The radioactive frogs escaped from the pond.

The prepositional phrase "from the pond" is an adverb modifying the verb *escaped,* telling where.

Editors should try to place prepositional phrases near what they modify, because the position of the modifier helps signal what it modifies.

> The woman saw the technician from the lab.

("From the lab" is an adjective telling which technician.)

> From the lab, the woman saw the technician.

("From the lab" is an adverb telling where she was when she saw the technician.)

The woman from the lab saw the technician.
("From the lab" is an adjective telling which woman.)

What at first appears to be a prepositional phrase may not be. In the sentence "The runners ran up the hill," *up* is a preposition, and the prepositional phrase "up the hill" modifies *ran.* However, in "The diners ran up a bill," *up* is an adverb used to change the meaning of *ran.* In "When the car reaches the top of the roller coaster, hold on to the bar," *on* and *to* are not prepositions; they are adverbs changing the meaning of *hold. Grasp* could be substituted for "hold on to."

Noun Phrases

A noun phrase is a group of words that acts as a noun. It consists of a noun and its modifiers.

the report
the book on my desk

Note that the modifiers of the noun in the noun phrase may be one-word adjectives or phrases functioning as modifiers. (The prepositional phrase "on my desk" modifies *book.*)

Verb Phrases

A verb phrase is a group of words with a verb at its center. A verb phrase may be a finite or nonfinite verb and its modifiers.

spoke loudly
slipped on the ice
to read aloud
tested yesterday

In the phrase "to read aloud," *to read* is an infinitive. In the phrase "tested yesterday," *tested* is a participle.

Participial phrases

A participial phrase contains a participle (a verb form usually ending in *-ing, -ed,* or *-en*) that may or may not have an object and modifiers. Participial phrases are used as adjectives.

playing
measuring samples
having tested
ended

A participial phrase provides more information about a noun, but it also suggests a verb's sense of action or being.

The surgeon performing the arthroscopy is Dr. Barnett.
("Performing the arthroscopy" tells which surgeon, so the phrase functions as an adjective.)

Having finished the report, he set it aside.
("Having finished the report" tells about him, so the phrase functions as an adjective even though it has a clear sense of action.)

Ms. Smith organized the meeting held yesterday.
("Held yesterday" tells which meeting, so the phrase functions as an adjective.)

A participial phrase should be placed near what it modifies, for clarity.

John saw Susan going into the meeting.
(Susan was going into the meeting.)
Going into the meeting, John saw Susan.
(John was going into the meeting.)

The present participle of a verb is formed by adding *-ing* to the infinitive form (without the *to*). The past participle (the third principle part of the verb) is listed in the dictionary definition; it may end in *-ed* or *-en,* or it may be an irregular form like *held* in "the meeting held yesterday."

Infinitive phrases

An infinitive phrase consists of an infinitive (the infinitive *to* and some form of a verb) and a modifier and/or an object. The most common infinitive form is the verb form found in the dictionary definition, with *to* before it to signal the infinitive. Infinitive phrases are used as nouns, adjectives, and adverbs.

to study
to have grown
to be examined

Infinitives and infinitive phrases can be used as the following parts of speech:

- Adjectives

The samples to test are in cooler 2.
(Infinitive *to test* tells which samples.)

Please bring in the slides to be examined.
(The infinitive *to be examined* tells which slides.)

- Adverbs

This laser is difficult to repair quickly.
(The infinitive phrase *to repair quickly* modifies the adjective *difficult,* telling how it is difficult.)

These seeds failed to germinate.
(The infinitive *to germinate* modifies the verb *failed,* telling how they failed.)

- Nouns

To be a corporate attorney was Susan's dream.
(The infinitive phrase "to be a corporate attorney" tells what Susan's dream was; the phrase is the subject of the sentence.)

He wanted to finish the report.
(The infinitive phrase "to finish the report" tells what he wanted; the phrase is the direct object of the verb *wanted.*)

Gerund phrases

Gerunds and gerund phrases are used as nouns. A gerund phrase consists of a gerund and a modifier and/or an object. The gerund may be a simple *-ing* verb form or a compound using *having* or *being* followed by a past participle.

Gerunds look like participles, but they are used as nouns, not modifiers.

> singing
> being seen
> playing squash

Gerunds can be used nearly every way a noun can. Gerund phrases can be subjects, subject complements, direct objects, appositives, and objects of prepositions, infinitives, and participles.

> His singing in the laboratory bothered his technicians.

(The gerund phrase "his singing in the laboratory" is the subject of the sentence.)

> On her vacations, she enjoys traveling in foreign countries.

(The gerund phrase "traveling in foreign countries" is the direct object of the verb *enjoys.*)

An infinitive phrase can often be substituted for the gerund phrase to create a simpler, smoother sentence.

> On her vacations, she likes to travel in foreign countries.

CLAUSES

Both kinds of clauses—main and subordinate—have a subject and a finite verb. Main clauses make complete statements and can stand alone as sentences. Subordinate clauses cannot stand alone; they are often called *dependent* clauses because they depend on a main clause for their meaning.

> Dr. Sullivan will move into her new lab when the wiring is completed.

In this sentence, "Dr. Sullivan will move into her new lab" is a main clause. It does not by itself tell when she will move in, but it can stand alone as a complete statement. "When the wiring is completed" is a dependent clause because it depends on the main clause for its meaning.

The distinction between main and dependent clauses is important to a technical editor. Statements made in main clauses get more emphasis than those in subordinate clauses. Being able to explain the difference between the two types of clauses can help writers follow two common guidelines for improvement: place important ideas in main clauses, to make sentences more forceful and emphatic, and avoid excessive subordination.

Main Clauses

A main clause is a statement or question that expresses a complete idea or piece of information.

The board of directors will meet Thursday.
Has Mr. Smith reviewed the proposal?
Although she wanted to work for IBM, she accepted a position with DavisCo.

In the third example, the subordinate clause "although she wanted to work for IBM" cannot stand alone; it is not complete in meaning. The main clause "she accepted a position with DavisCo" is complete in meaning and could function alone as a complete sentence.

When editors copyedit, they should identify the main clause in each sentence. By doing so, they can:

- Avoid a sentence fragment that does not make a complete statement.
- Avoid putting too many ideas or too much information in one sentence. (The old rule that a sentence is the statement of one idea is a good one.)
- Make sure the important idea or information is stated in a main clause, which the reader will focus on more than a phrase or a dependent clause.
- Punctuate the sentence correctly.

Subordinate Clauses

Subordinate clauses can be identified by subjects and finite verbs in clauses that cannot stand alone. Another way to identify most (but not all) dependent clauses is to watch for the words that introduce them: subordinating conjunctions and a variety of relative, indefinite, and interrogative pronouns, adjectives, and adverbs called *functional connectives.*

Subordinating conjunctions include:

after	that
although	unless
as	until
as if	when
because	where
before	whether
in order that	while
since	why

Subordinating conjunctions establish the logical connection between the dependent and the main clauses. That connection may be temporal ("I left the lab after he did"), causal ("I left the lab because he did"), conditional ("I'll leave the lab if he does"), spatial ("I left the key where he would find it"), among others. Whatever the connection, the subordinating conjunction establishes the relationship, but the word does not have a grammatical function in the dependent clause.

Functional connectives that link subordinate to main clauses may be:

Relative pronouns, such as *who* in "My brother, who lives in Missouri, teaches physiology." The relative pronoun *who* connects the subordinate to the main clause and has a grammatical function as the subject of the subordinate clause.

Indefinite and *interrogative pronouns,* such as *what* in "We are not sure what causes the rash." The pronoun *what* is indefinite (it doesn't replace anything in the main clause—it has no *antecedent*) and functions as the subject of the subordinate clause. There is little difference between indefinite and interrogative pronouns used to introduce subordinate clauses.

Relative adjectives, such as *whose* in "They are trying to find the employee whose car is parked in the lab director's space." The relative adjective *whose* (the antecedent of which is *employee*) functions as an adjective modifying *car* in the subordinate clause.

Indefinite relative adjectives, such as *whichever* in "Begin the test with whichever sample you want." The relative adjective *whichever* is indefinite (it has no antecedent in the main clause) and functions as an adjective modifying *sample* in the subordinate clause.

Indefinite and *interrogative adjectives,* such as *whose* in "I need to find out whose briefcase this is." The interrogative adjective *whose* introduces the indirect question (stated directly, "Whose briefcase is this?") and functions as an adjective modifying *briefcase* in the subordinate clause.

Relative adverbs, such as *when* in "I'll finish labeling those cabinets one day when I have time." The relative adverb *when* has the referent (but not antecedent) *day* in the main clause and modifies *have* in the subordinate clause.

Indefinite relative adverbs, such as *when* in "I'll finish labeling those cabinets when I have time." The indefinite relative adverb *when* has no referent in the main clause and functions as an adverb modifying *have* in the subordinate clause.

Interrogative adverbs, such as *where* in "We wondered where John left the key to the lab." The interrogative adverb *where* has no antecedent in the main clause. It introduces the indirect question in the subordinate clause (stated directly, "Where did John leave the key to the lab?") and functions as an adverb modifying *left* in the subordinate clause.

As the examples below demonstrate, subordinate clauses can be used as nouns, adjectives, and adverbs. Therefore, subordinate (dependent) clauses are identified as noun, adjective, or adverb clauses.

> Whoever wrote this report deserves a promotion.

(The subordinate clause "whoever wrote this report" is the subject of the verb *deserves* in the main clause. So the subordinate clause is a noun clause.)

> Use the dictionary that is on my desk.

(The subordinate clause "that is on my desk" provides more information about the word *dictionary* in the main clause, telling which dictionary to use. So the subordinate clause is an adjective clause.)

> After the Doppler effect was established using sound waves, it became important in studying electromagnetic radiation as well.

(The subordinate clause "after the Doppler effect was established using sound waves" cannot stand alone as a complete statement. Instead, it provides more information about the main clause "it [the Doppler effect] became important in studying electromagnetic radiation as well," telling when the effect became important. So the subordinate clause is an adverb clause.)

Many writing guides suggest using *that* for adjective clauses that are necessary to the meaning of the main clause (restrictive clauses) and *which* for clauses that are not necessary (nonrestrictive). In the previous example, the clause "that is on my desk" is necessary to be able to tell which dictionary to use; therefore, the adjective clause begins with *that,* not *which.* In the next sentence, notice how the distinction between *that* and *which* clarifies the meaning of the sentence.

> The Celsius system, which was developed by the Swedish physicist Anders Celsius, uses a scale that ranges from 0 to 100 degrees.

In the discussion of the article on auroras in Chapter 6 this sentence appeared: "The editor has split several paragraphs into shorter paragraphs that are easier to read." That is, the new paragraphs are easier to read because they are shorter. If the sentence read, "The editor has split several paragraphs into shorter paragraphs, which are easier to read," a general statement is made that shorter paragraphs are easier to read. The first sentence focuses on specific paragraphs, and "that" restricts the focus, so the subordinate clause is not set off by a comma. Nonessential dependent clauses are set off from the main clause with commas, however. Rules for punctuating subordinate clauses are presented in Chapter 12.

TYPES OF SENTENCES

Many technical writers have trouble writing simple, straightforward sentences. They tend to combine phrases and clauses to produce more complex statements, including many that are unnecessarily complex. Technical editors with a good understanding of phrases, clauses, and the types of sentences can more easily revise such sentences to make them clear, straightforward, and appropriate for their audience. Also, those editors can have more success when they discuss sentence structure with writers who have problems with it. Understanding the basic sentence patterns of English helps editors with both tasks (as well as with punctuating sentences correctly).

Each complete sentence states an idea or a piece of information. Sometimes, however, the meaning of a grammatically complete sentence may not seem clear. Sentences with pronouns may seem incomplete if taken out of context. For example, "He is an attorney" is a grammatically complete sentence, but the meaning is not clear until we substitute the name of the person for the pronoun *he.* Other types of sentences also rely on their context. "Absolutely not" on its own may not seem grammatically complete, but its meaning may be very clear in context: "Does this evidence suggest we should build a new test range? Absolutely not."

Editors should assume that nearly all readers will see contextual clues to meaning and not be puzzled by a sentence that seems incomplete. But watch out for true sentence fragments (see below).

Simple Sentence

A simple sentence states one complete idea or piece of information with only one main clause and no subordinate clauses.

> Galileo discovered the parabolic motion of projectiles in 1608.
> Was Einstein born in 1879?
> The report presents three solutions to the problem.

A simple sentence may have two or more subjects. Likewise, it may have two or more verbs.

> Susan and John were responsible for providing press coverage.
> In Tom's absence, Jim checked the stock and ordered parts.
> Bill and Angela monitored the tests and wrote the report.

Simple sentences are usually the most emphatic and clear. They are the easiest to punctuate correctly.

Compound Sentence

A compound sentence contains two or more complete ideas or statements of information with two or more main clauses and no dependent clauses.

> I stayed late to finish the report, and I was able to complete it.

(The two main clauses "I stayed late to finish the report" and "I was able to complete it" are joined by the conjunction *and.*)

> The spilled fuel ignited, and the rocket exploded.

(The two main clauses "the spilled fuel ignited" and "the rocket exploded" are joined by the conjuction *and.*)

Clauses can be joined in a compound sentence by a comma and a coordinating conjunction *(and, but, for, or, nor, so, yet),* and by a semicolon with or without a conjunctive adverb (see Chapter 12).

Sometimes editors choose to combine related simple sentences into compound sentences, to add variety or to elevate the level of a discussion so that it sounds less like an elementary school primer. More often, however, editors use a compound sentence to join syntactically related sentences, emphasizing the connection in meaning.

This is not to suggest, however, that editors should necessarily limit the frequency of simple sentences in a document. In documents intended for those who read English poorly because of limited education or language barriers, writers and editors should use a higher proportion of simple sentences to increase readability (see Chapter 4). Also, a higher proportion of simple sentences is appropriate in certain types of documents, such as instructions.

Complex Sentence

A complex sentence contains at least one main clause and at least one subordinate clause.

Send the report to Mr. Walker, who requested it.

(The adjective clause "who requested it" modifies *Mr. Walker* in the main clause.)

After you test the unit, take it to be packaged.

(The adverb clause "after you test the unit" modifies *take* in the main clause.)

Whoever broke this scale is in trouble.

(The noun clause "whoever broke this scale" is the subject of the verb *is* in the main clause.)

The idea or information in a sentence's main clause receives more emphasis than that in the subordinate clause, so key points should be stated in main clauses. When those statements require modification, the additional information may appear in a phrase or a dependent clause.

Some grammar texts identify a fourth type of sentence—the compound–complex sentence—which is defined as having at least two main clauses (making it "compound") and at least one subordinate clause (making it "complex"). However, a distinction between the complex and the compound–complex sentence is not particularly useful.

EXERCISE: SENTENCE STRUCTURE

Identify the phrases and clauses in the exercise below. Phrases are used within clauses, so put parentheses around each phrase and square brackets around each subordinate clause. Identify each main clause. For each phrase and dependent clause, state its type, the part of speech it is used as, and its function in the sentence. Finally, state the type of sentence. An answer key follows.

Sample

He wanted to work in New York because he enjoyed life in a big city.

"He wanted to work in New York" is a main clause.

"To work in New York" is an infinitive phrase used as a noun; it is the direct object of *wanted,* telling what he wanted.

"In New York" is a prepositional phrase used as an adverb to modify the verb *to work,* telling where he wanted to work.

"Because he enjoyed life in a big city" is a subordinate clause used as an adverb to modify the verb *wanted,* telling why he wanted to work there.

"In a big city" is a prepositional phrase used as an adjective to modify the noun *life,* telling what kind of life.

The sample is a complex sentence.

Exercise

1. In 1924, when he was a graduate student, de Broglie decided that particles must have wave properties.
2. Mounted on the back is a circuit breaker to control the power supply.
3. Mendel published the theories of inherited traits that he derived from studying garden peas; however, scientists failed to recognize the significance of his observations.
4. Jones told Allen to add whatever he wanted to the report.

Answer Key

1. In 1924, when he was a graduate student, de Broglie decided that particles must have wave properties.

"In 1924" is a prepositional phrase telling when de Broglie decided.

"When he was a graduate student" is a subordinate clause used as an adjective to describe the noun *1924.*

"De Broglie decided that particles must have wave properties" is the main clause with a subordinate clause within it.

"That particles must have wave properties" is a subordinate clause used as a noun, the direct object telling what de Broglie decided. It is introduced by the subordinating conjunction *that.*

The sentence is complex, with one main and one dependent clause.

2. Mounted on the back is a circuit breaker to control the power supply.

"Mounted on the back is a circuit breaker to control the power supply" is the main clause.

"Mounted on the back" is a participial phrase used as an adjective to describe the subject, *circuit breaker.*

"On the back" is a prepositional phrase used as an adverb to modify the participle *mounted,* telling where.

"To control the power supply" is an infinitive phrase used as an adjective to describe the subject, *circuit breaker.* (For the infinitive phrase, a subordinate clause could be substituted: "that controls the power supply.")

This is a simple sentence with only one main clause and no subordinate clause.

3. Mendel published the theories of inherited traits that he derived from studying garden peas; however, other scientists failed to recognize the significance of his observations.

"Mendel published the theories of inherited traits that he derived from studying garden peas" is a main clause with a subordinate clause within it.

"Other scientists failed to recognize the significance of his observations" is a main clause, linked to the first main clause with a semicolon and the conjunctive adverb *however.*

"Of inherited traits" is a prepositional phrase used as an adjective modifying *theories,* telling which theories.

"That he derived from studying garden peas" is a subordinate clause used as an adjective to modify *theories,* telling which theories. The subordinate clause is joined to the main clause by the relative pronoun *that,* which has *theories* as its antecedent and functions as the direct object of *derived* in the subordinate clause.

"From studying garden peas" is a prepositional phrase used as an adverb to modify *derived,* telling how he derived the theories.

"Studying garden peas" is a gerund phrase used as the object of the preposition *from.*

"To recognize the significance of his observations" is an infinitive phrase used as a noun, the direct object of *failed.*

"Of his observations" is a prepositional phrase used as an adjective to modify *significance,* specifying what is significant.

The sentence is complex; it has one main and one subordinate clause.

4. Jones told Allen to add whatever he wanted to the report.

"Jones told Allen to add whatever he wanted to the report" is the main clause, with a subordinate clause within it.

"To add" is an infinitive phrase used as a noun, the direct object of *told.*

"To the report" is a prepositional phrase used as an adverb to modify the infinitive *to add,* telling where to add.

"Whatever he wanted" is a subordinate clause used as a noun, the object of the infinitive *to add.* The subordinate clause is joined to the main clause by the indefinite relative pronoun *whatever,* which functions as the direct object of *wanted* in the subordinate clause.

The sentence is complex because it has a subordinate clause.

If you correctly identified the phrases and clauses in the exercise sentences, proceed with the discussion of grammar in this chapter and the discussion of punctuation in Chapter 12. If you had trouble, reread the sections on parts of speech, phrases, and clauses before you go on.

COMMON GRAMMATICAL ERRORS

Readers usually understand the meaning of a sentence with grammatical errors. For example, "Them's good apples" communicates clearly the intended meaning, but the errors in pronoun form and subject–verb agreement make the statement inappropriate for a technical document or presentation. Many less obvious grammatical errors do not interfere with meaning but create a negative impression and can distract readers who are accustomed to correct grammar. Also, some errors preclude clear communication, especially pronoun reference problems.

Most grammatical errors found frequently in technical documents can be grouped into three categories: sentence structure errors, verb agreement errors, and pronoun errors.

Sentence Structure Errors

Sentence structure errors include fragments, comma splices, and fused (run-on) sentences, as well as structural problems such as faulty parallelism, dangling or misplaced constructions, and split infinitives.

Sentence fragment

According to the definition of a sentence presented earlier in this chapter, a complete sentence has a subject and finite verb and can stand alone as a statement. Sentence fragments are phrases or subordinate clauses punctuated as though they were complete sentences but lacking the information needed to stand alone. Sentence fragments often result from incorrect typing of revisions.

> Errors: On account of bad weather down range.
> Whereas all the evidence suggests that it would be difficult to meet those design specifications.

If the context of the sentence fragment provides the rest of the necessary information, it might be allowed to stand as a deliberate fragment. Editors should use deliberate fragments rarely, however.

Comma splice

Joining main clauses with only a comma creates a comma splice, a major sentence error.

> Error: Mendel reported his findings in 1865, however most scientists were unaware of them for over twenty years.

If a comma joins the main clauses in a compound sentence, a coordinating conjunction should follow it. Main clauses can also be joined by a semicolon with a conjunctive adverb or by a semicolon alone.

Dangling or *misplaced modifier*

Separating modifiers from what they modify can create ambiguity.

> Error: The lab assistant was observed by the investigator removing samples from the laboratory.

To make a sentence clear, editors sometimes add a statement of what the modifier describes.

> Error: Stirring vigorously, the solution will change color.

When a modifier isn't clearly "attached" to what it modifies, the sentence can be confusing. In this example, the stirrer should be stated in the sentence, as in "When you stir the solution vigorously, it will change color," or the passive voice could be used: "When the solution is stirred vigorously, it will change color."

Split infinitive

A split infinitive occurs when the *to* and the verb of an infinitive are divided unnecessarily by a modifier.

> Error: We need to purchase this equipment soon to accurately test the prototype.

There is no need here to place *accurately* between *to* and *test,* so the phrase should read "to test accurately." Sometimes, however, it is best not to try to "unsplit" an infinitive.

> Error: He was instructed to merely discuss the contract, not sign it.

Placing *merely* after *discuss* focuses it on *the contract,* giving it the sense of "solely." Placing *merely* after *instructed* also undercuts the desired connection between *merely* and *discuss* and reduces the contrast between *discuss* and *sign.* "To merely discuss" is a split infinitive, but unsplitting it would alter the writer's intended emphasis. In this case, the split infinitive should stand.

Verb Agreement Errors

Verb agreement errors involve the wrong choice in number (singular or plural) because of problems with *false subjects,* "problem" nouns, and expletive constructions. When checking for verb agreement, also make sure the main clause has a subject and finite verb and is grammatically complete.

False subject

Errors in subject–verb agreement often result from making the verb agree with a word that is not its true subject. False subjects are usually objects of prepositions or other nouns coming between the subject and the verb.

> Error: Each of these sets of directions for starting the burners are hard to read.

The subject of *are* is neither *burners* (the object of the gerund *starting*) nor *directions* (the object of the second preposition *of*) nor *sets* (the object of the first preposition *of*). The subject is *each,* which is singular, so the verb must be *is.*

Problem noun

Nouns naming a collective group, such as *committee* and *class,* take a singular verb. If the members of the group are emphasized, a plural verb is used.

Errors: The committee on pricing guidelines meet in Room 602 at 3:00.
The committee votes at the end of their meetings.

In the first example, the committee as a group is emphasized, so the singular form *meets* should be used. In the second, voting by members should be emphasized with the plural verb *vote.*

Nouns ending in *-ics* are either singular or plural. Those that name a field such as *mathematics* or *physics* are normally singular. Nouns such as *acoustics* and *tactics,* referring to actions or characteristics, take plural verbs.

Certain pronouns and other expressions call for either singular or plural verbs. The pronouns *each, every, everybody, everyone, anybody, anyone, somebody, someone, nobody,* and *no one* take singular verbs. Expressions such as *either, neither,* and *the number of* take singular verbs, as do units of quantity such as 1 million dollars or ten tons. *Data* is either singular or plural; be consistent (and be prepared to argue, whichever usage you choose). A singular verb with *data* is common, treating it as a collective noun. The singular "datum" is rarely used. *None* takes a singular verb when used to indicate amount, as in "None of the coating darkens during testing." *None* takes a plural to indicate number, as in "None of the tests were conclusive."

Conjunctions affect verb agreement. When singular and plural subjects are joined by *or, either . . . or,* or *neither . . . nor,* the verb agrees with the subject *closer to it,* as in "Either aluminum or aluminum alloys are used to manufacture the part." Editors can solve the problem by placing the plural subject after the conjunction and making the verb plural. When *as well as* or *in addition to* is used to join subjects, the number of the noun following the conjunction does not affect the agreement of the verb with the noun or pronoun preceding the conjunction, as in "The president as well as the managers is going to the meeting."

Expletive construction errors

Expletives are used to place the subject after the verb. Using expletives occasionally leads to errors in subject–verb agreement. (This sentence is a *squinting* construction—*occasionally* could modify *using* or *leads,* so *occasionally* should be moved to follow *leads.*)

Error: There is a variety of ways to solve the problem.

In this sentence, *variety* has the sense of more than one way to solve the problem. It is not a collective noun calling for the singular subject *is,* so *are* should be used.

It is commonly used as an expletive to allow the subject of a sentence to follow the verb in a more natural sequence. The inverted word order in "It is unfortunate that Dr. Salk did not develop the polio vaccine sooner" is more natural than "That Dr. Salk did not develop the polio vaccine sooner is unfortunate," in which the subject (the noun clause) precedes the verb.

Pronoun Errors

Pronoun errors include lack of agreement with antecedents, vague pronoun reference, misuse of pronoun forms, and case errors.

Pronoun–antecedent agreement errors

Pronouns must agree with their antecedents (the words they take the place of). The pronouns listed above under problem noun verb agreement errors contribute to most agreement errors.

> Errors: Each of the documents are ready for printing.
> Everybody who use this program have trouble with it.

As *each* calls for a singular verb, *are* should be changed to *is* despite the fact that the sentence addresses more than one document. *Everybody* also calls for a singular verb, so *use* and *have* must be changed to *uses* and *has.*

Vague pronoun reference errors

The antecedent of a pronoun is sometimes unclear.

> Error: The test results indicated that the new composite will withstand rapid heating but become brittle. This prompted further study.

What prompted further study is not clear. Was it the test results or the brittleness? *This* and *that* frequently lead to vague pronoun reference.

> Error: A speaker can use their voice's tone and volume to emphasize a point.

This statement focuses on only one speaker. However, the plural "their" suggests more than one speaker. "Their" should be replaced by "his or her" to avoid the vague pronoun reference, or "a speaker" could be changed to "speakers" to solve the problem.

> Error: Lincoln did not have much formal education, but his Gettysburg Address is one of the world's great orations. That proves that you don't need a college degree to be a good speaker.

In the second sentence, *that* does not have a definite antecedent. A reader might not be sure what proves that you don't need a college degree, the idea that the Gettysburg Address is one of the world's great orations, or the fact that a man who had little formal education wrote it. The passage could be revised to read: "You don't need a college degree to be a good speaker. Lincoln did not have much formal education, but his Gettysburg Address is one of the world's great orations."

Editors can best check pronoun reference by asking: "Will readers know automatically what this pronoun refers to?"

Pronoun form errors

Compound forms of pronouns are sometimes substituted for simple pronouns.

Error: Marcia said she would send the report to President Gates and myself.

Many writers are awkward when they refer to themselves and have a (possibly unconscious) desire to make themselves seem more important. The same awkwardness or "puffing up" may lie behind the use of *hopefully* for *I hope,* which leaves the speaker out of a statement, making it sound objective rather than subjective.

Pronoun case errors

Personal pronouns are occasionally used in the wrong case, making a writer or speaker appear ignorant to a literate audience. Common errors include problems with prepositions ("between you and I"), gerunds ("He was upset with me being late"), subject complements ("She is older than him"), and subordinate clauses ("I met Smith, who the board chose president"). An increasingly common error is misusing the object of a preposition in the nominative case rather than the objective, especially when the object is *me.*

Error: Marcia said she would send the report to you and I.

A subject complement is sometimes put in the objective case rather than in the nominative (the subject case).

Error: The authors of the article are John Smith, Anita Towers, and me.

Pronouns should be examined carefully for correct case, because these problems jump out at literate readers and listeners (although verbal errors are usually judged more leniently). Notice how substituting *I* for *me* in the last example creates an awkward ending for the sentence. The sentence would be better restructured.

GRAMMAR-CHECKING SOFTWARE

A number of grammar-checking programs have been developed to help editors and writers correct grammatical errors in text. Such programs as Right Writer, Writer's Helper, VAX Grammar Checker, Grammatic Mac, and MacProof perform spelling checks; calculate readability scores; identify uncommon words, passive constructions, gender-biased diction, and long sentences; and search text for errors in sentence structure, grammar, and punctuation.

Use grammar-checking software when it is available, but never rely on it. The complexities of English sentence structure and the vagaries of English grammar have so far confounded software developers' attempts to construct grammar-checking software that can catch every problem. Some grammar-checking programs have troublesome designs; for example, many identify every passive verb form in the text and ask if the user would like to substitute an active form. As we have seen, passive voice is at times the best way to phrase a statement. In this situation, an editor would waste time reading and then skipping each passive-voice error message generated.

Grammar-checkers are becoming increasingly valuable, however, as new ones are developed and existing programs are improved. Readability indexing will continue to be one of their most useful features, but it will be some time before such software can check for problems in usability as well.

Editors should familiarize themselves with grammar-checkers because they can weed out some of the problems in a text after it is typed and before a micro-edit begins. However, you should never expect a grammar-checker to catch more than some of the errors in a document.

CHAPTER TWELVE

Punctuation

Checking the punctuation in a document is an important part of editing and proofreading, so editors must be able to recognize and correct errors in punctuation. Editors also need to be able to explain rules of punctuation and discuss why and how correct punctuation can make a document clearer, simpler, and more emphatic.

Speakers can use their voice's tone and volume, gestures, posture, and other rhetorical techniques in their delivery of their words. Writers cannot use those techniques for emphasis and clarity. They must rely on the words themselves. Punctuation allows writers to clarify sentence structure and therefore facilitate meaning.

Punctuation is often viewed as a way to represent orthographically the pauses a speaker would use if the text were read aloud. This is not the approach taken in this text. Rather, punctuation is considered a system of visible clues that clarify the structure and emphases of a sentence.

Correct punctuation is an editorial goal because, as with grammar and spelling, errors can suggest that the writer didn't know any better or didn't care enough about the document to make sure it was right. Either impression is a bad one.

Punctuation usage changes, and many punctuation rules are open to interpretation, making it much harder to master than spelling. A dictionary can resolve questions of spelling simply, but punctuation depends on the situation. Some grammar texts present so many examples of punctuation that it is difficult to see the rules that underlie them. Some style guides (such as *Chicago*) try to cover all of the rules and possible exceptions, making the discussion thorough but complicated. To many writers and editors, there seem to be too many conflicting rules.

Writing teachers and punctuation guides occasionally advise against "unnecessary" punctuation. The word "unnecessary" is unfortunate, because it suggests that writers can use punctuation arbitrarily. This is not the case. In technical documents, the tendency today is to use punctuation only when it is needed and to use as little punctuation as possible. Therefore, only necessary punctuation should be used. This discussion focuses on a limited number of general punctuation rules that cover most common situations. For a more comprehensive treatment, refer to *The Chicago Manual of Style* or one of the books on punctuation listed in the bibliography.

COMMAS

Writers have more trouble with commas than other marks of punctuation, especially if they don't have a good sense of sentence structure. Your study of grammar in Chapter 11 should have helped you see how phrases and clauses make up sentences. With that background, it will be relatively easy to determine where commas should be used. Some general guidelines follow.

- Use a comma to set off material not essential to the meaning of the sentence.

    Tuesday's meeting, which will be held in the main auditorium, will begin at 2:00.
    Please send a copy of the report to Ms. Sanders, the personnel director.
    Mr. Miller, who recently worked for Federal Dynamics, has joined the engineering department of Hopkins Electric.

- Use a comma to set off an introductory phrase or clause.

    Although sales were up, increased costs resulted in lower profits.
    Since last week, thirty units have been completed.
    While the shipping staff were at lunch, Mr. Clemmons called to find out when the order would be sent.

- Use a comma in front of a coordinating conjunction used to join main clauses.

    Ms. Smith does not need the report until Monday, but I finished it yesterday.
    Someone left the window open last night, and it's cold in the office this morning.
    Susan has applied for a transfer to Phoenix, and I think she'll get it.

- Use a comma to separate items in a series.

    The company has offices in Atlanta, Boston, and Chicago.
    This seems to be an effective, inexpensive solution to a very difficult problem.
    A bachelor's degree in business, three years' experience, and programming experience are listed as requirements for the position.

- Use a comma to separate parts of dates and addresses.

    February 7, 1983, was her first day at work.
    The manufacturing facilities are located in Tulsa, Oklahoma, and Jackson, Mississippi.
    His correct address is 5317 Curry Ford Road, M-201, Orlando, Florida 32812.

- Use a comma to prevent misreading.

  Inside, the technician was running tests.

  Frequently, depressed people need to be hospitalized.

Commas for Nonessential Material

Many sentences have phrases and clauses that clarify or add to the meaning but are not essential to its meaning. Set off such phrases and clauses with commas.

> Madelyn Anderson, the director of sales, earned an MBA at Harvard.

The phrase "the director of sales" tells us more about Anderson, but it is not essential to the meaning of "Madelyn Anderson earned an MBA at Harvard."

When you set off a nonessential word, phrase, or clause, make sure you do it fully. If the sentence were punctuated "Madelyn Anderson, the director of sales earned an MBA at Harvard," it would appear Madelyn Anderson is addressed in the statement. Such *direct address* calls for a comma (or two, if the name is not the first or last word of the sentence) to set off the noun in direct address. The director of sales would appear to be someone other than Anderson.

If a nonessential phrase or clause comes at the beginning of a sentence, it is called an *introductory* phrase or clause and is usually set off by a comma, as discussed below.

> The message read: "When you get back from lunch, please bring the Adams file to my office."

Sometimes it is difficult for editors to decide whether material is essential to the meaning of a sentence. Read it without the phrase or clause in question and see if the sentence makes sense.

> Give the file to Mr. Johnson, who is sitting by the door.
> Give the file to the man who is sitting by the door.
> Move Ms. Sullivan into the office with a window.
> SamData, of Newton, Massachusetts, is the primary supplier of that software.

In the first example, "who is sitting by the door" is set off by a comma, so the reader will assume that the clause provides additional but not vital information. The dependent clause could be left out, and the sentence would make sense: "Give the file to Mr. Johnson." In the second sentence, the meaning would be incomplete if the subordinate clause "who is sitting by the door" were left out, so no comma is used.

In the third sentence, *with* is not preceded by a comma. Therefore, the reader is to understand that the phrase is essential. The definite article *the* requires the information to tell which office. Without the phrase, the reader would not know into which office to move Ms. Sullivan.

In the final example, the phrase "of Newton, Massachusetts," is set off by commas to indicate that the additional information about SamData is not essential to the meaning of the main clause. If the phrase were necessary to distinguish SamData from another firm with the same name, no commas would be used.

> The fuel system, designed by Rota and Associates, has proved very effective.

The participial phrase *designed by Rota and Associates* is set off by commas, indicating that it is not necessary to the meaning of the sentence. It does not tell which fuel system; the reader will assume there is only one. However, if commas were not used the meaning would be very different.

> The fuel system designed by Rota and Associates has proved very effective.

Not using commas around *designed by Rota and Associates* indicates that the participial phrase is essential to the meaning of the main clause. That is, the phrase tells *which* fuel system has proved very effective. Readers will assume that there are other systems in addition to Rota's.

Consider the difference that commas around *SketchIt* would make in the following sentence.

> SamData's graphics program SketchIt is easy to learn to use.

If *SketchIt* is not set off by commas, the reader will assume that *SketchIt* tells which of SamData's graphics programs is easy to learn to use. The reader assumes that SamData has more than one graphics program, and *SketchIt* tells which, so *SketchIt* is essential to the meaning of the main clause.

If *SketchIt* is set off by two commas, the reader will assume that SamData has only one graphics program, which does not need to be named in the sentence because SamData has only one. The program's name could be left out.

A comma can also set up a quick contrast between two words or short phrases. At the beginning or end of the sentence, the contrast phrase is set off with a single comma. When the word or phrase comes elsewhere in the sentence, a pair of commas surrounds it.

> The reflecting telescope was invented by Newton, not Galileo.
> Despite the announcements in the papers, Truman, not Dewey, was declared the winner of the election.

The comma after *papers* sets off the introductory phrase "despite the announcements in the papers." The commas around "not Dewey" set off that contrasting phrase.

Phrases or subordinate clauses that tell when, where, or why at the *end* of the sentence are usually not set off by a comma. However, if the question of when, where, or why has already been answered by another phrase, that phrase or clause is set off.

I shall finish the report after I have talked to the auditors.
The shipment was delayed because of bad weather in Dallas.
I plan to have the test results by Friday, when I am supposed to report to Ms. Gorman on the tests.

Titles and degrees are usually not necessary to distinguish a person from another of the same name, so they are set off as nonessential information.

Constance Silverson, MD
Charles Henry, PhD

Exercise 1

Provide commas as needed according to the guidelines in this section.

1. Mr. Harris's son David who lives in Los Angeles will be coming to visit next weekend.
2. The main problem with the system temporary losses of power should be solved by Wednesday.
3. I plan to study the report while I'm on the plane to Seattle.
4. The software that is most appropriate for this project is ATAFALQUE.
5. Ms. Sanders has already left for Building 12 where the meeting will be held.

Answer Key

1. Mr. Harris's son David, who lives in Los Angeles, will be coming to visit next weekend.

Commas are needed after *David* and *Angeles* to set off the nonessential dependent clause "who lives in Los Angeles" from the main clause. The reader would assume that Mr. Harris has only one son named David, so the reader does not need the clause to tell which son. If the writer wished to indicate that Mr. Harris has only one son, commas are needed after *son* and *David* to set off the name.

2. The main problem with the system, temporary losses of power, should be solved by Wednesday.

The information in the phrase "temporary losses of power" is not essential to the meaning of the main clause; therefore, the phrase is set off by commas. If the writer wanted to emphasize the nature of the problem, the phrase should be removed from the appositive construction and emphasized, perhaps by restating the sentence: "The system's temporary losses of power should be fixed by Wednesday."

3. I plan to study the report while I'm on the plane to Seattle.

No commas are needed because the clause "while I'm on the plane to Seattle" establishes when the report will be studied and is at the end of the sentence.

4. The software that is most appropriate for this project is ATAFALQUE.

No commas are needed. The clause "that is most appropriate for this project" specifies which software and is essential to the meaning of the main clause.

5. Ms. Sanders has already left for Building 12, where the meeting will be held.

A comma should be used after *12* to set off the dependent clause "where the meeting will be held." Because *Building 12* establishes where the meeting will be held, the sentence is complete without the dependent clause.

Commas for Introductory Phrases and Clauses

Most sentences begin with the subject of the main clause followed by the verb. Sometimes, however, the main clause is postponed until after an introductory phrase or clause that introduces the main clause but is not part of it. An introductory phrase or clause is separated from the main clause with a comma.

> After I have lunch, I'll finish reading the report and evaluate it.
> Having named the new director of sales, Ms. Warren announced that she would also select a new director of finance.
> Sometime after 6:00 PM on the evening of March 28, someone tried to enter the secure area.

If the introductory phrase is short, no comma is needed.

> In June she will assume her new position.
> Thus we had to change the design.

Certain longer words that link a sentence to the previous one require a comma to emphasize the connection.

> Therefore, he had to change the design.

The longer word *therefore* takes a comma; the shorter *thus* does not. These comma use rules apply also in compound sentences, at the beginning of the second main clause.

> The building committee would not approve the architect's plan; therefore, he had to change the design.
> The new security system has been effective; since it was installed, no one has been able to enter the secure area without authorization.

If the two main clauses in a compound sentence are joined by a comma and a coordinating conjunction (as discussed below), an introductory phrase or clause at the beginning of the second main clause is often not set off.

> The manufacturing staff ordered an N/C drill press, but after it arrived they discovered that no one had been trained to set it up.

The clause "after it arrived" is introductory, but it is not set off because then the comma separating the clauses would not delineate the two main parts of the sentence as clearly.

Exercise 2

Provide commas as needed according to the guidelines in this section.

1. Although changes to the original design were proposed the head of Engineering chose to stick with the original design.
2. Some day I'll finish my course work and have my degree.
3. When Hopkins found out that Schwarz had resigned he organized a farewell party.
4. The meeting was to be held at the Mason Hotel; however we had to change the site to a larger hotel.
5. Aware of the commercial potential of the device the Smith Company bought the rights to it.

Answer Key

1. Although changes to the original design were proposed, the head of Engineering chose to stick with the original design.

A comma after *proposed* will separate the introductory clause "although changes to the original design were proposed" from the main clause.

2. Some day I'll finish my course work and have my degree.

No comma is needed because *some day* is a short introductory phrase.

3. When Hopkins found out that Schwarz had resigned, he organized a farewell party.

A comma after *resigned* separates the introductory clause "when Hopkins found out that Schwarz had resigned" from the main clause.

4. The meeting was to be held at the Mason Hotel; however, we had to change the site to a larger hotel.

The comma after *however* separates that conjunctive adverb from the second main clause of the sentence.

5. Aware of the commercial potential of the device, the Smith Company bought the rights to it.

A comma is needed after *device* to set off the introductory phrase "aware of the commercial potential of the device" from the main clause.

Commas for Joining Main Clauses

Two main clauses can be combined in one sentence in a number of ways, most commonly by a comma and a coordinating conjunction. (Main clauses can also be joined with a semicolon, as discussed below.) A comma with a coordinating conjunction suggests a close, simple relationship between the two clauses.

> The House passed the bill, but the Senate defeated it.
> The report must be submitted by June 15, or the contracting officer can declare us noncompliant with the contract.

A comma is placed in front of the coordinating conjunction. The coordinating conjunctions are: *and, but, or, nor, for, so,* and *yet.* If the two main clauses are short and closely balanced, you can join them with a comma.

> John is 34, Susan 27. (The verb *is* is understood in the second clause.)
> Interest rates rose, stock prices fell.

Conjunctions used as coordinating conjunctions can be used in other ways too. Be careful not to confuse the use of coordinating conjunctions to join main clauses with the other possible uses: to join subjects, verbs, objects, and so on.

> I gave the copies of the report to John, and Jim later asked me for one.
> I gave the copies of the report to John and Jim.

In the first sentence, *and* joins the two main clauses, so it is preceded by a comma. If the conjunction joins sentence parts that are not main clauses, a comma is not used except in a series. In the second sentence, *and* joins the two objects of a preposition, so no comma is used before it.

If a coordinating conjunction is used to join two main clauses but the comma is omitted, a *fused* or *run-on* sentence is formed.

> I gave the copies of the report to John and Jim later asked me for one.

Such sentences can be confusing. Without a comma after *John,* the sentence suggests on first reading that John and Jim received copies of the report.

If a coordinating conjunction is not used with a comma between two main clauses, a *comma splice* is formed.

> I gave the copies of the report to John, Jim later asked me for one.

On first reading, the comma after *John* may make it seem like the first item in a series. When readers reach *later,* however, they will realize that *John* and *Jim* are not items in a series. The editor's goal is to make every sentence clear on first reading so readers won't have to reread the sentence or stop midway through it and figure out the first part. In this sentence, there should be a stronger, clearer break between the two clauses. A comma after *John* is not enough, but a semicolon would be.

Exercise 3

Provide commas as needed according to the guidelines in this section.

1. The company had expected record profits in 1991 but the recall of its largest selling product cut sales drastically.
2. He must not have read the report or he would not have asked that question.
3. He wanted to be assigned to the Santa Barbara office however he was sent to Chicago.
4. The longer he works the more mistakes he makes.
5. The new design labs and the proposed office complex will be completed in 1993 and the manufacturing plant will be finished by 1995.

Answer Key

1. The company had expected record profits in 1991, but the recall of its largest selling product cut sales drastically.

A comma after *1991* is needed, before the coordinating conjunction *but* joining the two main clauses.

2. He must not have read the report, or he would not have asked that question.

A comma should be used after *report,* before the coordinating conjunction *or* joining the two main clauses.

3. He wanted to be assigned to the Santa Barbara office; however, he was sent to Chicago.

A semicolon, not a comma, should be used after *office* because *however* is not a coordinating conjunction. A comma is needed after *however* to separate that introductory conjunction from the second main clause in the sentence, "He was sent to Chicago."

4. The longer he works, the more mistakes he makes.

A comma without a conjunction should follow *works* to divide the two closely related, balanced main clauses.

5. The new design labs and the proposed office complex will be completed in 1993, and the manufacturing plant will be finished by 1995.

A comma is needed after *1993* because the coordinating conjunction *and* joins the two main clauses. No comma should be used after *labs* because *and* joins the two subjects of the first main clause, *labs* and *complex,* not two main clauses.

Commas for Items in a Series

Items in a series of nouns, verbs, adjectives, adverbs, prepositions, and even phrases and clauses should be separated by commas.

> John, James, and Anne attended the meeting.
> The office was decorated in beige, brown, and gold.
> We edited, typed, and proofread the report yesterday.
> His hobbies include playing golf, reading science fiction, and going to the movies.

Nearly all technical writers remember having learned that the comma after the next-to-the-last item in a series (the *serial comma*) is optional. But the decision whether to use a comma should be controlled by the style guide or the sense of the sentence, not the preference or habit of the writer, editor, or proofreader. In the following examples, notice how no comma after the penultimate item in the series can make the sentence harder to understand on the first reading.

> This article reports a perception of the state of veterinary medicine in China based on visits to two veterinary schools, a zoo veterinary hospital, two agricultural communes, and a veterinary research institute and meetings with educators, government officials, veterinarians, and animal scientists.

One way to solve the confusion ("visits . . . to meetings"?) is to clarify that the perception is based on two sources of information: (1) visits, and (2) meetings.

> This article reports a perception of the state of veterinary medicine in China based on visits to two veterinary schools, a zoo veterinary hospital, two agricultural communes, and a veterinary research institute, and on meetings with educators, government officials, veterinarians, and animal scientists.

The longer a sentence gets, the more an editor must try to make it easier for readers to handle. A second possibility here is to make two shorter sentences out of one long, confusing one.

The serial comma can be especially useful when listing items joined by conjunctions, such as names of firms or people. Without the comma, confusion may result.

> The comptroller examined prospectuses from Arthur Anderson, Price Waterhouse, Ernst and Young, Deloitte Haskins & Sells, and Coopers and Lybrand.

In this example, a comma is needed after "Sells" to indicate simply the separate items in the series.

Adjectives preceding a noun can form a different type of series. In general, two or more adjectives before a noun are separated by commas, but editors must be careful not to mistake an adverb for an adjective. Consider the following sentences.

> She is a bright, hard-working analyst.
> She drives a bright red car.

In the first sentence, *bright* and *hard-working* are adjectives modifying *analyst.* They could be joined by *and,* so a comma should be used. In the second sentence, *bright* is an adverb modifying *red.* No comma should be used.

Exercise 4

Provide commas as needed according to the guidelines in this and the previous sections.

1. Last week he flew to Chicago Minneapolis and Atlanta.
2. Wallace's enthusiastic dedicated staff completed the project on time not three days late.
3. Too many managers too few workers and too much paperwork characterize many American businesses.
4. To save the file press the function key type a file name and press the Enter key.
5. Revising editing and proofreading are important steps in creating technical documents.

Answer Key

1. Last week he flew to Chicago, Minneapolis, and Atlanta.

Commas are needed after *Chicago* and *Minneapolis* to separate the items in the series.

2. Wallace's enthusiastic, dedicated staff completed the project on time, not three days late.

A comma is needed after *enthusiastic* to indicate that it is one of a series of adjectives (with *dedicated*) modifying *staff.* Also, a comma is needed after *time* to set off the nonessential phrase "not three days late."

3. Too many managers, too few workers, and too much paperwork characterize many American businesses.

Commas should be used after *managers* and *workers* to set off the series of phrases.

4. To save the file, press the function key, type a file name, and press the Enter key.

A comma is needed after *file* to set off the introductory phrase "to save the file," as discussed earlier. Commas after *key* and *name* separate the items in the series of phrases that makes up the main clause's verb phrase.

5. Revising, editing, and proofreading are important steps in creating technical documents.

Commas after *revising* and *editing* separate the series of gerunds that makes up the subject of the main clause verb *are*.

Commas for Dates and Addresses

Commas are used to separate parts of dates and addresses. In general, editors should impose as few commas as possible, following the guidelines below.

Dates

When a month, day, and year are stated in the traditional date format, place a comma after the day of the month.

> Einstein died on April 18, 1955.

When you state the month, day, and year in the international or "military" format, do not use a comma.

> Einstein died on 18 April 1955.

If the date is followed by other text in the sentence, place a comma after the year as well. (In a sense, the year is appositive to the month and day.)

> Einstein died on April 18, 1955, and was buried in Princeton.

The comma after *1955* concludes the date; it does not signal the beginning of a new main clause.

When you do not state the year after a day, do not use a comma.

> November 13 is his birthday.

When you do not state the day of the month, do not use a comma.

> He visited Moscow in July 1989.

Addresses

Separate the name of a city from its county, state, or country by a comma. Likewise, a comma follows the county, province, state, or country after the name of a city or other municipality.

> The 1990 International Technical Communication Conference was held in Santa Clara, California.
> Cheyenne, Wyoming, is the windiest city in the United States.
> He was born in Raleigh, North Carolina, where his parents had been living for two years.

In the third sentence, note that the comma after *Carolina* concludes the address but also separates the nonessential dependent clause "where his parents had been living for two years" from the main clause.

The United States Postal Service now requests mailing addresses be written on envelopes in all caps, with no punctuation.

DR. DONALD SAMSON
5317 CURRY FORD ROAD M-201
ORLANDO FL 32812-8857

When the address is written out in a sentence, use a comma after the end of each line: Dr. Donald Samson, 5317 Curry Ford Road M-201, Orlando, FL 32812-8857.

On envelopes and in inside addresses, a one-word job title should be placed after the person's name, separated by a comma. If the title is two or more words, place it on the line below the name.

MR. ROGER BACON, PRESIDENT
TECHNICAL COMMUNICATIONS CONSULTANTS
P. O. BOX 206
ORLANDO FL 32857-0206

MS. MARY LAWRENCE
DIRECTOR OF PERSONNEL
TECHNICAL COMMUNICATIONS CONSULTANTS
P. O. BOX 206
ORLANDO FL 32857-0206

Use the two-letter Postal Service abbreviations for state names, and use the nine-digit ZIP codes if possible.

Exercise 5

Provide commas as needed, following the guidelines in this and the preceding sections.

1. Send your response by September 12 1993 to Mr. Carl Andrews 40 Hogan Street Suite 5 Knoxville TN 37994-2283.
2. On 10 March 1876 Bell was granted his patent for the telephone.
3. Albuquerque New Mexico would be a good site for the test facility.
4. The shipment was to go to Columbus Ohio not Columbus Georgia.
5. Punctuate the following address:
   MR. RUDOLPH ZIMMERMAN
   MARKETING DIRECTOR
   JONES BABCOCK STEELE AND COMPANY
   1400 WASHINGTON AVENUE SUITE 20
   CHICAGO IL 60606-0493

Answer Key

1. Send your response by September 12, 1993, to Mr. Carl Andrews, 40 Hogan Street, Suite 5, Knoxville, TN 37994-2283.

A comma should follow *12* to separate the day of the month from the year. A comma follows *1992,* as it is part of the date. Because Andrews's address is written out in a sentence, commas should follow *Andrews, Street, 5,* and *Knoxville.* No comma should follow *TN.*

2. On 10 March 1876, Bell was granted his patent for the telephone.

A comma is needed after *1876* to set off the introductory phrase "on 10 March 1876" from the main clause. (The date "10 March 1876" would not otherwise require punctuation.)

3. Albuquerque, New Mexico, would be a good site for the test facility.

Albuquerque and *Mexico* should be followed by commas.

4. The shipment was to go to Columbus, Ohio, not Columbus, Georgia.

Commas are needed after *Columbus, Ohio,* and *Columbus* to separate the names of the cities and states. The comma after *Ohio* also sets off the contrasting phrase "not Columbus, Georgia."

5. Punctuate the following address:
   MR. RUDOLPH ZIMMERMAN
   MARKETING DIRECTOR
   JONES, BABCOCK, STEELE, AND COMPANY
   1400 WASHINGTON AVENUE SUITE 20
   CHICAGO IL 60606-0493

Regarding the company name, check the company letterhead or a business directory for whether or not commas are used after Jones, Babcock, and Steele. No commas are added after *AVENUE* or *CHICAGO.*

Commas for Setting Off Quotations

One of the most difficult punctuation decisions for most editors is introducing quotations. Sometimes quotations are worked into sentences with only the quotation marks to delineate them; sometimes a colon or a comma introduces a quotation.

In general, use a comma to divide a short quotation from its attribution (the name of the speaker or writer).

> "Let's meet on Thursday instead," John said.
> "A better time to meet," John said, "would be Thursday."
> John said, "Let's meet on Thursday instead."

In these sentences, *John said* is the attribution or source tag. The commas after *instead, meet, said,* and *said* separate the attribution from the quoted statement.

If a quotation is longer or set up more formally, a colon is often used before the attribution.

> Of his landing on the moon, Neil Armstrong said: "This is one small step for a man, one giant leap for mankind."

Often a quotation in a sentence is not set off by punctuation, especially when the attribution does not come next to the quotation. If the structure of the sentence does not call for a comma before the quotation, editors should not use one.

> Neil Armstrong called his landing on the moon "one small step for a man, one giant leap for mankind."

The structure of this sentence is similar to "They named the star Xybos 3." *Xybos 3* completes the statement about naming the star just as "one small step for a man, one giant leap for mankind" completes the statement about what Armstrong called his landing on the moon. If a quotation is worked tightly into the structure of the sentence, it is usually not set off.

Sometimes the sentence structure requires a comma before or after the quotation.

> Neil Armstrong's landing on the moon was, as he put it, "one small step for a man, one giant leap for mankind."

The commas after *was* and *it* separate the nonessential dependent clause "as he put it" from the main clause. The comma after *it* does not introduce the quotation.

Exercise 6

Provide commas as needed according to the guidelines in this and the preceding sections.

1. "The monitor won't work until you plug it in Dave" Mary said.
2. Anne replied "I'm sure we can finish on time."
3. Mr. Hoffman argued that the company ought to "get out of the construction business and into fast food."
4. "In the back of the report" Amy noted "there is an appendix with the computer printouts."
5. The awards committee chose Jane Thompson the supervisor of customer service "Most Helpful Employee."

Answer Key

1. "The monitor won't work until you plug it in, Dave," Mary said.

A comma after *in* separates the noun in direct address, *Dave,* as discussed earlier in this chapter. Also, a comma is needed after *Dave* to separate the quoted statement from the attribution *Mary said.*

2. Anne replied, “I’m sure we can finish on time.”

A comma should be used after *replied* to separate the attribution *Anne replied* from the quotation.

3. Mr. Hoffman argued that the company ought to “get out of the construction business and into fast food.”

No comma should be used in this sentence, where the quoted statement is worked into the structure.

4. “In the back of the report,” Amy noted, “there is an appendix with the computer printouts.”

Commas should be used after *report* and *noted* to set off the attribution *Amy noted* from the quotation.

5. The awards committee chose Jane Thompson, the supervisor of customer service, “Most Helpful Employee.”

Commas are needed after *Thompson* and *service* to set off the nonessential phrase “the supervisor of customer service.” No comma is needed to set off the quotation. If “the supervisor of customer service” were dropped from the sentence, no comma would be needed: “The awards committee chose Jane Thompson ‘Most Helpful Employee.’ ”

Commas to Prevent Misreading

On rare occasions, commas might be needed to keep readers from misinterpreting a sentence.

> Inside, Dr. Sanders was setting up equipment.

The comma after the adverb *inside* prevents the reader from first thinking that “inside Dr. Sanders” is a prepositional phrase. The same problem exists in the following sentence.

> As you know, the problems facing the farm manager today are difficult and growing.

Without the comma, this sentence might be thought to mean “because you know the problems facing the farm manager today are difficult and growing,” and a main clause would be expected to follow. However, *problems* is not the direct object of *know;* it is the subject of *are,* and “as you know” is an introductory clause that should be set off by a comma.

Avoid inserting commas that are not needed to prevent misreading, as they can cloud the meaning of a sentence.

> Unusually, large doses of the drug can induce sedation in cattle and mice or excitement in horses, dogs, and cats.

In this sentence, the comma after *unusually* is incorrect, not just unnecessary. *Unusually* is not an introductory adverb that should be set off by a comma; the sentence does not say that the drug rarely has that effect. *Unusually* modifies the adjective *large,* indicating the type of dose that has that effect.

Notice how the next sentence can be interpreted two ways, depending on whether a comma is used.

> Frequently depressed people need to be hospitalized.
> Frequently, depressed people need to be hospitalized.

In the first example, *frequently* is an adverb modifying *depressed,* and no comma should separate the adverb from the adjective it modifies. The sentence is about people who are frequently depressed. In the second example, *frequently* is an introductory adverb in the sentence, and it should be set off by a comma. The sentence says that depressed people frequently need to be hospitalized. Notice how the meaning of the sentence would change again if *frequently* were the last word in the sentence.

Exercise 7

Provide commas as needed according to the guidelines in this and the preceding sections.

1. Most important people like to be left alone.
2. In January weather can delay shipments.
3. Often fallen trees block the road to the test range.
4. Ms. Jackson the head of Purchasing is calling on line 1.
5. The painting supervisor wants four new spray guns split between the first and second shift.

Answer Key

1. Most important people like to be left alone.

A statement about "most important people" would not require a comma. But if the statement is a general one about all people—"people like to be left alone"—a comma would be needed to set off *most important* from the main clause and prevent misreading.

2. In January, weather can delay shipments.

A comma is needed after *January* to prevent it from seeming to describe *weather.* The introductory phrase *in January* is a short one that would not normally be set off by a comma, as discussed earlier. However, in this sentence, a comma, is needed to prevent possible misunderstanding.

3. Often, fallen trees block the road to the test range.

A comma should follow the introductory word *often* to prevent the suggestion that some trees are often fallen. The awkwardness could also be solved by moving the word *often* closer to what it modifies: "Fallen trees often block the road to the test range."

4. Ms. Jackson, the head of Purchasing is calling on line 1. *Or:* Ms. Jackson, the head of Purchasing, is calling on line 1.

The punctuation depends on whether Ms. Jackson is being told the head of Purchasing is calling on line 1 (the first answer) or whether Ms. Jackson is the head of Purchasing and is calling someone. In the first reading, the comma sets off the direct address of Ms. Jackson. In the second, the two commas set off the nonessential appositive "the head of Purchasing."

5. The painting supervisor wants four new spray guns, split between the first and second shift.

This sentence needs a comma after *guns* to prevent the illogical suggestion that the supervisor wants the spray guns *split* rather than used whole. The problem could also be solved by choosing a synonym for *split* such as *distributed.*

See the exercise on all uses of the comma at the end of this chapter for more practice sentences.

SEMICOLONS

Semicolons have two primary uses: to join main clauses and to separate items in a series. They are also used in some reference systems.

Semicolons for Joining Main Clauses

Placing two main clauses in the same sentence emphasizes the connection between them, as discussed above. A semicolon after the first clause establishes more of a break than does a comma; however, the semicolon does unite the clauses and suggest a close connection between them.

Main clauses can be joined by a semicolon alone or by a semicolon and the proper conjunction.

> Tesla espoused alternating current; Edison attacked it.
> Until the seventeenth century, people thought maggots developed spontaneously in decaying meat; however, Francesco Redi demonstrated that they developed from the eggs of flies.

The conjunctions used after a semicolon to join the main clauses are called *conjunctive* or *transitional adverbs.* These include:

accordingly	in addition
also	indeed
as a result	instead

besides	moreover
consequently	nevertheless
furthermore	then
hence	therefore
however	thus

Remember that if a coordinating conjunction *(and, but, or, nor, for, so, yet)* joins the clauses, a comma should be used in front of it.

Again, most of the conjunctive adverbs listed above would be followed by a comma if used with a semicolon to join two main clauses. However, short conjunctive adverbs such as *hence, thus,* and *then* are rarely followed by a comma. In general, monosyllabic conjunctions are run into the text when they appear after a semicolon.

Semicolons for Separating Items in a Series

Items in a series should be separated by punctuation, usually a comma (see above). But when the items in the series are clauses or long phrases, or when the items have commas within them, use semicolons.

> The primary researchers were Dr. Thomas Kent, the director of the Medical School at the University of Biloxi; Dr. Patricia Lancaster, professor of physiology at the University of Bristol; and Dr. William Simpson, of Kellogg Pharmaceuticals.
>
> "Other gross observations were distributed between the age groups and included multiple, firm, elevated mucosal nodules 2- to 4-cm in diameter in the second stomach compartment; occasional encysted parasites in blubber and skeletal muscle; focal ulcerations in the first and second stomach; and petechial and ecchymotic hemorrhage of serosal surfaces of abdominal viscera." (From the *Journal of the American Veterinary Medical Association,* Vol. 187, No. 11 (1985), p. 1138.)

If commas were used instead of semicolons in the first example, a reader might think there were five primary researchers: (1) Dr. Thomas Kent, (2) the director of the Medical School at Biloxi, (3) Dr. Patricia Lancaster, (4) the professor of physiology at Bristol, and (5) Dr. William Simpson of Kellogg Pharmaceuticals. In the second example, the semicolons separate the observations more clearly than would commas.

Occasionally a semicolon should be used between numbers that contain commas.

> In the U.S. in 1985, deaths from all causes totaled 2,086,400; 96,000 more than the year before.

Semicolons for Separating References

Semicolons are sometimes used in multiple references to works cited. The system recommended by many style guides lists the name of the author and the year of publication in parentheses. If more than one source is cited, a semicolon separates them.

> Other researchers (Smith 1982; Lee 1984) concur.

Exercise

Provide semicolons and commas as needed according to the guidelines presented here.

1. The first round of tests proved inconclusive however we were able to eliminate some candidate materials.
2. Setting up business as a consultant involves several legal requirements: drawing up incorporation papers filing them with the state obtaining a business license from the city county and perhaps state and filing a notice of the company name.
3. We tried to convince the supervisor to postpone the meeting but she refused.
4. Mr. Burns suggested an overhead rate of 30 percent Mr. Fletcher argued for a much higher rate.
5. The shipment from the Dayton plant will be delayed five days thus we won't be able to start assembling until next week.

Answer Key

1. The first round of tests proved inconclusive; however, we were able to eliminate some candidate materials.

Because the conjunctive adverb *however* joins the two main clauses, a semicolon is needed after *inconclusive* and a comma after *however.*

2. Setting up business as a consultant involves several legal requirements: drawing up incorporation papers; filing them with the state; obtaining a business license from the city, county, and perhaps state; and filing a notice of the company name.

The series of requirements introduced by the colon has four separate items. Semicolons are needed after each because the third requirement, "obtaining a business license from the city, county, and perhaps state," is itself a series. Commas would not be enough to differentiate the items in the two series.

3. We tried to convince the supervisor to postpone the meeting, but she refused.

A comma should follow the word *meeting* because the coordinating conjunction *but* is used to join the two main clauses.

4. Mr. Burns suggested an overhead rate of 30 percent; Mr. Fletcher argued for a much higher rate.

This sentence has two main clauses but no conjunction joining them. Therefore, a semicolon should be used after *30 percent* to join the clauses correctly. The clauses are too long and not balanced enough for only a comma to join them.

5. The shipment from the Dayton plant will be delayed five days; thus we won't be able to start assembling until next week.

The conjunctive adverb *thus* joins the main clauses; thus a semicolon should be used after *days*. No comma should be used after *thus* because it is short and not as disruptive as a conjunction like *therefore* or *moreover*.

COLON

The colon has two main uses: to introduce a quotation formally and to introduce and emphasize a word or phrase. It also has a function in references, book titles, and business letters.

Colons to Introduce a Quotation

To emphasize a quotation at the end of a complete sentence, place a colon before it.

> John stated: "I saw no one enter Building 19."
> He quoted Ambrose Bierce's definition of patriotism: "Patriotism is the last refuge of a scoundrel."

If you do not want to set up the quotation this formally, or if you want to vary the way you introduce quotations, work it into your sentence in one of the following ways.

> John stated that he saw no one enter Building 19.
> He disliked Ambrose Bierce's definition of patriotism as "the last refuge of a scoundrel."

In the first example, no quotation marks are used for the indirect discourse; the quotation becomes a noun clause telling what John stated. In the second example, quotation marks are used to indicate that Bierce's original wording is being used, but it is introduced less formally (without a colon) and is worked into the structure of the sentence.

Colons to Emphasize a Word or Phrase

To lead up to a word or phrase that explains or summarizes the preceding part of the sentence, place a colon in front of the word or phrase. The colon indicates that you are about to state your point.

> Connelly decided there was only one solution to the problems in the personnel department: fire the supervisor.

> The water from the burst pipe damaged several items: four word processors, one typewriter, and two calculators.

A colon is a major break in the sentence that emphasizes what follows. It is often used after a statement that ends with "the following." The second example above could read, "The water from the burst pipe damaged the following: four word processors, one typewriter, and two calculators." However, a colon would not be used if the sentence read "The water from the burst pipe damaged several items, including four word processors, a typewriter, and two calculators." When the word or phrase to be emphasized is short, an integral part of the sentence, a colon is not used. In "The first-shift supervisors are William Catesby and Alice Jones," "William Catesby and Alice Jones" completes the identification of the supervisors by naming them. No colon is needed.

If the emphasized word or phrase comes at the beginning of the sentence instead of the end, use a dash.

> Lower corporate and property tax rates, lower utility costs, no state income tax—all these made Jonesville an attractive site for the new plant.

Other Uses of Colons

Colons are used to state the full title of a book with a subtitle and to separate the name of the publisher from the place of publication in references.

> Peters, Thomas J. and Robert H. Waterman, Jr. *In Search of Excellence: Lessons from America's Best Run Companies.* New York: Harper and Row, 1982.

Colons are used in business letters and memoranda following the salutation or subject lines.

> Dear Mr. Smith:
> Subject: Request for rate information

Exercise

Provide colons as needed according to the guidelines in this section.

1. Mrs. Montague introduced the two new marketing specialists Jane Amodio and Jack Bridges.
2. Just before the downpour ended his dedication speech, Mr. Armbruster said "We will remember this day for many years."
3. Schell, John and John Stratton, *Writing on the Job A Handbook for Business and Government,* New York New American Library, 1984.
4. When the supervisor said we'd have to work on the analysis some more tomorrow, I remembered Macbeth's speech about "tomorrow, and tomorrow, and tomorrow."

5. Mrs. Montague introduced Jane Amodio and Jack Bridges, the two new marketing specialists.

Answer Key

1. Mrs. Montague introduced the two new marketing specialists: Jane Amodio and Jack Bridges.

A colon should be used after *specialists* to indicate that they will be listed in the sentence.

2. Just before the downpour ended his dedication speech, Mr. Armbruster said: "We will remember this day for many years."

A colon is needed after *said* to introduce the quotation.

3. Schell, John and John Stratton, *Writing on the Job: A Handbook for Business and Government,* New York: New American Library, 1984.

A colon after *Job* separates the subtitle *A Handbook for Business and Government* from the main title of the book. Also, a colon is needed after *New York* to separate the name of the publisher from the place of publication. (Other formats might call for periods rather than commas after *Stratton, Government,* and *Library.*)

4. When the supervisor said we'd have to work on the analysis some more tomorrow, I remembered Macbeth's speech about "tomorrow, and tomorrow, and tomorrow."

No colon should be used to introduce the quotation, which has been worked into the structure of the sentence.

5. Mrs. Montague introduced Jane Amodio and Jack Bridges, the two new marketing specialists.

No colon should be used because the names are not led up to and emphasized as in the first sentence. Instead, the nonessential phrase "the two new marketing specialists" is set off from the main clause by a comma.

APOSTROPHE

Twenty years from now, apostrophes may have disappeared from all but the most formal (and best edited) writing. Apostrophes are used incorrectly more often than any other mark of punctuation, even by educated writers. Editors must understand their two main uses: to indicate possession and to indicate contraction. Apostrophes also form the the plurals of some letters, numbers, and acronyms.

Apostrophes to Indicate Possession

To indicate that some person or thing is possessed by (or closely related to or associated with) another person or thing, use an apostrophe to indicate possession. For either singular or irregular plural nouns, add *'s* to form the possessive.

> Saturn's rings
> Mars's diameter
> the oscilloscope's power source
> the women's room

If the possessor is a regular plural (formed by *-s* or *-es*), add just the apostrophe. Make sure to form plurals properly before you add the apostrophe. The house in which the Jones family lives is "the Joneses' house." "Jones's" is the singular possessive, as in "Deliver the package to Dr. Jones's lab."

When two or more people possess something, be careful to use the apostrophe correctly to avoid confusion. For example, "I read Tom and John's reports" means I read the reports written by Tom and John. But "I read Tom's and John's reports" implies I read two or more reports, at least one written by Tom and one by John.

The possessive form in gerund phrases describes someone's actions.

> Mendel's experimenting with peas led to important discoveries in genetics.

With a personal pronoun, the possessive form (without an apostrophe) indicates the possession.

> Mendel contributed significantly to the field of genetics through his experimenting with peas.
> Her careful proofreading uncovered several errors in the final draft of the report.

Sometimes the apostrophe is not used when there is a close association between the possessor and the possession, especially when the possessor is not a person.

> I've already read most of the NIH report.

Apostrophes to Indicate Contraction

Contractions (phrases shortened into words) are formed with an apostrophe to indicate that one or more letters have been dropped.

> can't
> he's
> won't
> I'm

Editors should avoid contractions in formal reports and limit them in correspondence. However, contractions are acceptable in in-house documents such as memos, announcements, instructions, and drafts of reports. Some contractions are so common that to say *cannot* instead of *can't* is to express special emphasis that may be inappropriate.

One of the most common apostrophe problems involves *its* and *it's*. *It's* is a contraction for *it is* or *it has*. *Its* is the possessive form of the pronoun *it*.

> It's easy to recognize the operator's manual by its yellow cover.

In this example, *it's* is a contraction for *it is*, and *its* is a possessive pronoun taking the place of *the manual's*. The sentence could read: "It is easy to recognize the operator's manual by the operator manual's yellow cover."

> We can make the printer work by disabling the paper-out switch, but it's not the best way to repair its problem.

In this sentence, *it's* is the contraction for *it is*, and *its* is substituted for *the copier's*.

Apostrophes to Indicate Plurals

Apostrophes are frequently used to form the plurals of numbers, letters, and acronyms. Style guides vary on this use of the apostrophe.

> His 8's look like S's.
> For a research and development organization, Smith and Peters has very few physicists with PhD's.

Exercise

Provide apostrophes as needed according to the guidelines in this section.

1. Smiths report was written much better than Jones.
2. "I cant finish the project today," she said. "Its going to take several hours, and Ive got too much to do."
3. By the look on their faces, especially the women, John knew he had said something wrong.
4. This typewriter needs cleaning; the As and Ps have blurred centers.
5. Its not surprising that his report isnt as well organized as hers.

Answer Key

1. Smith's report was written much better than Jones's.

An apostrophe and *s* are needed after *Smith* to indicate possession, as they are after *Jones*.

2. "I can't finish the project today," she said. "It's going to take several hours, and I've got too much to do."

Apostrophes are needed in the contractions *can't* in the first sentence and in *it's* and *I've* in the second.

3. By the look on their faces, especially the women's, John knew he had said something wrong.

An apostrophe and an *s* are needed after *women* to form the possessive referring to their faces. Like *children,* or *geese, women* is an irregularly formed plural.

4. This typewriter needs cleaning; the A's and P's have blurred centers.

Apostrophes may be necessary to form the plurals of the letters A and P, depending on the editorial style guide.

5. It's not surprising that his report isn't as well organized as hers.

Apostrophes are needed in the contractions "it's" and "isn't," but no apostrophes are needed with the personal pronouns "his" and "hers."

QUOTATION MARKS

Quotation marks are used primarily to reproduce another person's words exactly as written or spoken. Whether the quotation is introduced by a comma or a colon or worked into the structure of the sentence and not set off by punctuation, begin and end it with quotation marks.

> Internal Revenue Service guidelines state: "You may deduct what you actually gave to organizations that are religious, charitable, educational, scientific, or literary in purpose."

Try to keep quotations brief and work them in smoothly so that if the sentence were read aloud, a listener could not immediately identify the quotation.

Quotation marks are also used to draw attention to a word as an example rather than for what it means.

> The word "hopefully" is nearly always misused.

Quotation marks can set off an ironic, colloquial, or proverbial word or phrase.

> That company's strategic planning is mostly of the "seat of the pants" school.

Finally, quotation marks identify titles of report sections, book chapters, articles, and other works published as parts of a larger work.

For a quotation within a quotation, single quotation marks are used. To punctuate a quotation at the end of a sentence, do the following:

- Put a comma or period inside the quotation marks

  "Profits should be up next quarter," said Ms. Smith, "if sales continue at the present rate."

- Put a colon or semicolon outside the final quotation mark

  The president disagreed with the comment that the plan was "arbitrary"; he thought it was carefully designed.

- Put a question mark inside the quotation marks only if you are quoting a question.

  At the meeting, Ms. Sullivan asked: "Why not give Thompson Industries the subcontract, given their performance on the Astro project?"

  Did she say: "I move for adjournment"?

Exercise

Provide single and double quotation marks as needed according to following the guidelines in this section. Provide commas and colons as needed for attributions.

1. Mr. Perkins said You should go see what they did to the cafeteria.
2. Mr. Perkins said that you should go see what they did to the cafeteria.
3. I don't know why the word joystick came to be used for this control Tim said.
4. I'm tired of asking what I can do for the company James said and I want to ask instead what the company has done for me.
5. In his conclusion Mr. Addison said I want to make sure that everyone recognizes my desire to accept full responsibility for lab operations. As long as I am lab manager, the buck stops here.

Answer Key

1. Mr. Perkins said, "You should go see what they did to the cafeteria."

Quotations marks should be used before *You* and after the period following *cafeteria* to indicate that Mr. Perkins is being quoted. A comma should follow *said* to set off the attribution "Mr. Perkins said."

2. Mr. Perkins said that you should go see what they did to the cafeteria.

No punctuation is needed. Mr. Perkins's statement is being reproduced indirectly in the clause beginning with *that.*

3. "I don't know why the word 'joystick' came to be used for this control," Tim said.

Quotation marks should precede *I* and follow *control* to indicate that Tim is being quoted. Single quotation marks around joystick draw attention to the use of the word as a term. A comma inside the quotation marks is needed after *control,* to set off the attribution.

4. "I'm tired of asking what I can do for the company," James said, "and I want to ask instead what the company has done for me."

Quotation marks should be used before *I'm,* after *company,* before *and,* and after *me* to indicate that James is being quoted. Commas are needed after *company* and *said* to set off the attribution *James said.*

5. In his conclusion, Mr. Addison said: "I want to make sure that everyone recognizes my desire to accept full responsibility for lab operations. As long as I am lab manager, 'the buck stops here.' "

Quotation marks should be used before *I* and after *here* to indicate that Addison is being quoted. The colon after *said* sets off the attribution and introduces the quotation formally. The phrase borrowed from Truman is in single quotation marks to indicate the quotation within the quotation. The period should go inside the single and double quotation marks.

HYPHENS AND DASHES

Hyphens and dashes look similar but have very different uses. Generally, hyphens are used within words, and dashes are used between words and numbers.

Hyphens

Use a hyphen to divide a word when you must carry the end of the word over to the next line. Break the word according to the syllable division in a good dictionary or a word division list such as Silverthorn's *Word Division Manual.*

Certain compound words, especially adjectives such as *long-lived, decision-making, well-dressed,* and *low-frequency,* require a hyphen, depending on the applicable standard.

If two compounds share the same base word, that word is written only once.

> Please get me a 100- or 150-watt light bulb from the cabinet.

Technical terms such as the names of chemical compounds frequently use hyphens.

> 2-hydroxy-para-cymene 4,5,6,7-tetrachloro-2-(2-dimethylaminoethyl) isoindoline dimethylchloride

Hyphens are also used in some abbreviations, such as *ft-lb* and *g-cm;* in names of theories and equations, such as *Ginzburg-Landau theory;* and in compounds naming devices and mechanisms, such as "shell-and-tube exchanger." Again, check your company or agency's style guide or recommended dictionary to see if such terms should be hyphenated.

Hyphens are also used when numbers from 21 to 99 are written out.

> The screw machines in Building 4 are twenty-four years old.

Dashes

There are two main kinds of dashes: *em dashes* (as wide as the typeset capital letter M) and *en dashes* (as wide as the capital N).

The en dash means "to" or "and" in a range or group of dates, places, and times.

> He'll be on the New York–Boston shuttle today.
> The discussion is on pages 204–211.
> Treatment of gangrene became more effective during the War of 1914–1918.
> She worked on the SmithCo–Davis project last year.

The em dash has more varied uses; one is to set off a series at the beginning of a sentence.

> Floor space, age, lot size, location—all these factors affect the value of a house.

This sentence could be turned around and punctuated with a colon: "Several factors affect the value of a house: floor space, age, lot size, and location." To state it more concisely, eliminate the emphasis on factors: "Floor space, age, lot size, and location affect the value of a house."

The em dash can also set off material for emphasis. If the material is in the middle of the sentence, be sure dashes surround it.

> The staff I chose for the project—Tom Brown, Ann Jones, and Dick Thompson—are the best people we have in radar-absorbing structures.

In this example, an editor could use parentheses at the beginning and end of the series of names. Parentheses would deemphasize the names; using dashes draws attention to them. Using commas might suggest that Brown, Jones, and Thompson are not among the people chosen for the project. Often, such appositive series that contain commas must be set off by dashes. (Note the hyphenated compound adjective "radar-absorbing.")

Dashes are also used in references and in dialogue. A long dash in a bibliography indicates that that item was written by the same author as that immediately above it. In this use, most standards call for a 3-em dash followed by a period. In dialogue, em dashes are sometimes used to indicate a new speaker. En and em dashes should be marked by the editor (see Chapter 3).

Exercise

Provide hyphens and dashes as needed according to the guidelines in this section.

1. This is a time tested method for determining conductivity.
2. Ward, Bleeker, Thomas, Quincy, and Stevens the entire group came in Saturday to finish the tests.
3. When she got to the hotel in Salt Lake City, Ms. Walker realized imme diately that she had left her briefcase on the plane.
4. The advantages of working as an independent consultant flexibility, self determination, higher pay attract engineers away from the company every year.
5. Thirty two controllers passed the test, including the youngest applicant, who was only twenty three years old.

Answer Key

1. This is a time-tested method for determining conductivity.

A hyphen is needed in the compound adjective *time-tested.*

2. Ward, Bleeker, Thomas, Quincy, and Stevens—the entire group came in Saturday to finish the tests.

A dash should be used after *Stevens* to set off the nonessential phrase listing the people in the group. A dash after "group" would emphasize that the five people form the entire group.

3. When she got to the hotel in Salt Lake City, Ms. Walker realized imme-diately that she had left her briefcase on the plane.

Because the word *immediately* is carried over to the next line, a hyphen should follow *imme* to indicate the word division.

4. The advantages of working as an independent consultant—flexibility, self-determination, higher pay—attract engineers away from the company every year.

A hyphen should be used in *self-determination.* Also, dashes should be used before *flexibility* and after *determination* to set off the nonessential phrase "flexibility, self-determination, higher pay" from the main clause.

5. Thirty-two controllers passed the test, including the youngest applicant, who was only twenty-three years old.

The numbers *thirty-two* and *twenty-three* need to be hyphenated.

PARENTHESES

Parentheses can:

- Set off nonessential material
- Define acronyms and abbreviations
- Refer to books, articles, or reports
- Call out graphics
- State scientific names and terms

Parentheses to Set Off Nonessential Material

Parentheses can enclose material that is not essential to the meaning of the sentence. Parentheses are frequently used instead of commas to set off nonessential words or phrases that the writer wants to deemphasize but still include as explanation.

> The meetings were attended by several company directors: Ms. Brown (marketing), Mr. Fox (sales), Mr. King (finance), and Ms. Williams (planning).

Material in parentheses is less significant than that set off by commas. Like dashes, parentheses provide information that could not be set off with commas without possible confusion.

> Three of the supervisors (Mr. Ball, Ms. Selvidge, and Mr. Wenger) are opposed to the new production quotas.

In this example, dashes could be used to emphasize who opposed the quotas. If commas were used after *supervisors* and *Wenger* instead of parentheses to set up the appositive phrase with the three names, the sentence could be misinterpreted to mean six people in all oppose the quotas.

Parentheses often enclose numbers or letters that list the items in a series. Parentheses are not used in a displayed list.

> The directors have three items on the agenda: (1) choice of a site for the new manufacturing facility, (2) selection of a new vice president of finance, and (3) response to the shareholders' proposals.

Parentheses to Define Acronyms and Abbreviations

Use parentheses to define acronyms—a series of letters that stands for an organization or other phrase—and abbreviations.

> Just-In-Time (JIT) inventory techniques have been used successfully by many companies.
> In the equations, force is measured in dynes (d).

On the first use of an acronym that will appear frequently in the document, write out the phrase fully, followed by the acronym in parentheses. Editors should ensure that all acronyms in the document are defined when first used, except commonly known ones (e.g., NASA, NIH).

Parentheses in Reference Citations

Many companies, agencies, and publications use parentheses to refer to books, articles, and other materials. Many handbooks on style, including *The Chicago Manual of Style,* the *APA Style Sheet,* and the *MLA Style Manual,* recommend using parenthetical references rather than the traditional (and now old-fashioned) system of footnotes.

> Recent studies of the syndrome (Smith 1982; Walker 1984; Albert 1985) have proved inconclusive.
>
> Jones's study of hypothermia (4) was recommended highly by the panel.

The complete references, with author's name, title, place and date of publication, and page numbers, would be listed at the end of the article. Numbered references appear in the order cited, while the name–date system lists entries in alphabetical order by the author's last name, depending on the format recommended.

Parentheses for Graphics Callouts

Callouts for graphics are often placed in parentheses.

> The results (Table 1) suggest that we should use a different aluminum alloy.
>
> The ablative coating showed signs of deterioration after the tests (Figure 1).

Many writers prefer to draw more attention to the graphics by not using parentheses:

> The results shown in Table 1 suggest that we should use a different aluminum alloy.
>
> Figure 1 illustrates deterioration of the ablative coating caused by the tests.

Either system of callouts is usually acceptable, but they should not be mixed in a document.

Parentheses to State Scientific Names and Terms

In technical documents, parentheses are used to state genus and species when added for clarification, and other scientific information such as complete chemical composition.

> Autopsies were performed on 23 pygmy sperm whales *(Kogia breviceps)* and 6 dwarf sperm whales *(Kogia simus).*

Tris (2-hydroxyethyl) tallow ammonium chloride is used in cleaning and disinfecting solutions for contact lenses.

In scientific notation, genus and species names are italicized. Genus names are initial cap, and species names lower case.

Parentheses in Math

Parentheses are also used in equations and formulas.

The formula for the present value of $1 is:

$$PV = \frac{S}{(1+i)^n}$$

where *PV* is the present value at the beginning, *S* is the amount in the future, *i* is the interest rate, and *n* is the number of periods.

BRACKETS

Square brackets appear mainly in scientific and mathematical notation and in quotations. Brackets can insert a word or phrase of clarification in a passage being quoted.

In his testimony, Mr. Robert Stoddard said: "I saw Wallace and Meyers go into the office, and he [Meyers] was carrying a large briefcase."

A common use of square brackets for clarification in a quotation is with the Latin term *sic,* which indicates that the wording is exact despite an obvious error.

According to the contract, "the managing firm will be paid $4,000 *[sic]* a year for overseeing operations."

Exercise

Provide parentheses, brackets, italics, and commas as needed according to the guidelines presented here.

1. As she got ready for her vacation, she packed two magazines the novel she was reading On the Run two bathing suits and her new summer clothes.
2. The National Aeronautics and Space Administration NASA was established to develop the United States' space program.
3. The food pellets West and Company manufacture for rainbow trout Salmo gairdneri are very different from those they prepare for brown trout Salmo trutta.
4. Utility costs for the last ten years are shown in Figure 4.
5. "During the meeting," Janice said, "Jenson argued against Healey's proposal so convincingly that the president asked him Healey to prepare a new proposal."

Answer Key

1. As she got ready for her vacation, she packed two magazines, the novel she was reading *(On the Run)*, two bathing suits, and her new summer clothes.

Assuming the novel was titled *On the Run, On the Run* should be placed in parentheses so it does not seem to be an additional item in the list. Commas are needed after *magazines, Run)*, and *suits* to separate the items in the series. The title *On the Run* should be italicized.

2. The National Aeronautics and Space Administration (NASA) was established to develop the United States' space program.

NASA should be in parentheses to set up the acronym in a document designed for lay readers.

3. The food pellets West and Company manufacture for rainbow trout *(Salmo gairdneri)* are very different from those they prepare for brown trout *(Salmo trutta)*.

Salmo gairdneri and *Salmo trutta* should be in parentheses because they add for clarification the scientific names of two types of trout (salmonoids, more accurately). As genus and species names, they should be underlined for italics.

4. Utility costs for the last ten years are shown in Figure 4.

No parentheses should be used because the reference to Figure 4 has been worked into the sentence.

5. "During the meeting," Janice said, "Jenson argued against Healey's proposal so convincingly that the president asked him [Healey] to prepare a new proposal."

Healey should be in square brackets to indicate that it is clarification added to Janice's statement.

To help editing students master the punctuation guidelines presented here, self-tests of punctuation follow. In each passage, insert only the punctuation called for by this chapter. Answer keys with explanations follow each passage.

SELF-TEST 1: USES OF COMMAS

Insert commas where needed in the following passage. All other necessary punctuation has been provided.

One important way to write better is to control your sentences. Don't let sentences get too long. The longer your sentences get the harder they will be to understand. Proper sentence length depends on your audience. A college-educated audience that is familiar with your subject can handle longer sentences than an educated audience unfamiliar with your subject or a less educated audience.

Keep your sentence structure simple and your audience will have less trouble understanding each sentence. Rely on the sentence structures we use when we speak: subject–verb subject–verb–object and subject–verb–complement. Whenever you use a different structure you make the sentence more difficult. Don't try to impress readers with long or complicated sentences designed to make them think you're intelligent; rather impress them with how clearly and simply you can present your material.

Keep your sentences straightforward. Focus on who (or what) does what. The sentence "That the test results were faulty not valid was apparent to the engineer" is more complex than "The engineer knew that the test results were faulty not valid." Both sentences say the same but the second version has a simpler more familiar structure.

Avoid unnecessary subordinate clauses which can make a sentence longer as well as more complex. Use subordinate clauses only when necessary. For example "Send the report that was written by Smith to Jessica Locke who is the supervisor of training" can be stated more concisely and simply as "Send Smith's report to Jessica Locke the supervisor of training." The second version is shorter containing fewer words for the reader to process and simpler having only one set of subject and verb.

Try to begin each sentence with the subject followed soon after by the verb. Don't have too many words in front of the subject. Also don't split the subject and verb unnecessarily. Sentences are easier for readers to process when they can see clearly and easily what is being said. Don't oversimplify to the point of "Dick loves Jane" but do avoid sentences that are not straightforward. For example the sentence "Michael Smith having graduated in May 1992 went to work for Wheeler and Smith in San Anselmo California" could be restated as "After graduating in May 1992 Michael Smith went to work for Wheeler and Smith in San Anselmo California."

Rely on active voice sentences and use passive voice when the doer of an action is unimportant. In active voice the subject of the verb performs the action. For example active voice is used in the sentences "Robots can perform manufacturing operations that are dangerous to humans" and "Ms. Stevens reported the results of the audit completed January 15 1992." In passive voice the subject of the verb receives the action. The passive versions read: "Manufacturing operations that are dangerous to humans can be performed by robots" and "The results of the audit completed January 15 1992 were reported by Ms. Stevens."

Passive voice is sometimes more appropriate. The active voice sentence "Robots can perform manufacturing operations that are dangerous to humans" has more focus on *robots* than on the kind of operations they can perform. However the passive voice sentence "Manufacturing operations that are dangerous to humans can be performed by robots" has more focus on *manufacturing operations. Manufacturing operations* is the subject of the sentence and the real doer of the action *robots* is deemphasized by being tucked away in a prepositional phrase. Focus the sentence on what you want to emphasize. But remember that active voice sentences are more forceful direct and economical.

Answer Key

One important way to write better is to control your sentences. Don't let sentences get too long. The longer your sentences get, the harder they will be to understand.

> A comma is needed after *get* to join the two balanced main clauses. The sentence is not a comma splice because the clauses are short and balanced in structure.

Proper sentence length depends on your audience. A college-educated audience that is familiar with your subject can handle longer sentences than an educated audience unfamiliar with your subject or a less educated audience.

Keep your sentence structure simple, and your audience will have less trouble understanding each sentence.

> The comma after *simple* is needed with the coordinating conjunction *and* to join the two main clauses. (The imperative verbs in the main clauses have "you" understood as their subjects.)

Rely on the sentence structures we use when we speak: subject–verb, subject–verb–object, and subject–verb–complement.

> Commas are needed after *verb* and *object* to separate the items in the series.

Whenever you use a different structure, you make the sentence more difficult.

A comma is needed after *structure* to set off the introductory dependent clause from the main clause.

Don't try to impress readers with long or complicated sentences designed to make them think you're intelligent; rather, impress them with how clearly and simply you can present your material.

The comma after *rather* sets off the introductory conjunctive adverb from the second main clause in the sentence.

Keep your sentences straightforward. Focus on who (or what) does what. The sentence "That the test results were faulty, not valid, was apparent to the engineer" is more complex than "The engineer knew that the test results were faulty, not valid."

Commas are needed after *faulty, valid,* and *faulty* to set off both uses of the contrasting nonessential phrase *not valid.*

Both sentences say the same, but the second version has a simpler, more familiar structure.

The comma after *same* and the coordinating conjunction *but* join the main clauses. The adjective *simpler* is used in a series with the adjective *more familiar* to modify *structure.* "And" could be used between the adjectives, so a comma is used after *simpler.*

Avoid unnecessary subordinate clauses, which can make a sentence longer as well as more complex.

The nonessential subordinate clause *which can make a sentence longer as well as more complex* should be set off from the main clause by a comma after *clauses.*

Use subordinate clauses only when necessary. For example, "Send the report that was written by Smith to Jessica Locke, who is the supervisor of training" can be stated more concisely and simply as "Send Smith's report to Jessica Locke, the supervisor of training."

A comma is needed after *example* to set off the introductory phrase *for example.* Commas should be used after *Locke* in both uses, to set off the nonessential clause *who is the supervisor of training* and the nonessential phrase *the supervisor of training.*

The second version is shorter, containing fewer words for the reader to process, and simpler, having only one set of subject and verb.

The commas after *shorter* and *process* set off from the main clause the nonessential participial phrase *containing fewer words for the reader to process.* The comma after *simpler* sets off the nonessential participial phrase *having only one set of subject and verb.*

Try to begin each sentence with the subject followed soon after by the verb. Don't have too many words in front of the subject. Also, don't split the subject and verb unnecessarily.

The comma after *also* sets off the introductory transition.

Sentences are easier for readers to process when they can see clearly and easily what is being said. Don't oversimplify to the point of "Dick loves Jane," but do avoid sentences that are not straightforward.

The comma after *Jane* and the coordinating conjunction *but* join the main clauses.

For example, the sentence "Michael Smith, having graduated in May 1992, went to work for Wheeler and Smith in San Anselmo, California" could be restated as "After graduating in May 1992, Michael Smith went to work for Wheeler and Smith in San Anselmo, California."

A comma is needed to set off the introductory phrase *For example.* The commas after *Smith* and *1992* set off the nonessential phrase *having graduated in May 1992* from the main clause. The comma after the second *1992* sets off the nonessential introductory phrase *After graduating in May 1992* from the main clause. The comma after *Anselmo* is used to separate the city and state in the address.

Rely on active voice sentences, and use passive voice when the doer of an action is unimportant.

The coordinating conjunction *and* used to join the main clauses requires a comma after *sentences.*

In active voice, the subject of the verb performs the action of the verb.

The comma after *voice* sets off the introductory phrase *in active voice.*

For example, active voice is used in the sentences "Robots can perform manufacturing operations that are dangerous to humans" and "Ms. Stevens reported the results of the audit completed January 15, 1992."

Commas are used to set off the introductory phrase *for example* and in the date *January 15, 1992.* No comma is called for after *sentences* because the two quoted sentences that follow are necessary to tell which sentences. No comma would be used after *humans* because the two quoted sentences are not main clauses in this sentence.

In passive voice, the subject of the verb receives the action.

The comma after *voice* sets off the introductory phrase.

The passive versions read: "Manufacturing operations that are dangerous to humans can be performed by robots" and "The results of the audit completed January 15, 1992, were reported by Ms. Stevens."

Commas are needed after *15* and *1992* to present the date. No comma should be used after *robots* because the *and* that follows *robots* does not join main clauses.

Passive voice is sometimes more appropriate. The active voice sentence "Robots can perform manufacturing operations that are dangerous to humans" has more focus on *robots* than on the kind of operations they can perform. However, the passive voice sentence "Manufacturing operations that are dangerous to humans can be performed by robots" has more focus on *manufacturing operations.*

> The comma after *however* sets off the introductory transition. No commas should be used after *sentence* and *robots* because the quoted sentence tells which passive voice sentence and is necessary to the meaning of the sentence.

Manufacturing operations is the subject of the sentence, and the real doer of the action, *robots,* is deemphasized by being tucked away in a prepositional phrase.

> A comma is needed after *sentence* with the coordinating conjunction *and* to join the main clauses. Commas are needed after *action* and *"robots"* to set off the nonessential appositive *"robots"*.

Focus the sentence on what you want to emphasize. But remember that active voice sentences are more forceful, direct, and economical.

> Commas should be used after *forceful* and *direct* to separate the items in the series. No comma should be used after the short transition *but.*

If you had any trouble with this exercise, review the appropriate section(s) of the discussion of commas in this chapter before you go on to the self-test on punctuation.

SELF-TEST 2: MARKS OF PUNCTUATION

In the following, insert punctuation where it is called for by the guidelines in this book.

English has been written since 900 AD or so but only over the centuries did our present system of punctuation develop. In Old English which is best known to readers today as the language of the epic poem Beowulf the scribes did not skip spaces between words nor did they use punctuation marks. Even when a reader learns the vocabulary and grammar of Old English which are very different from Modern English writing in Old English is much harder to read because it lacks punctuation.

Modern punctuation developed from ancient writers attempts to make it easier to read written material aloud they used points marks to indicate where and how long to pause. Greek writers of the third century BC used marks to separate phrases. Three of our terms for punctuation comma colon and period derive from the Greek words for three different lengths of phrases which were followed by a dot at the middle bottom or top of the space after the last letter.

Latin writers of the first century BC to the second century AD used a point between each word to aid the reader and sometimes they signaled a new paragraph by placing its first two letters in the left margin. Extra space was sometimes left at the end of a sentence and a larger letter a capital might begin the next sentence.

However Greek and Latin punctuation systems were maintained by few writers in the early Middle Ages when knowledge of Greek and Latin declined. Irish English and German scribes developed their own systems of punctuation. They began to leave a space after each word and gradually writers began to mark the end of a sentence with a space and the beginnings of sentences and paragraphs with a larger letter. By the tenth century the hyphen came to indicate word division between lines. By the twelfth century the period and the question mark had developed from symbols used in music.

The sixteenth century Italian printer Aldus Manutius who died in 1515 and his grandson also named Aldus Manutius developed a regularized system of punctuation which included the comma the colon and the period. And the younger Manutius the Encyclopaedia Britannica says stated plainly for the first time the view that clarification of syntax sentence structure is the main object of punctuation page 276.

The first English writer to argue that punctuation should be used to indicate the syntax of the sentence rather than how the sentence should be read aloud was Ben Jonson the famous poet and playwright 1572 1637. Jonsons influence in England and the Manutiuses throughout Europe was considerable and by 1700 English punctuation as we know it today was established. It has certain characteristics the comma semicolon quotation mark dash question mark exclamation mark and period have specific uses a space separates words and a longer space separates sentences the beginnings of paragraphs are indented and a capital letter begins a sentence a proper noun or a title.

Punctuation of writing in English continues to change however. In the eighteenth century punctuation was used very heavily but twentieth century usage is characterized by only as much punctuation as necessary to clarify the structure of the sentence. The present trend is toward less not more punctuation in academic and business writing.

Answer Key

English has been written since AD 900 or so, but only over the centuries did our present system of punctuation develop.

A comma is needed after *so* because the coordinating conjunction *but* is used to join the two main clauses. No periods are used in AD.

In Old English, which is best known to readers today as the language of the epic poem *Beowulf*, the scribes did not skip spaces between words, nor did they use punctuation marks.

Commas are needed after *English* and *Beowulf* to set off the nonessential clause *which is best known to readers today as the language of the epic poem Beowulf* from the main clause. No comma should be used after *poem* because the title of the poem is essential to know which epic poem. Because *Beowulf* is the title of an epic poem, it should be underlined. A comma should be used after *words* because the coordinating conjunction *nor* is used to join the two main clauses.

Even when a reader learns the vocabulary and grammar of Old English, which are very different from Modern English, writing in Old English is much harder to read because it lacks punctuation.

The introductory clause *even when a reader knows the vocabulary and grammar of Old English* and the nonessential clause *which are very different from Modern English* that is part of it should be set off from the main clause by a comma after the second *English.* The nonessential clause *which are very different from Modern English* should be set off from the introductory clause it modifies by a comma after the first *English.* In a sense, two commas are called for after the second *English:* one to set off the introductory clause, and another to finish setting off the second dependent clause, which is a nonessential part of the introductory clause. However, we use only one comma. No comma should be used after *read* because the following dependent clause tells why it is difficult to read and comes at the end of the sentence.

Modern punctuation developed from ancient writers' attempts to make it easier to read written material aloud; they used "points" (marks) to indicate where and how long to pause.

The apostrophe after *writers* indicates possession; the attempts were those of the writers. The semicolon after *aloud* joins the two main clauses. The two main clauses could be separate sentences, but they are closely related in content (the marks were their attempts), so they belong in one sentence. The quotation marks around "points" emphasize the word as a term rather than what it stands for, and the parentheses around "marks" indicate that "marks" is provided to explain the word "points."

Greek writers of the third century BC used marks to separate phrases.

No punctuation is needed. *Third century* would be hyphenated only if the phrase were used as an adjective as in "third-century writers." BC does not need periods.

Three of our terms for punctuation—"comma," "colon," and "period"—derive from the Greek words for three different lengths of phrases, which were followed by a dot at the middle, bottom, or top of the space after the last letter.

> Dashes should be used to set off the nonessential phrase *"comma," "colon," and "period"* from the noun they are in apposition to, *terms*. Commas would not set off the phrase sufficiently because the phrase has commas in it. Commas are needed after *comma* and *colon* to separate the items in the series. The names of the three marks of punctuation should be in quotation marks because the words that stand for the marks are being emphasized, not the marks themselves. The comma after *phrases* is needed to set off the nonessential dependent clause *which were followed by a dot at the middle, bottom, or top of the space following the last letter*. (The word *which* used instead of "that" helps the reader see that the clause is not essential to the meaning of the main clause.) The commas after *middle* and *bottom* are needed to separate the items in the series.

Latin writers of the first century BC to the second century AD used a point between each word to aid the reader, and sometimes they signaled a new paragraph by placing its first two letters in the left margin.

> A comma is needed after *reader* because the coordinating conjunction *and* is used to join the two main clauses. No commas should be used around *sometimes*. Also, no hyphen should be used in *second century* because it is not used as an adjective to modify *AD*. No commas should be used after *century* in *first century BC* and *second century AD* because *BC* and *AD* are adjectives telling which first and second century.

Extra space was sometimes left at the end of a sentence, and a larger letter (a capital) might begin the next sentence.

> A comma should be used after *sentence* because the coordinating conjunction *and* is used to join the two main clauses. Parentheses are needed around *a capital* to indicate that the phrase is additional, nonessential information about *a larger letter*. Setting off the appositive *a capital* with commas after *letter* and *capital* might lead a reader to think that *letter* and *capital* are the first two items in a series. The phrase *a capital* could be set off by dashes, but that would emphasize the phrase more than needed. The reader of the sentence would assume that the larger letter is a capital.

However, Greek and Latin punctuation systems were maintained by few writers in the early Middle Ages, when knowledge of Greek and Latin declined.

> The introductory adverb *however* should be set off from the main clause by a comma. A comma should be used after *ages* to set off the nonessential clause *when knowledge of Greek and Latin declined*. Some style guides might specify initial caps for *middle ages*.

Irish, English, and German scribes developed their own systems of punctuation.

Commas are needed after *Irish* and *English* to separate the items in the series.

They began to leave a space after each word, and gradually writers began to mark the end of a sentence with a space and the beginnings of sentences and paragraphs with a larger letter.

A comma should be used after *word* with the coordinating conjunction *and* is used to join the two main clauses. No comma should be used after *space* because the *and* that follows it does not join another main clause to the main clause; it joins the infinitives *to mark* and *[to, understood] use.* The adverb *gradually* should not be set off by a comma from the verb *began,* which it modifies.

By the tenth century, the hyphen came to indicate word division between lines.

The comma after *century* is needed to set off the introductory phrase from the main clause.

By the twelfth century, the period and the question mark had developed from symbols used in music.

Again, the comma after *century* sets off the introductory phrase. No comma should be used after *developed* because the phrase *from symbols used in music* provides essential information about how they developed.

The sixteenth-century Italian printer Aldus Manutius, who died in 1515, and his grandson, also named Aldus Manutius, developed a regularized system of punctuation, which included the comma, the colon, and the period.

A hyphen is needed in *sixteenth-century* because it is used as an adjective. Commas are needed after *Manutius* (first use) and *1515* to set off the nonessential clause *who died in 1515* from the main clause. Commas are needed after *grandson* and the second *Manutius* to set off the nonessential phrase *also named Manutius* from the main clause. Commas should be used after *comma* and *colon* to separate the items in the series. No comma should be used after *printer* because for the meaning of the sentence to be clear we need to know which printer. The nonessential dependent clause *which included the comma, the colon, and the period* should be set off from the main clause by a comma after *punctuation.*

And the younger Manutius, the *Encyclopaedia Britannica* says, "stated plainly for the first time the view that clarification of syntax [sentence structure] is the main object of punctuation" (page 276).

Commas should be used after *Manutius* and *says* to set off the attribution *the Encyclopaedia Britannica says. Encyclopaedia Britannica* is the title of a book and therefore must be underlined. Quotation marks are needed for the quoted material from the *Encyclopaedia Britannica.* From the sentence it is not apparent that the material is quoted exactly as it appeared in the original. The material could have been paraphrased—that is, put into the words of the writer of the sentence. In either case, the attribution and the page reference would be used. If you are not sure whether the material is quoted or paraphrased, check the source. Parentheses should be used for the page-number reference *page 276*. Square brackets are needed to set off the phrase *sentence structure* that was added to the quotation. If the statement about Manutius were paraphrase rather than quotation, parentheses instead of brackets would be used around *sentence structure*.

The first English writer to argue that punctuation should be used to indicate the syntax of the sentence, rather than how the sentence should be read aloud, was Ben Jonson, the famous poet and playwright (1572–1637).

Commas are needed after *sentence* and *aloud* to set off the contrasting nonessential clause *rather than how the sentence should be read aloud* from the main clause. A comma should be used after *Jonson* to set off the nonessential phrase *the famous poet (1572–1673)*. The dates of Jonson's birth and death should be enclosed in parentheses and separated by an en dash.

Jonson's influence in England and the Manutiuses' throughout Europe was considerable, and by 1700 English punctuation as we know it today was established.

A comma must be used after *considerable* because the coordinating conjunction *and* joins the two main clauses. The dependent clause *as we know it today in English* should not be set off by commas; it is essential to the meaning of the second main clause, telling which punctuation. No comma should be used after *England* because the *and* after *England* is not used to join two main clauses.

It has certain characteristics: the comma, semicolon, quotation mark, dash, question mark, exclamation mark, and period have specific uses; a space separates words, and a longer space separates sentences; the beginnings of paragraphs are indented; and a capital letter begins a sentence, a proper noun, or a title.

A colon is needed after *characteristics* to set up and emphasize the following naming of the characteristics. Semicolons are needed after *uses, sentences,* and *indented* to separate the items in the series. Commas would not be sufficient to separate the items because at least one item has a comma in it. Commas must be used after *comma, semicolon, mark, dash, mark,* and *mark* in the first series to separate the items in the series. A comma is needed after *words* in the second item in the series, because in that item the two main clauses *a space separates words* and *a longer space separates sentences* are joined by the coordinating conjunction *and*. Commas are needed after *sentence, noun,* and *title* to separate the items in the series.

Punctuation of writing in English continues to change, however.

A comma is needed after *change* to set off the nonessential transitional word *however* from the main clause.

In the eighteenth century, punctuation was used very heavily, but twentieth-century usage is characterized by only as much punctuation as necessary to clarify the structure of the sentence.

> A comma is needed after *century* to set off the introductory phrase *in the eighteenth century*. No hyphen is needed in *eighteenth century* because it is used as a noun, not an adjective. A comma should be used after *heavily* with the coordinating conjunction *but* to join the two main clauses. A hyphen should be used in the adjective *twentieth-century*. No comma should be used after *necessary* because the phrase that follows *necessary* is essential to the meaning of the sentence, telling why the punctuation is necessary.

The present trend is toward less, not more, punctuation in academic and business writing.

> Commas should be used after *less* and *more* to set off the nonessential contrasting phrase *not more* from the main clause.

Bibliography

STYLE MANUALS

American Chemical Society. *Handbook for Authors of Papers in American Chemical Society Publications.* 1978.

American Institute of Physics, Publication Board. *Style Manual for Guidance in the Preparation of Papers for Journals published by the American Institute of Physics and Its Member Societies.* 3rd ed. American Institute of Physics, 1978.

American Mathematical Society. *A Manual for Authors of Mathematical Papers.* 8th ed. 1990.

———. *Typing Guide for Mathematical Expressions.* 1986.

American National Standards Institute. *Scientific and Technical Reports: Organization, Preparation, and Production.* 1987.

American Physical Therapy Association. *Style Manual.* 1985.

American Psychological Association. *Publication Manual of the American Psychological Association.* 3rd ed. 1983.

American Society of Civil Engineers. *ASCE Authors' Guide to Journals, Books, and Reference Publications.* 1986.

Associated Press Stylebook and Libel Manual. Addison Wesley, 1982.

Austin, Mike, and Ralph Dodd. *The ISTC Handbook of Technical Writing and Publication Techniques: A Practical Guide for Managers, Engineers, Scientists, and Technical Publications Staff.* Heinemann, 1983.

Barclay, William R., M. Therese Southgate, and Robert Mayo, comp. *Manual for Authors & Editors: Editorial Style & Manuscript Preparation.* Lange Medical Publications, for the American Medical Association, 1981.

Council of Biology Editors, Style Manual Committee. *Style Manual: A Guide for Authors, Editors, and Publishers in the Biological Sciences.* 5th ed. 1983.

Dodd, Janet, ed. *The ACS Style Guide: A Manual for Authors and Editors.* American Chemical Society, 1985.

Howell, John B. *Style Manuals of the English-Speaking World.* Oryx Press, 1983.

Reynolds, Helen L., ed. *The Association of Analytical Chemists Style Manual.* 1972.

Silverthorn, J. E., and Devern J. Perry. *Word Division Manual for the Basic Vocabulary of Business Writing.* 3rd ed. Southwestern, 1984.

Simmons, Barbara, ed. *Typing Guide for Mathematical Expressions.* Society for Technical Communication, 1976.

Swanson, Ellen. *Mathematics into Type.* Rev. ed. American Mathematical Society, 1987.

United States Government Printing Office. *Style Manual.* 1984.

———. *Word Division: Supplement to the United States Government Printing Office Style Manual.* 1987.

University of Chicago Press. *The Chicago Manual of Style.* 13th ed. 1982.
Warren, Thomas. *Words into Type.* 4th ed. Prentice-Hall, 1993.
Webb, Robert. ed. *The Washington Post Deskbook on Style.* McGraw-Hill, 1978.

EDITING

American National Standards Institute. *Proof Corrections.* 1989.
Beach, Mark. *Editing Your Newsletter: How to Produce an Effective Publication using Traditional Tools and Computers.* 3rd ed. Coast to Coast Books, 1988.
Bennett, John B. *Editing for Engineers.* Wiley Interscience, 1970.
Bishop, Claude. *How to Edit a Scientific Journal.* Williams and Wilkins, 1989.
Bohle, Robert H. *Publication Design for Editors.* Prentice-Hall, 1989.
Boston, Bruce O., ed. *Stet! Tricks of the Trade for Writers and Editors.* Editorial Experts, 1986.
Brittain, Robert. *A Pocket Guide to Correct Punctuation.* 2nd ed. Barron's, 1990.
Brooks, Brian, and James Pinson. *Working with Words: A Concise Handbook for Media Writers and Editors.* St. Martin's, 1989.
Butcher, Judith. *Copy-Editing: The Cambridge Handbook.* 3rd ed. Cambridge University Press, 1991.
Cheney, Theodore. *Getting the Words Right.* Writer's Digest Books, 1983.
Clements, Wallace, and Robert G. Waite. *Guide for Beginning Technical Editors.* Society for Technical Communication, 1983.
Duffy, Thomas M., and Robert Waller, eds. *Designing Usable Texts.* Academic Press, 1985.
Eisenberg, Anne. *Guide to Technical Editing.* Oxford University Press, 1992.
England, R. Breck. *Proofreading, Editing, and Writing.* Shipley, 1989.
Farkas, David. *How to Teach Technical Editing.* Society for Technical Communication, 1986.
Freedman, George. *The Technical Editor's and Secretary's Desk Guide.* McGraw-Hill, 1985.
Garst, Robert E., and Theodore Bernstein. *Headlines and Deadlines: A Manual for Copy Editors.* 4th ed. Columbia University Press, 1981.
Gibson, Martin. *Editing in the Electronic Era.* 3rd ed. Iowa State University Press, 1991.
Gong, Gwendolyn, and Sam Dragga. *Editing: The Design of Rhetoric.* Baywood, 1989.
Hart, Horace. *Hart's Rules for Compositors and Readers at the University Press, Oxford.* 39th ed. Oxford University Press, 1983.
Haverty, John R., ed. *Webster's Medical Office Handbook.* G & C Merriam, 1979.
Jack, Judith. *Editing for Engineers: A Workbook in Technical Editing for Students.* Atwater, 1987.
Judd, Karen. *Copy Editing.* Crisp Publications, 1989.
Lee, Jo Ann. *Proofreading for Wordprocessing.* Harcourt Brace Jovanovich, 1988.
Morgan, Peter. *An Insider's Guide for Medical Authors and Editors.* ISI Press, 1986.
Mullins, Carolyn. *The Complete Manuscript Preparation Style Guide.* Prentice-Hall, 1982.
O'Connor, Maeve. *How to Copyedit Scientific Books and Journals.* Williams and Wilkins, 1989.
———. *The Scientist as Editor: Guidelines for Editors of Books and Journals.* Krieger, 1979.
Oxford University Press. *Oxford Dictionary for Writers and Editors.* 1981.
Paxson, William C. *The Mentor Guide to Punctuation.* New American Library, 1986.
Pickens, Judy. *The Copy-to-Press Handbook: Preparing Words and Art for Print.* John Wiley, 1985.

Plotnick, Arthur. *The Elements of Editing: A Modern Guide for Editors and Journalists.* Macmillan, 1982.

Rude, Carolyn, ed. *Teaching Technical Editing.* Association of Teachers of Technical Writing, 1985.

———. *Technical Editing.* Wadsworth, 1991.

Seraydarian, Patricia E. *Proofreading for Information Processing.* Science Research Associates, 1988.

Shaw, Harry. *Punctuate It Right!*. Harper and Row, 1986.

Smith, Peggy. *Mark My Words: Instruction and Practice in Proofreading.* Editorial Experts, 1987.

———. *Proofreading Manual and Reference Guide* and *Proofreading Workbook.* Editorial Experts, 1981.

Stoughton, Mary. *Substance and Style: Instruction and Practice in Copyediting.* Editorial Experts, 1989.

Stovall, James G., Charles C. Self, and L. Edward Mullins. *On-Line Editing.* Prentice-Hall, 1984.

Strunk, William, and E. B. White. *The Elements of Style.* 3rd ed. Macmillan, 1979.

Stultz, Russell A. *The Business Side of Writing.* Wordware, 1989.

Van Buren, Robert, and Mary Fran Buehler, *The Levels of Edit.* 2nd ed. California Institute of Technology, Jet Propulsion Laboratory, 1980. Reprinted by Society for Technical Communication, 1991.

Visual Education Corporation Staff. *Proofreading Skills for Business.* Prentice-Hall, 1986.

Weiss, Edmond H. *Writing Remedies.* Oryx Press, 1990.

Zook, Lola, ed. *Technical Editing: Principles and Practices.* Society for Technical Communication, 1975.

GRAPHICS, PRINTING, AND GRAPHIC ARTS

Biegeleisen, J. I. *Handbook of Type Faces and Lettering.* 4th ed. Prentice-Hall, 1982.

Bove, Tony, Cheryl Rhodes, and Wes Thomas. *The Art of Desktop Publishing; Using Personal Computers to Publish It Yourself.* 2nd ed. Bantam Doubleday Dell, 1987.

Cleveland, William S. *The Elements of Graphing Data.* Wadsworth, 1985.

Craig, James. *Designing with Type.* Rev. ed. Watson-Guptill, 1983.

———. *Phototypesetting: A Design Manual.* Watson-Guptill, 1978.

Crow, Wendell C. *Communication Graphics.* Prentice-Hall, 1986.

Demoney, Jerry, and Susan Meyer. *Paste-up and Mechanicals: A Step-by-Step Guide to Preparing Art for Reproduction.* Watson-Guptill, 1982.

Dodt, L. *Graphic Arts Production.* American Technical Publications, 1990.

Duff, Jon M. *Industrial Technical Illustration.* Van Nostrand Reinhold, 1982.

Field, Janet N., ed. *Graphic Arts Manual.* Ayer Company Publications, 1980.

Grant, Hiram. *Engineering Drawing.* McGraw-Hill, 1962.

Hickman, Dixie E. *Teaching Technical Writing: Graphics.* Association of Teachers of Technical Writing, 1985.

Holmes, Nigel. *Designer's Guide to Creating Charts and Diagrams.* Watson-Guptill, 1991.

———. *Designing Pictorial Symbols.* Watson-Guptill, 1990.

International Paper Company. *Pocket Pal: A Graphic Arts Production Handbook.* 13th ed. 1983.

Jensen, Cecil, and Jay Helsel. *Fundamentals of Engineering Drawing and Design.* 3rd ed. McGraw-Hill, 1989.

Karsnitz, John R. *Graphic Arts Technology*. Delmar, 1984.
Labuz, Ronald. *How to Typeset from a Word Processor*. Bowker, 1984.
MacGregor, A. J. *Graphics Simplified: How to Plan and Prepare Effective Charts, Graphs, Illustrations, and Other Visual Aids*. University of Toronto Press, 1979.
Magnan, George. *Using Technical Art: An Industry Guide*. John Wiley, 1970.
Meyerowitz, Michael, and Sam Sanchez. *The Graphic Designer's Basic Guide to the Macintosh*. Allworth Press, 1990.
Parker, Roger C. *Looking Good in Print: A Guide to Basic Design for Desktop Publishing*. Ventana Press, 1988.
Rice, Stanley. *Type-Caster: Universal Copyfitting*. Van Nostrand Reinhold, 1980.
Rowbotham, George E. *Engineering and Industrial Graphics Handbook*. McGraw-Hill, 1982.
Ruggles, Philip K. *Printing Estimating*. 3rd ed. Delmar, 1991.
Silver, Gerald A. *Professional Printing Estimating*. 2nd ed. Van Nostrand Reinhold, 1984.
Spence, William P. *Engineering Graphics*. 2nd ed. Prentice-Hall, 1988.
Stone, Charles. *Xplaining Macintosh: The Complete Primer of Macintosh Computing*. Multisoft Resources, 1990.
Thomas, T. A. *Technical Illustration*. 3rd ed. McGraw-Hill, 1978.
Tufte, Edward. *The Visual Display of Quantitative Information*. Graphics Press, 1983.
White, Alex. *How to Spec Type*. Watson-Guptill, 1986.
White, Jan. *Editing By Design*. 2nd ed. Bowker, 1982.
———. *Graphic Design for the Electronic Age*. Watson-Guptill, 1989.
———. *Mastering Graphics: Design and Production Made Easy*. Bowker, 1983.
———. *Using Charts and Graphs*. Bowker, 1984.
Wilde, Richard. *Information Graphics: Visual Thinking for Graphic Communications*. Van Nostrand Reinhold, 1987.
———. *Visual Literacy: A Conceptual Approach to Graphic Problems*. Watson-Guptill, 1991.
Wood, Phyllis. *Scientific Illustration: A Guide to Biological, Zoological, and Medical Rendering Techniques, Design, Printing, and Display*. Van Nostrand Reinhold, 1982.

TECHNICAL WRITING

Alley, Michael. *The Craft of Scientific Writing*. Prentice-Hall, 1987.
Alred, Gerald J., Carles T. Brusaw, and Walter E. Oliu. *Handbook of Technical Writing*. 2nd ed. St. Martin's, 1982.
Andrews, Deborah C., and Margaret Blickle. *Technical Writing: Principles and Forms*. 2nd ed. Macmillan, 1982.
Barnum, Carol M. *Prose and Cons: The Do's and Don'ts of Business and Technical Writing*. National Publishers, 1986.
Barrass, Robert. *Scientists Must Write: A Guide to Better Writing for Scientists, Engineers and Students*. Chapman and Hall/Wiley, 1978.
Barnett, Marva. *Writing for Technicians*. 3rd ed. Delmar, 1987.
Barrett, Edward, ed. *Text, Context, and HyperText: Writing with and for the Computer*. MIT Press, 1988.
Bates, Jefferson. *Writing with Precision: How to Write So That You Cannot Possibly Be Misunderstood*. 3rd ed. Acropolis, 1990.
Bazerman, Charles. *Shaping Written Knowledge: The Genre and Activity of the Experimental Article in Science*. University of Wisconsin Press, 1988.

Beason, Pamela S., and Patricia Williams. *Technical Writing for Business and Industry: A Practical Guide*. Scott Foresman, 1989.

Beene, LynnDianne, and Peter White. *Solving Problems in Technical Writing*. Oxford University Press, 1988.

Bjelland, Harley. *Writing Better Technical Articles*. TAB Books, 1990.

Blicq, Ron. *Technically—Write!: Communicating in a Technological Era*. 3rd ed. Prentice-Hall, 1986.

———. *Writing Reports to Get Results: Guidelines for the Computer Age*. IEEE Press, 1987.

Bly, Robert W., and Gary Blake. *Technical Writing: Structure, Standards, and Style*. McGraw-Hill, 1982.

Bolsky, Morris. *Better Scientific and Technical Writing*. Prentice-Hall, 1988.

Booher, Dianna. *Let's Get Technical about Writing: Guide for Scientists, Engineers, and Data Processors*. John Wiley, 1989.

———. *To the Letter: A Handbook of Model Letters for the Busy Executive*. Lexington Books, 1988.

———. *Writing for Technical Professionals*. John Wiley, 1989.

Brockmann, R. John. *The Case Method in Technical Communication: Theories and Models*. Association of Teachers of Technical Writing, 1985.

Brown, John F. *Engineering Report Writing*. 3rd ed. United Western Press, 1989.

Browning, Christina. *Guide to Effective Software Technical Writing*. Prentice-Hall, 1984.

Bruffee, Kenneth. *A Short Course in Writing*. 3rd ed. Little, Brown, 1985.

Bruner, Ingrid, et al. *The Technician as Writer: Preparing Technical Reports*. Macmillan, 1980.

Burnett, Rebecca E. *Technical Communication*. 2nd ed. Wadsworth, 1990.

Cain, B. Edward. *The Basics of Technical Communicating*. American Chemical Society, 1988.

Carter, Sylvester P. *Writing for Your Peers: The Primary Journal Paper*. Praeger, 1987.

Chapman, Charles F., and Barbara Lynch. *Writing for Communication in Science and Medicine*. Van Nostrand Reinhold, 1980.

Clements, Wallace, and Robert Berlo. *The Scientific Report: A Guide for Authors*. Society for Technical Communication, 1984.

Conway, William D. *Essentials of Technical Writing*. Macmillan, 1987.

Cook, Claire K. *Line by Line: The MLA's Guide to Improving Your Writing*. Houghton Mifflin, 1986.

Couture, Barbara, and Jone R. Goldstein. *Cases for Technical and Professional Writing*. Scott Foresman, 1985.

Dagher, Joseph P. *Technical Communication: A Practical Guide*. Prentice-Hall, 1978.

Damerst, William A., and Arthur H. Bell. *Clear Technical Communication: A Process Approach*. 3rd ed. Harcourt Brace Jovanovich, 1989.

Davis, Michael. *Manuals that Work: A Guide for Writers*. Nichols, 1990.

Day, Robert. *How to Write and Publish a Scientific Paper*. 3rd ed. Oryx Press, 1988.

DeGeorge, James, Gary A. Olson, and Richard Ray. *Style and Readability in Technical Writing: A Sentence-Combining Approach*. Random House, 1984.

Dobrin, David. *Writing and Technique*. National Council of Teachers of English, 1989.

Ebel, H. F., et al. *The Art of Scientific Writing: From Student Reports to Professional Publications in Chemistry and Related Fields*. VCH Publishers, 1987.

Eisenberg, Anne. *Writing Well for Technical Professions*. Harper Collins, 1988.

Elliott, Stephen P., ed. *The Complete Book of Contemporary Business Letters*. Round Lake Publishing, 1988.

Emerson, Frances B. *Developing Technical Writing Skills.* Houghton Mifflin, 1987.
Ewing, David W. *Writing for Results in Business, Government, and the Professions.* 2nd ed. John Wiley, 1979.
Farr, Alfred D. *Scientific Writing for Beginners.* Blackwell Scientific, 1985.
Feinberg, Susan. *Components of Technical Writing.* Holt, Rinehart and Winston, 1989.
Flaherty, Stephen M. *Technical and Business Writing: A Reader-Friendly Approach.* Prentice-Hall, 1990.
Forbes, Mark. *Writing Technical Articles, Speeches, and Manuals.* John Wiley, 1988.
Fourdrinier, Sylvia, and Henrietta J. Tichy. *Effective Writing for Engineers, Managers, Scientists.* 2nd ed. John Wiley, 1988.
Gibson, Walker. *Tough, Sweet, and Stuffy: An Essay on Modern American Prose Styles.* Indiana University Press, 1966.
Gillman, Leonard. *Writing Mathematics Well: A Manual for Authors.* Mathematical Association of America, 1987.
Grimm, Susan. *How to Write Computer Documentation for Users.* 2nd ed. Van Nostrand Reinhold, 1986.
Haines, Roger W., and Donald R. Bahnfleth. *Effective Communications for Engineers.* TAB Books, 1989.
Harkins, C., and D. L. Plung, eds. *A Guide for Writing Better Technical Papers.* Institute of Electrical and Electronics Engineers, 1982.
Harris, John S. *Teaching Technical Writing: A Pragmatic Approach.* Association of Teachers of Technical Writing, 1989.
Harty, Kevin J. *Strategies for Business and Technical Writing.* 3rd ed. Harcourt Brace Jovanovich, 1988.
Helgeson, Donald V. *Handbook for Writing Technical Proposals that Win Contracts.* Prentice-Hall, 1985.
Holtz, Herman. *The Complete Guide to Writing Readable User Manuals.* Business 1 Irwin, 1988.
Horton, William. *How to Write User-Seductive Documentation.* Society for Technical Communication, 1991.
Houp, Kenneth W., and Thomas E. Pearsall. *Reporting Technical Information.* 6th ed. Macmillan, 1988.
Huckin, Thomas N. *Technical Writing and Professional Communication: A Handbook for Non-native Speakers.* McGraw-Hill, 1989.
———, and Leslie Olsen. *Principles of Communication for Science and Technology.* McGraw-Hill, 1983.
Huth, Edward J. *How to Write and Publish Papers in the Medical Sciences.* 2nd ed. Williams and Wilkins, 1990.
Jordan, Michael P. *Fundamentals of Technical Description.* Krieger, 1984.
Katz, Michael J. *Elements of the Scientific Paper.* Yale University Press, 1985.
Katzin, Emanuel. *How to Write a Really Good User's Manual.* Van Nostrand Reinhold, 1985.
King, Lester S. *Why Not Say it Clearly?: A Guide to Expository Writing.* 2nd ed. Little, Brown, 1991.
Kolin, Philip C., and Janeen L. Kolin. *Models for Technical Writing.* St. Martin's Press, 1985.
Krull, Robert. *Word Processing for Technical Writers.* Baywood Press, 1988.
Lannon, John M. *Technical Writing.* 4th ed. Scott Foresman, 1988.
Louth, Richard, and Ann Martin Scott, eds. *Collaborative Technical Writing: Theory and Practice.* Association of Teachers of Technical Writing, 1989.

Mair, David, and Nancy Roundy. *Strategies for Technical Communication.* Little, Brown, 1985.

Markel, Michael. *Technical Writing Essentials.* St. Martin's, 1988.

———. *Technical Writing: Situations and Strategies.* 2nd ed. St. Martin's, 1987.

Mathes, J. C., and Dwight W. Stevenson. *Designing Technical Reports: Writing for Audiences in Organizations.* 2nd ed. Macmillan, 1991.

McGehee, Brad. *The Complete Guide to Writing Software User Manuals.* Writer's Digest Books, 1984.

McGuire, Peter J., and Sara M. Putzell. *A Guide to Technical Writing.* Harcourt Brace Jovanovich, 1987.

Michaelson, Herbert B. *How to Write and Publish Engineering Papers and Reports.* 3rd ed. Oryx Press, 1990.

Miller, Diane F. *Guide for Preparing Software User Documentation.* Society for Technical Communication, 1988.

Miller, Ryle L. *How to Write for the Professional Journals: A Guide for Technically Trained Managers.* Quorum Books, 1988.

Mills, Gordon H., and John A. Walter. *Technical Writing.* 5th ed. Holt, Rinehart and Winston, 1986.

Mitchell, Joan. *The New Writer: Techniques for Writing Well with a Computer.* Microsoft Press, 1987.

Odell, Lee, and Dixie Goswami, eds. *Writing in Nonacademic Settings.* Guilford Press, 1985.

Pauley, Steven E., and Daniel G. Riordan. *Technical Report Writing Today.* 3rd ed. Houghton Mifflin, 1987.

Pearsall, Thomas, and Donald Cunningham. *How to Write for the World of Work.* 2nd ed. Dryden Press, 1990.

Pfeiffer, William S. *Technical Writing: A Practical Approach.* Macmillan, 1990.

Pickett, Nell Ann, and Ann A. Laster. *Technical English: Writing, Reading, and Speaking.* 5th ed. Harper and Row, 1988.

Pinchuck, Isadore. *Scientific and Technical Translation.* Andre Deutsch, 1977.

Price, Jonathan. *How to Write a Computer Manual: A Handbook of Software Documentation.* Benjamin/Cummings, 1984.

Riney, Larry. *Technical Writing for Industry.* Prentice-Hall, 1989.

Robinson, Patricia A. *Fundamentals of Technical Writing.* Houghton Mifflin, 1985.

———, and Gretchen H. Schoff. *Writing and Designing Operator Manuals.* 2nd ed. Lewis Publishers, 1991.

Rutherford, Andrea. *Basic Communication Skills for Electronics Technology.* Prentice-Hall, 1988.

Ryan, D. L. *Graphical Displays for Engineering Documentation.* Dekker, 1987.

Sageev, Pneena P. *Helping Researchers Write—So Managers Can Understand.* Battelle Press, 1986.

Samuels, Marilyn S. *The Technical Writing Process.* Oxford University Press, 1989.

Sanders, Norman. *Photographing for Publication.* Bowker, 1983.

Sandman, Peter, Carl Klompus, and Betsy Yarrison. *Scientific and Technical Writing.* Holt, Rinehart and Winston, 1985.

Schoenfeld, Robert. *The Chemist's English.* 2nd ed. VCH Publishers, 1986.

Sherman, Theodore A., and Simon S. Johnson. *Modern Technical Writing.* 5th ed. Prentice-Hall, 1990.

Sides, Charles. *How to Write and Present Technical Information.* 2nd ed. Oryx Press, 1991.

Stratton, Charles R. *Technical Writing: Process and Product.* Holt, Rinehart and Winston, 1984.
Stuart, Ann. *The Technical Writer.* Holt, Rinehart and Winston, 1988.
Sullivan, David, William L. Sullivan, and J. Wesley Sullivan. *Desktop Publishing: Writing and Publishing in the Computer Age.* Houghton Mifflin, 1989.
Sullivan, Frances J., ed. *Basic Technical Writing.* Society for Technical Communication, 1987.
Tench, Ann Gregson, and Isabelle Kramer Thompson. *Communication for Technicians: Reading, Writing, and Speaking on the Job.* Prentice-Hall, 1988.
Tryzna, Thomas N., and Margaret W. Batschelet. *Writing for the Technical Professions.* Wadsworth, 1987.
Turk, Christopher, and John Kirkman. *Effective Writing: Improving Scientific, Technical, and Business Communication.* 2nd ed. Routledge/Chapman and Hall, 1989.
Van Alstyne, Judith S. *Professional and Technical Writing Strategies.* 2nd ed. Prentice-Hall, 1990.
Warren, Thomas L. *Technical Writing.* Wadsworth, 1985.
Weiss, Edmond H. *How to Write Usable User Documentation.* 2nd ed. Oryx Press, 1991.
———. *The Writing System for Engineers and Scientists.* Prentice-Hall, 1982.
Whalen, Timothy. *Writing and Managing Winning Technical Proposals.* Artech House, 1987.
Wilkinson, Antoinette M. *The Scientist's Handbook for Writing Papers and Dissertations.* Prentice-Hall, 1991.
Wischerth, G. E. *Interim Standards: Technical Manual and Report Formats.* Society for Technical Communication, 1981.
Young, Matt. *Technical Writer's Handbook.* University Science Books, 1988.
Zimmerman, Donald. *The Random House Guide to Technical and Scientific Communication.* Random House, 1987.

PRESENTATIONS

Aslett, Don. *Is There a Speech Inside You?* Writer's Digest Books, 1989.
Booth, Vernon. *Communicating in Science: Writing and Speaking.* Cambridge University Press, 1985.
Gray, James G., Jr. *Strategies and Skills of Technical Presentations.* Greenwood Press, 1986.
Manko, Howard H. *Effective Technical Speeches and Sessions: A Guide for Speakers and Program Chairmen.* McGraw-Hill, 1969.
Meilach, Dona Z. *Dynamics of Presentation Graphics.* 2nd ed. Business 1 Irwin, 1990.
Turner, Barry T. *Effective Technical Writing and Speaking.* 2nd ed. Business Books, 1978.

GRAMMAR AND USAGE

Bernstein, Theodore. *Dos, Don'ts, and Maybes of English Usage.* New York Times Books, 1977.
———. *The Careful Writer: A Modern Guide to English Usage.* Macmillan, 1977.
———. *Miss Thistlebottom's Hobgoblins.* Simon and Schuster, 1984.
———. *Watch Your Language.* Macmillan, 1976.
Blumenthal, Joseph C. *English 3200: A Programmed Course in Grammar and Usage.* 3rd ed. Harcourt Brace Jovanovich, 1981.

Booher, Dianna. *Good Grief, Good Grammar.* Fawcett, 1989.
Delton, Judy. *The 29 Most Common Writing Mistakes and How to Avoid Them.* Writer's Digest Books, 1991.
Freeman, Morton S. *A Handbook of Problem Words and Phrases.* ISI Press, 1987.
———. *The Wordwatcher's Guide to Good Writing and Grammar.* Writer's Digest Books, 1990.
Kaplan, Jeffrey P. *English Grammar: Principles and Facts.* 2nd ed. Prentice-Hall, 1988.
Liles, Bruce L. *A Basic Grammar of Modern English.* 2nd ed. Prentice-Hall, 1987.
Master, Peter A. *Science, Medicine, and Technology: English Grammar and Technical Writing.* Prentice-Hall, 1986.
Morris, William and Mary, eds. *Harper Dictionary of Contemporary Usage.* 2nd ed. Harper and Row, 1985.
Pence, R. W., and D. W. Emery. *A Grammar of Present-day English.* 2nd ed. Macmillan, 1963.
Randall, Bernice. *Webster's New World Guide to Current American Usage.* Simon and Schuster, 1988.
Rook, Fern. *Slaying the English Jargon.* Society for Technical Communication, 1983.
Rozakis, Laurie. *The Random House Guide to Usage and Punctuation.* Random House, 1991.
Safire, William. *Coming to Terms.* Doubleday, 1991.
———. *Fumblerules.* Doubleday, 1990.
———. *I Stand Corrected: More on Language.* New York Times Books, 1984.
———. *Language Maven Strikes Again.* Doubleday, 1990.
———. *On Language.* New York Times Books, 1980.
———. *Take My Word for It.* Henry Holt, 1986.
———. *What's the Good Word?* New York Times Books, 1982.
———. *You Could Look It Up.* New York Times Books, 1988.
Schwager, Edith. *Medical English Usage and Abusage.* Oryx Press, 1991.
Strumpf, Michael, and Auriel Douglas. *Painless, Perfect Grammar.* Monarch Press, 1985.
Turner, Rufus P. *Grammar Review for Technical Writers.* Krieger, 1981.
Walpole, Jane. *A Writer's Guide: Easy Ground Rules for Successful Written English.* Prentice-Hall, 1980.
Webster's Dictionary of English Usage. Merriam-Webster, 1989.
Weiss, Edmond H. *One Hundred Writing Remedies: Practical Exercises for Technical Writing.* Oryx Press, 1990.
Westheimer, Patricia W., and Julie Wheatcroft. *Grammar for Business.* Scott Foresman, 1989.
Yarber, Robert E. *Reviewing Basic Grammar.* 2nd ed. Scott, Foresman, 1986.

DICTIONARIES AND OTHER REFERENCE WORKS

American Men and Women of Science 1992–1993. 18th ed. Bowker, 1992.
Ballentyne, D. W. G., and D. R. Lovett. *A Dictionary of Named Effects and Laws in Chemistry, Physics, and Mathematics.* 4th ed. Chapman and Hall, 1980.
Belzer, Jack, ed. *Encyclopedia of Computer Science.* 21 vols. Dekker, 1975–1989.
Black, William Campbell. *Black's Law Dictionary.* 6th ed. West Publishing, 1990.
Chen, Ching-chih. *Scientific and Technical Information Sources.* 2nd ed. MIT Press, 1987.
Clayton, L. Thomas, ed. *Taber's Cyclopedic Medical Dictionary.* 16th ed. F. A. Davis, 1989.
Daintith, John, ed. *The Facts on File Dictionary of Physics.* Facts on File, 1990.

Edmunds, Robert A. *The Prentice-Hall Encyclopedia of Information Technology.* Prentice-Hall, 1987.

Freedman, Alan. *The Computer Glossary Sourcebook.* 5th ed. AMACOM, 1990.

Gillispe, Charles C., ed. *Dictionary of Scientific Biography.* 8 vol. Macmillan, 1981.

Glenn, J. A. *Dictionary of Mathematics.* Barnes and Noble, 1984.

Graf, Rudolf F. *The Encyclopedia of Electronic Circuits.* TAB Books, 1985.

———. *Modern Dictionary of Electronics.* 5th ed. H. W. Sams, 1977.

———, and George J. Whalen. *The Reston Encyclopedia of Biomedical Engineering Terms.* Reston Publishing, 1977.

Jay, Frank, ed. *IEEE Standard Dictionary of Electrical and Electronic Terms.* Rev. ed. IEEE Standards, 1988.

Karlin, Leonard and Muriel S. *Medical Secretary's and Assistants' Encyclopedic Dictionary.* Prentice-Hall, 1984.

Parker, Sybil, ed. *McGraw-Hill Dictionary of Scientific and Technical Terms.* 4th ed. McGraw-Hill, 1989.

———. *McGraw-Hill Encyclopedia of Physics.* McGraw-Hill, 1983.

Rosenberg, Jerry M. *Dictionary of Artificial Intelligence and Robotics.* John Wiley, 1986.

———. *Dictionary of Business and Management.* John Wiley, 1985.

———. *Dictionary of Computers, Information Processing, and Telecommunications.* 2nd ed. John Wiley, 1987.

Rothenberg, Mikel A., and Charles F. Chapman. *Dictionary of Medical Terms for the Nonmedical Person.* 2nd ed. Barron's, 1989.

Rothwell, William S. *The Vocabulary of Physics.* National Book Company, 1987.

Sippl, Charles J. *Computer Dictionary.* 4th ed. Howard Sams, 1985.

The Software Encyclopedia 1991. Bowker, 1991.

Stearn, William T. *Botanical Latin: History, Grammar, Syntax, Terminology, and Vocabulary.* 3rd ed. David and Clark, Newton Abbot, Devon, 1983.

Thrush, Paul W. *A Dictionary of Mining, Mineral, and Related Terms.* Government Printing Office, 1968.

Waldman, Harry. *Dictionary of Robotics.* Macmillan, 1985.

OTHER USEFUL TITLES

Davies, P. C. W. *The Forces of Nature.* 2nd ed. Cambridge, 1986.

Downing, Douglas. *A Dictionary of Mathematical Terms.* Barron's, 1987.

Graf, Rudolf F., and George J. Whalen. *How It Works, Illustrated: Everyday Devices and Mechanisms.* Popular Science, 1974.

———. *The TAB Handbook of Hand & Power Tools.* TAB Books, 1984.

Gribbin, J. *In Search of Schroedinger's Cat.* Corgi Books, 1985.

Hayslett, H. T. *Statistics Made Simple.* Doubleday, 1968.

Herrington, Donald E. *How to Read Schematics.* 4th ed. Howard W. Sams, 1986.

Kline, Morris. *Mathematics for the Nonmathematician.* Dover, 1985.

King, William J. *The Unwritten Laws of Engineering.* American Society of Mechanical Engineers, 1944.

Schumacher, Michael. *Creative Conversations: The Writer's Guide to Conducting Interviews.* Writer's Digest Books, 1990.

Sawyer, W. W. *What Is Calculus About?* Mathematical Association of America, 1961.

Index